D0231277

D. G. Laing R. L. Doty W. Breipohl (Eds.)

The Human Sense of Smell

With 74 Figures

Springer-Verlag
Berlin Heidelberg New York London
Paris Tokyo Hong Kong Barcelona
Budapest

Dr. DAVID G. LAING
CSIRO Sensory Research Center
P.O. Box 52
North Ryde, NSW 2113
Australia

Dr. RICHARD L. DOTY
Smell and Taste Center
School of Medicine
University of Pennsylvania
3400 Spruce Street
Philadelphia, PA 19104, USA

Assoc. Professor WINRICH BREIPOHL
University of Queensland
Department of Anatomy
St. Lucia, Brisbane
Queensland 4067, Australia

ISBN 3-540-53355-9 Springer-Verlag Berlin Heidelberg New York
ISBN 0-387-53355-9 Springer-Verlag New York Berlin Heidelberg

Library of Congress Cataloging-in-Publication Data. The Human sense of smell / D.G. Laing, R.L. Doty, W. Breipohl (eds.). p. cm. Includes bibliographical references and index. ISBN 3-540-53355-9 (alk. paper). – ISBN 0-387-53355-9 (alk. paper) 1. Smell. I. Laing, David G. (David George) II. Doty, Richard L. III. Breipohl, W. [DNLM: 1. Smell. 2. Smell – physiology. WV 301 H918] QP458.H86 1991 612.8'6–dc20 DNLM/DLC for Library of Congress 91-5102

Printed in Germany

Typesetting: Camera ready by author
31/3145-543210 – Printed on acid-free paper

PREFACE

The human sense of smell is vastly underrated, many regarding it as a sense of the past, a sense that has outlived its usefulness. Anecdotal comments berate the inability of humans to detect, discriminate, or identify odors as proficiently as dogs or insects, yet such commentators forget that no single instrument can sense many chemicals at the low concentrations achieved by the nose, or discriminate and identify odors with the speed of the nose. Forgotten too is the role of smell in food acceptance, fragrance appreciation and as a warning device for spoiled food, toxic gases and the presence of fire. Encouragingly, the number of scientists studying various aspects of the sense increases year by year as evidenced by the number of journal publications and swelling attendances at smell related conferences, belying the notion that this exquisite sense should be disregarded. How the sense of smell works, why it is so efficient in the detection, discrimination and identification of chemicals, and its role and influence on the physiology and behavior of humans has provided the *raison de etre* for the significant increase in interest in the science of smell.

In essence, the production of this book is a consequence of the rapid expansion of research on the sense of smell and the need for a text that deals comprehensively with the many facets of the human sense of smell. Indeed, it was the great diversity of information gained in recent years from scientists in many disciplines that prompted us to produce a multi-author text, the expertise needed for the task far exceeding the capabilities of a single author.

Accordingly, the book consists of five sections; Anatomy, Physiology and Chemistry (Part 1); Measurement Methods (Part 2); Development and Senescence (Part 3); General Operating Characteristics (Part 4), and Clinical and Health Aspects (Part 5). The contributing authors were asked to provide an up-to-date review of the specific area relevant to their expertise rather than solely a historical perspective or description of their latest research, the aim being to provide readers with a state of the art text.

Despite the specific expertise needed to write each Chapter, there is some overlap of contents from time to time as might be expected in a multidisciplinary book. For example, contributors describe the use of similar encephalographic (EEG) techniques for different purposes in Chapters 6 and 17, whilst others have different reasons for studying anatomical structures such as the olfactory mucosa, as in Chapters 1, 3 and 14. For this reason, special emphasis has been given to methods for measuring human responses to odors (Chapter 5) which are relevant to many Chapters, particularly those dealing with olfactory adaptation, memory, mixture perception, comparison of the sense of smell of humans and animals, and olfactory dysfunction.

Two Chapters are devoted to Development and Senescence. Though relatively thin on information at present, they represent two rapidly expanding areas of research. Studies of development are particularly important for establishing the role of experience and genetics in our ability to smell and in the establishment of odor preferences. Senescence on the other hand has considerable implications for olfactory dysfunction, changes in appreciation of foods and fragrances, and the link to Alzheimer's and other diseases (Chapters 9 and 14).

Finally, in view of the emphasis currently given to clinical studies of smell, particularly in the United States, Chapters on olfactory dysfunction (14), epidemiological aspects (15) and the effects of drugs on smell (16) are included.

In conclusion, the contributors have done their best to produce a text that is informative, at times controversial, and at the very least provides the reader with a solid grasp of current knowledge of the human sense of smell and future directions for research.

D G Laing

R L Doty

W Breipohl

ACKNOWLEDGEMENTS

The Editors would like to thank all the contributors for the efforts they have made to produce this text. They also wish to acknowledge the enormous contribution of Mrs Pat Gould (CSIRO) who played a major role in the layout of the text and illustrations, and in its general production. Thanks also go to Mrs Lyn Keen (CSIRO) for her contributions during the typing and correction of drafts. Finally, the Editors thank the management of the CSIRO Food Research Laboratory for providing the facilities and staff required for the preparation of the book for camera ready publication.

CONTENTS

PART 1 ANATOMY, PHYSIOLOGY AND CHEMISTRY

3. PHYSIOLOGY OF OLFACTORY RECEPTION AND TRANSDUCTION: GENERAL PRINCIPLES

T V Getchell and M L Getchell

4. MOLECULAR STRUCTURE AND SMELL

M Chastrette and D Zakarya

PART 2 MEASUREMENT OF OLFACTORY RESPONSES

5. PSYCHOPHYSICAL MEASUREMENT OF ODOR PERCEPTION IN HUMANS

R L Doty

6. HUMAN ELECTRO-OLFACTOGRAMS AND BRAIN RESPONSES TO OLFACTORY STIMULATION

G Kobal and Th Hummel

CONTRIBUTORS

Numbers in parentheses indicate the pages on which the authors' contributions begin.

Gary K Beauchamp (165), Monell Chemical Senses Center, 3500 Market Street, Philadelphia, PA 19104, USA.

William S Cain (215), John B Pierce Foundation Laboratory, 290 Congress Avenue, New Haven, Connecticut 06519, USA.

M Chastrette (75), Laboratoire de Chimie Organique Physique, Université Claude Bernard, Lyon 1, 43 Boulevard du 11 Novembre 1918, 69622 Villeurbanne Cedex, France.

Rene A de Wijk (197), Psychonomie, Buys Ballot Laboratory, Princetonplein 5, NL-3584, CC Utrecht, The Netherlands.

Richard L Doty (93, 153, 179), Smell and Taste Center, Hospital of the University of Pennsylvania, 3400 Spruce Street, Philadelphia, PA 19104, USA.

Thomas V Getchell (59), Sanders-Brown Center on Aging, College of Medicine, University of Kentucky, Lexington, Kentucky 40536, USA.

Marilyn L Getchell (59), Sanders-Brown Center on Aging, College of Medicine, University of Kentucky, Lexington, Kentucky 40536, USA.

Loredana M Harrison (333), Department of Psychology, University of New Hampshire, Conant Hall, Durham, NH 03824, USA.

Thomas Hummel (133), Institut für Pharmakologie und Toxikologie, Universität Erlangen-Nürnberg, Universitätsstr, 22, D-8520 Erlangen, Germany.

Bruce W Jafek (1), Department of Otolaryngology and Head and Neck Surgery, University of Colorado School of Medicine, Box B-210, 4200 East Ninth Avenue, Denver, Colorado 80262, USA.

Roger A Jennings (259), Bowman Gray Technical Center, R J Reynolds Tobacco Company, Winston-Salem, NC 27102, USA.

Gerd Kobal (133), Institut für Pharmakologie und Toxikologie, Universität Erlangen-Nürnberg, Universitätstr 22, D-8520 Erlangen, Germany.

E P Köster (197), Psychological Laboratory, Utrecht University, Sorbonnelaan 16, 3584 CA Utrecht, The Netherlands.

David G Laing (239), CSIRO Sensory Research Centre, Food Research Laboratory, PO Box 52, North Ryde, NSW 2113, Australia.

Harry Lawless (359), Department of Food Science, NY State College of Agriculture and Life Sciences, Cornell University, Ithaca, NY 14853, USA.

Robert G Mair (333), University of New Hampshire, Department of Psychology, Conant Hall, Durham, New Hampshire 03824-3567, USA

Julia A Mennella (165), Monell Chemical Senses Center, 3500 Market Street, Philadelphia, PA 19104, USA.

David T Moran (1), Department of Otolaryngology and Head and Neck Surgery, University of Colorado School of Medicine, Box B-210, 4200 East Ninth Avenue, Denver, Colorado 80262, USA.

Patricio Reyes (27), Department of Neuropathology, Jefferson Medical College, 1025 Walnut Street, Philadelphia, PA 19107, USA.

J Carter Rowley III (1), Department of Otolaryngology and Head and Neck Surgery, University of Colorado School of Medicine, Box B-210, 4200 East Ninth Avenue, Denver, Colorado 80262, USA.

Frank R Schab (215), Perceptual and Cognitive Science Group, Operating Sciences Department, General Motors Research Laboratories, Warren, MI 48090-9055, USA.

Brian S Schwartz (305), John Hopkins University School of Hygiene and Public Health, Division of Occupational Health, Room 7041, 615 North Wolfe Street, Baltimore, MD 21205, USA.

Allen M Seiden (281), Department of Otolaryngology and Maxillo-facial Surgery, University of Cincinnati, 231 Bethesda Avenue, Cincinnati, OH 45267, USA.

Michael T Shipley (27), Department of Anatomy and Cell Biology, University of Cincinnati, Mail Location 521, Cincinnati, OH 45267, USA.

David V Smith (281), Department of Otolaryngology and Maxillofacial Surgery, College of Medicine, University of Cincinnati Medical Center, Mail Location 528, 231 Bethesda Avenue, Cincinnati, Ohio 45267-0528, USA

James C Walker (259), Bowman Gray Technical Center, R J Reynolds Tobacco Company, Winston-Salem, NC 27102, USA.

D Zakarya (75), Université Mohammed V, Faculté des Sciences, Rabat, Morocco.

PART 1

ANATOMY, PHYSIOLOGY AND CHEMISTRY

1

THE ULTRASTRUCTURE OF THE HUMAN OLFACTORY MUCOSA.

DAVID TAYLOR MORAN

BRUCE W JAFEK

J CARTER ROWLEY III

I. INTRODUCTION

The purpose of this Chapter is to present a detailed description of the ultrastructure of the human olfactory mucosa. To this end, the Chapter is divided into four major sections which describe the biopsy technique; the ultrastructure of the normal olfactory mucosa in humans; the microvillar cell, and histopathologic changes of the olfactory mucosa in patients with olfactory dysfunctions.

In 1980, an electron-microscopic investigation of the histopathology of olfactory dysfunction was initiated in this laboratory. A prerequisite to this was a thorough understanding of the fine structure of the normal human olfactory mucosa. Although the olfactory mucosae of many vertebrates had been described in the literature, little electron-microscopic information was available on that of humans. Consequently, we set out to define the ultrastructure of the human olfactory mucosa, and promptly discovered why this had not yet been done: it is physically quite difficult to obtain biopsies of human olfactory epithelium. In humans, the olfactory epithelium is restricted to a very small area in the roof of the nasal cavity, sits 7 cm deep to the nostril, and is impossible to reach with conventional biopsy instruments. Our first task was to design an instrument that would permit safe removal of small pieces of olfactory mucosa while a patient was under local anesthesia. Since the primate olfactory epithelium is known to regenerate its neuronal elements after section of the olfactory nerve (Graziadei et al. 1980), it was considered that such a biopsy would not compromise the sense of smell in tissue donors.

II. THE BIOPSY TECHNIQUE

An instrument has been designed in this laboratory that permits the safe removal of small biopsies of olfactory mucosa from normal persons under local anesthesia. An overview of the procedure is presented here; details are available in the literature (Lovell et al. 1982).

First, the subjects are given an appropriate physical and intranasal examination by an otorhinolaryngologist, and psychophysically evaluated for olfactory function using the University of Pennsylvania Smell Identification Test (UPSIT: Doty et al. 1984b). After explaining the procedure to the subject, and obtaining informed, written consent, the nasal cavity is sprayed with a 4% aqueous cocaine solution that acts as a local anesthetic and vasoconstrictor. Next, the supine subject's nasal cavity is inspected with an Olympus SES-1711D selfoscope. An arthroscope of small caliber is essential here, since the passage to the olfactory epithelium, bounded laterally by the turbinates and medially by the septum, is extremely narrow. The biopsy instrument is gently advanced to the olfactory region near the roof of the nasal cavity (Fig 1). The instrument itself (Fig 2) is a long, thin stainless steel rod tipped by a flat, sharp knife-blade reflected back in the direction of the handle. When in position, the instrument is rotated so that the blade faces the wall of the superior septum, and a thin slice of olfactory mucosa is shaved off. When done properly, the knife-blade passes through the lamina propria, yielding a small, 1 mm-square piece of olfactory epithelium underlain by a thin sheet of lamina propria. Although quite safe in the hands of an experienced surgeon, this procedure should only be performed by a trained otorhinolaryngologist, since the olfactory region is near the cribriform plate of the ethmoid bone, a thin region which, if perforated, could result in leakage of cerebrospinal fluid and possible infection.

Tissues are immediately immersed in fixative, and prepared for electron microscopy (see Moran et al. 1982a, 1987, for details). In order to obtain large fields of view in the electron microscope, thin sections are mounted on slot grids using the Domino Rack technique described by Moran and Rowley (1987).

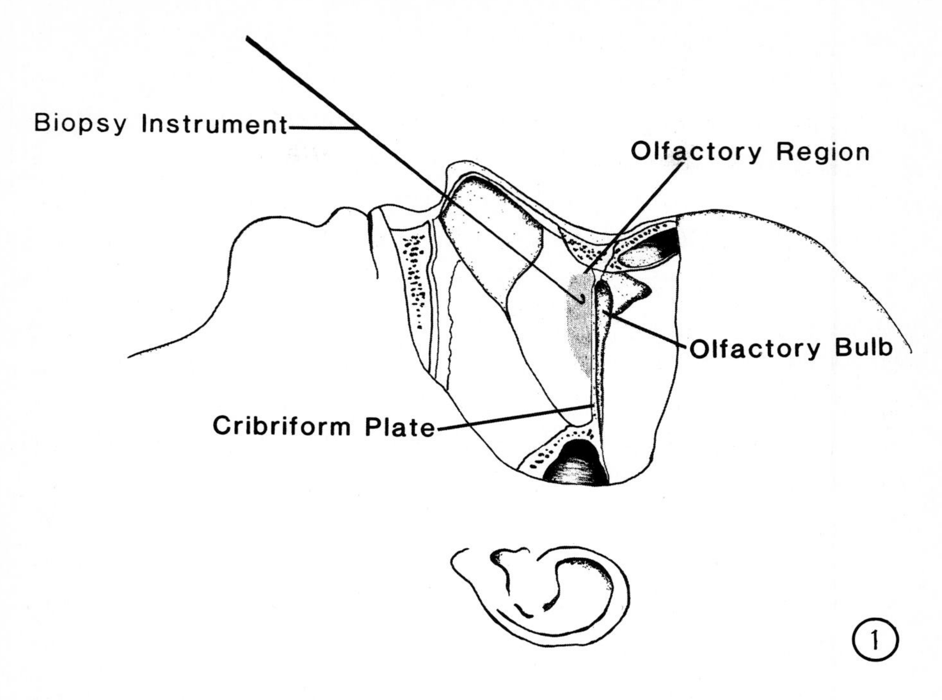

Fig 1. Diagram showing position of instrument when obtaining a small biopsy of human olfactory mucosa.

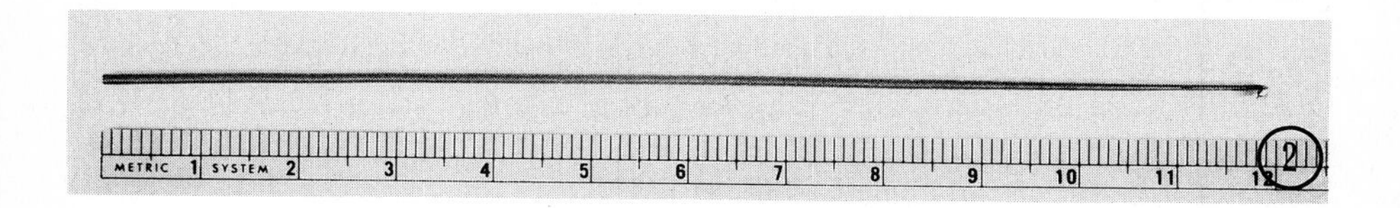

Fig 2. The biopsy instrument, shown to metric scale.

A. Location and Distribution of Olfactory Epithelium

The precise location and distribution of the olfactory epithelium in humans of different ages is not well documented. In order to find a suitable region to biopsy, serial sections through the putative olfactory region of several cadavers from persons 60-70 years of age were taken. Examination of these thick sections by light microscopy showed that the olfactory epithelium covers an area of about 1 square centimeter per nostril, and exists on the cribriform plate of the ethmoid bone on part of the upper surface of the superior turbinate, and on the superior aspect of the nasal septum. Given that smell function seems to decrease with age (Doty et al. 1984a, Schiffman 1979), one would suspect the olfactory region of a younger person to cover a larger area. Unfortunately, such data are not yet available. Both light and scanning electron microscopic observations, however, show that the olfactory epithelium is not a continuous sheet, but is interspersed with patches of respiratory epithelium (Nakashima et al. 1984, Morrison and Costanzo 1990). This fact may account for our 50% success rate in obtaining biopsies of olfactory epithelium from the superior nasal septum.

III. ULTRASTRUCTURE OF THE HUMAN NASAL MUCOSA

A. General Morphology of Respiratory Mucosa

The vast majority of the human nasal cavity is lined by respiratory mucosa. Presumed to be devoid of olfactory receptors, the respiratory mucosa within the nasal cavity is similar to that of the trachea, bronchi, and bronchioles. Lined by a ciliated pseudostratified columnar epithelium supported by a highly vascular lamina propria, the respiratory mucosa helps to warm and moisten incoming air and remove inhaled particulate matter, thereby rendering the inspired air suitable for passage into the delicate pulmonary alveoli, where gas exchange occurs.

Depicted in Figure 3, the respiratory epithelium contains three major cell types: goblet cells, ciliated cells, and basal cells. The goblet cells, which secrete mucus onto the epithelial

---

Fig 3. Low magnification electron micrograph of the respiratory epithelium that lines the vast majority of the human nasal cavity. B, basal cell; C, ciliated epithelial cell; G, goblet cell, N, nasal cavity; arrow, motile cilia. 2,550X

Fig 4. Low magnification electron micrograph of the human olfactory epithelium showing ciliated olfactory receptor cells (R), their dendrites (D), and olfactory vesicles (arrows) at the dendrite tips that project into the nasal cavity (N). A microvillar cell (M) and supporting cells (S) are present, as are basal cells (B). 1,200X

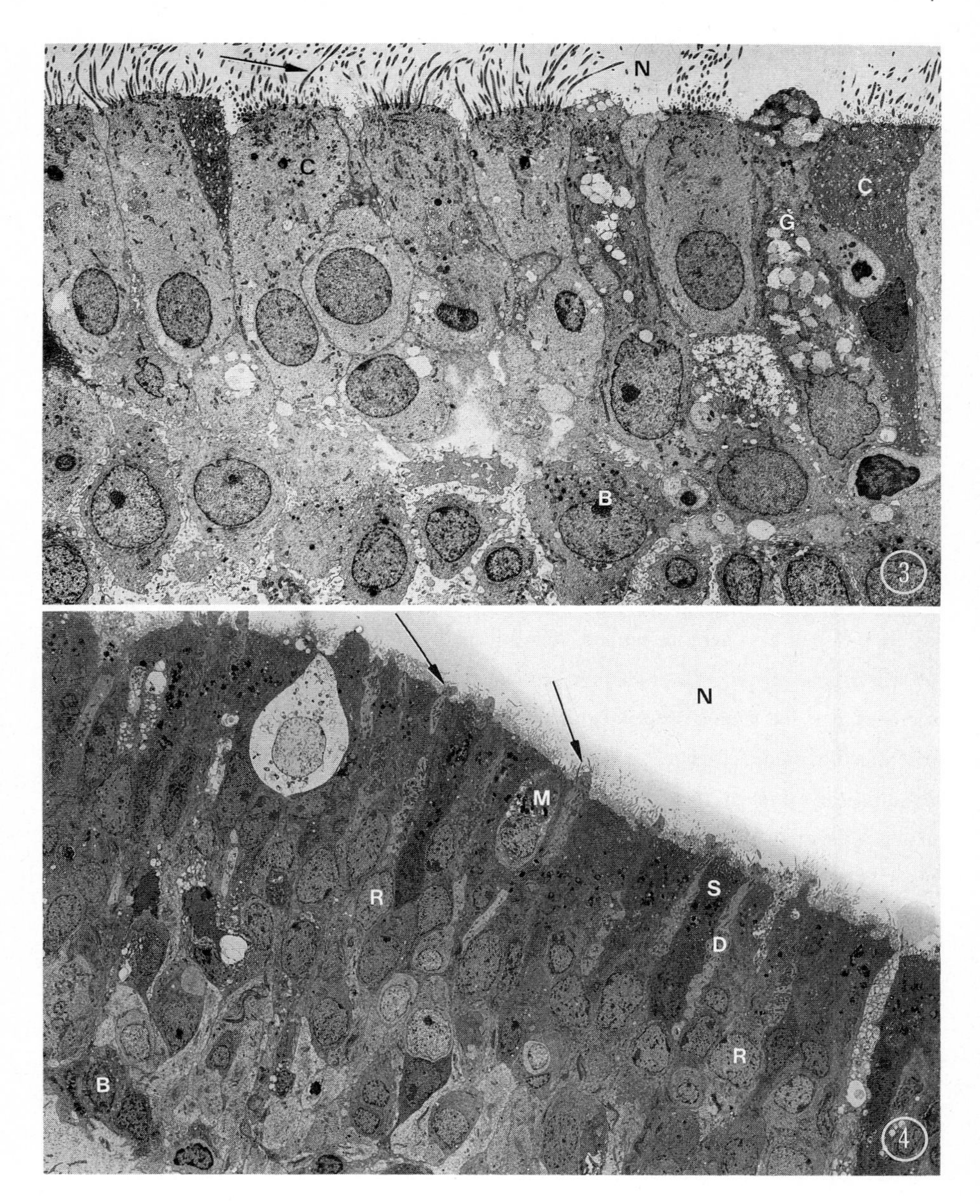

N
C
C
G
B
3
N
M
R
S
D
R
B
4

surface, have an apical cytoplasm filled with mucigen droplets synthesized by a dense cytoplasm rich in mitochondria, rough endoplasmic reticulum, and Golgi complexes. The biosynthetic components of the cytoplasm, along with the nucleus, are often compressed into the basal region of the cell by the mass of stored mucigen that is the cell's own product.

Tall ciliated columnar cells, with nuclei located centrally in the epithelium, have many long, motile cilia and short microvilli at the apical surface of each cell. The cilia, which grow out from a row of basal bodies, beat in metachronal waves that sweep the mucus blanket and its entrapped particles toward the pharynx, where the material can be swallowed or expectorated.

Both ciliated and goblet cells eventually degenerate and die, and are subsequently replaced by new cells derived from the mitotically-active basal cells. The basal cells, small, polygonal cells with centrally located nuclei, are spread in a monolayer on top the well developed basement membrane. Beneath the basement membrane is connective tissue of the lamina propria. In addition to collagen and elastic fibers and the fibroblasts that secrete them, the lamina propria is highly vascular, so much so that the nasal mucosa is a true erectile tissue. This property of the lamina propria is manifest not only during the inflammatory process, but in the normal nasal cycle, in which each half of the nasal cavity alternately swells and deflates, making air flow mainly through one nostril or the other at any given time. The lamina propria is rich in nerve fibers and in cells of the immune series. Plasma cells, for example, are abundant. They secrete antibodies, many of which find their way into the surface mucus, where they can be used as a first line of immunologic defense against inhaled antigens.

B. General Morphology of Olfactory Mucosa

The microanatomy of the olfactory mucosa, when viewed by low magnification electron microscopy (Fig 4), appears quite different from the respiratory mucosa. Lacking both goblet cells and ciliated epithelial cells, the olfactory mucosa is rich in ciliated olfactory receptors cells and attendant supporting, or sustentacular, cells. In addition, it contains a morphologically distinct class of cells of uncertain function called microvillar cells. In the lower reaches of the epithelium, sitting atop the basement membrane, are the basal cells.

The latter are important stem cells that give rise to new olfactory receptor neurons following normal cell death or injury to the olfactory epithelium (Monti Graziadei and Graziadei 1979). This remarkable regenerative capacity is unique to the olfactory system, since most mammalian neurons, once lost, are gone for all time; the mammalian nervous system is considered to have a static cell population (Goss 1964).

The fine structure of the four major cell types of the olfactory epithelium, the ciliated olfactory receptor cells, microvillar cells, supporting cells, and basal cells, and the lamina propria which serves as their life-support system, will be described below.

C. Ciliated Olfactory Receptor Cells

The ciliated olfactory receptor cells illustrated in Figures 4-11, are primary sensory cells, slender bipolar neurons that send a dendrite to the site of stimulus reception at the epithelial surface and an axon directly to the olfactory bulb of the brain. Consequently, the ciliated olfactory receptor cells in the human nose are "naked neurons", which are directly exposed to the nasal cavity and forge an anatomic link between the atmosphere and the brain. Their position at the frontier of the animal, which suits them so well for the detection of odorants, puts the central nervous system at some risk: these "naked neurons" serve as a portal of entry for pathogens (Shipley 1985).

As shown in Figure 4, ciliated olfactory receptor cells are abundant, occurring at 3-5 μm intervals along the epithelial surface. Counts of receptor cells in cross-sectional electron images (taken from tissue donors 21-25 years of age) shows they have an approximate population density of 30,000 bipolar neurons per square millimeter of tissue. Given our cadaver calculations (from persons 60-70 years of age) of a total olfactory area of 2 square centimeters, the total receptor population is estimated to be of the order of 6 million per nose. By inference, a greater number would be present in a younger person with a larger total area of olfactory mucosa.

The nuclei of the ciliated olfactory receptor cells, which occupy a broad band in the lower third of the epithelium, are usually located below the nuclei of the supporting cells and above those of the basal cells. A typical receptor cell, cut in longitudinal section, is shown in Figure 5. Here, the small, fusiform cell body sends a thin dendrite some 40 μm

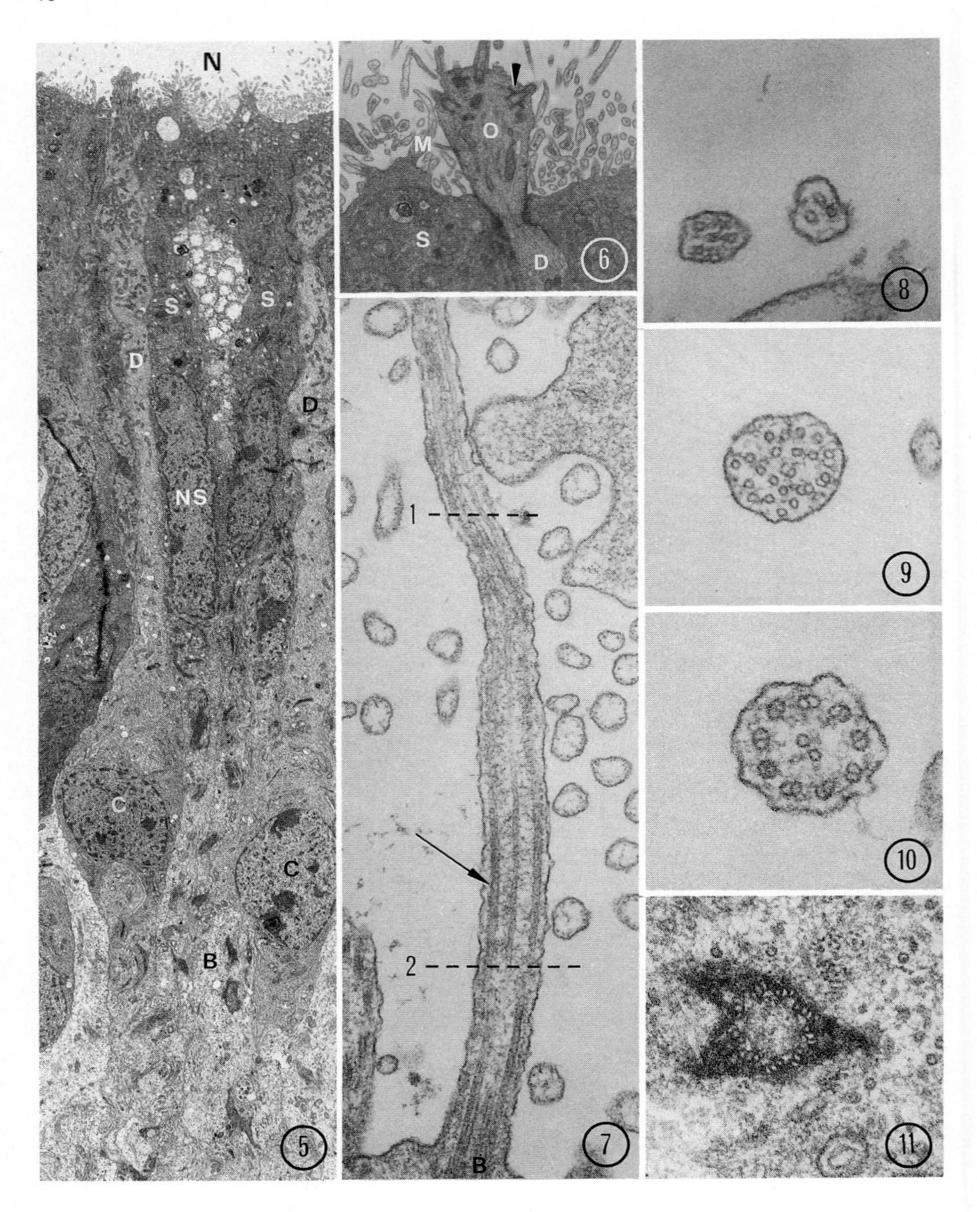
N
S
S
D
D
NS
C
C
B
5
M
O
S
D
6
1
2
B
7
8
9
10
11

to the epithelial surface. As it approaches the surface, the dendrite gains a large population of mitochondria, as do many sensory cells at the site of stimulus reception and sensory transduction. The dendrite is tipped by an olfactory vesicle, or olfactory knob (Fig 6), measuring about 1.5 μm in diameter that projects 2 μm from the epithelial surface. The olfactory vesicle, stabilized by an extensive skeleton of cytoplasmic microtubules, contains several mitochondria, and has a large number of basal bodies situated beneath the cell membrane. As seen in Figure 6, a group of olfactory cilia, usually numbering from 10 to 30, project from the basal bodies and extend along the sensory surface of the olfactory epithelium.

The ultrastructure of a single human olfactory cilium is illustrated in Figures 7-11. Figure 7, a longitudinal section, shows a single sensory cilium as it projects from a basal body near the lateral surface of an olfactory vesicle. Shown in cross-section in Figure 11, the

---

Fig 5. Low power electron image of human ciliated olfactory receptor cells, cut in longitudinal section, showing their cell bodies (C), dendrites (D), and an olfactory vesicle projecting into the nasal cavity (N). Supporting cells (S) encircle the receptor cells. NS, nucleus of supporting cell; B, basal pole of supporting cell. 3,550X

Fig 6. Longitudinal section through an olfactory vesicle (O) containing basal bodies (arrowhead) from which sensory cilia emerge. Nearby, the microvilli (M) of supporting cells (S) are evident. D, dendrite of olfactory receptor cell. 9,600X

Fig 7. Longitudinal section through part of a sensory cilium of an olfactory receptor cell. B, basal body; arrow, microtubules of outer doublet of ciliary axoneme; dotted lines 1 and 2 represent the levels through which ciliary cross-sections in Figures 9 and 10 were cut. 58,500X

Fig 8. Cross-section through the tip of an olfactory cilium, distal to the regions of the ciliary shaft depicted in Figure 7. The slender ciliary lash is supported by an axoneme containing few single microtubules. 90,000X

Fig 9. Cross-section through an olfactory cilium, cut at level 1 in Figure 7. Most of the outer doublets have given way to single microtubules. 89,000X

Fig 10. Cross-section through the base of an olfactory cilium cut at level 2 in Figure 7. Here, the axoneme displays the "9+2" pattern of organization; no dynein arms are present, suggesting the cilium is immotile. 89,000X

Fig 11. Cross-section through the basal body of an olfactory cilium, showing the 9 triplets of microtubules characteristic of basal bodies and centrioles. 80,500X

basal body contains nine triplets of microtubules. The base of the ciliary shaft, imaged in both long (Fig 7) and cross section (Fig 11), is supported by a "9+2" axoneme consisting of nine outer doublets of microtubules surrounding a single central pair. The dynein arms typical of motile cilia are absent. As the ciliary shaft proceeds distally, it becomes thinner. The "9+2" pattern of organization is lost; the axonemal doublets splay out, and proceed on as single microtubules. This is demonstrated in Figure 10, which shows a cross-section through an olfactory cilium cut at level 1 in Figure 7. As one approaches the distal tip of the cilium, it thins out even more, until the slender shaft is supported by a reduced axoneme of 1-3 single microtubules (Fig 8).

It is interesting to compare the ultrastructure of motile cilia from the respiratory mucosa with that of sensory cilia from the olfactory mucosa. We obtained a fortuitous biopsy from a normosmic 22 year old man in which respiratory and olfactory epithelia were present, side-by-side, in the same piece of tissue. When this tissue was prepared for electron microscopy, cut in cross-section, and viewed on end, examples of respiratory cilia (Fig 12) and olfactory cilia (Figs 13-14) were obtained that had been prepared in precisely the same way at the same time. Figure 12 shows a cross-section from four respiratory cilia. These cilia, known to be motile, contain a "9+2" axoneme with conspicuous dynein arms (arrows) projecting from the "A"-subfiber of each outer doublet toward the "B"-subfiber of the adjacent doublet. Dynein arms provide the motive force for ciliary beating; when they are removed, either biochemically in the laboratory (Gibbons 1967), or genetically, as in Kartagener's syndrome (Afzelius 1978), the cilium becomes immotile. The axonemes of the olfactory cilia, thought to be the site of sensory transduction in olfactory bipolar neurons (Lancet 1986), appear quite different. Of the three olfactory cilia shown in cross-section in Figures 13-14, one has a "9+2" axoneme, one is "9+3", and one is "9+4". Of perhaps greater significance, however, is the observation that the "A"-subfibers of all of the outer doublets lack dynein arms which indicates that the olfactory cilia in humans are immotile, and will move only passively in response to the flow of mucus which surrounds them.

D. The Microvillar Cell

In addition to the ciliated olfactory receptor cells, supporting cells and basal cells, the human olfactory epithelium contains a morphologically distinct class of cells called

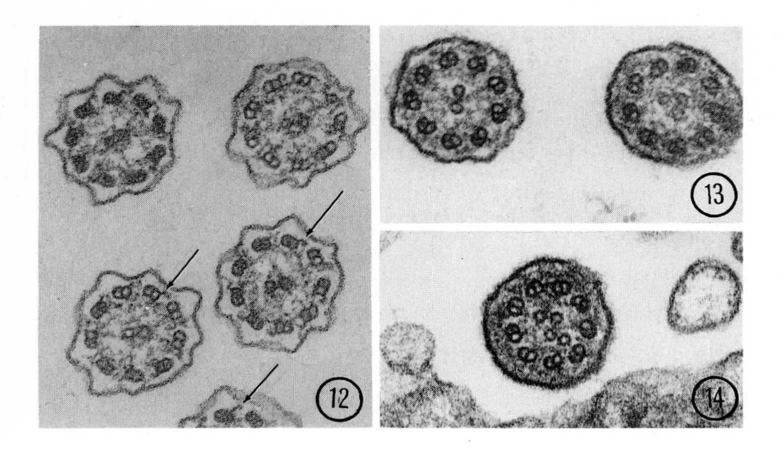

Figs 12, 13, and 14. Cross-sections through cilia taken from the respiratory (Fig 12) and olfactory (Figs 13,14) epithelia of a normosmic 22-year old male. These images come from the same thin section of a single block of tissue. Dynein arms, present on the motile, respiratory cilia (arrows), are absent in the olfactory cilia. The olfactory cilium in Figure 14 has an additional central pair of microtubules. 80,000X

microvillar cells. Described in detail in the literature (Moran et al. 1982a, 1982b), these flask-shaped cells usually have an electron-lucent cytoplasm, and of all the cells it is their nuclei that are situated closest to the surface of the olfactory epithelium. The number of microvillar cells seems to vary slightly between biopsies of olfactory epithelium taken from different persons. No pattern related to age or sex has yet emerged. In cases where a census of cell populations has been taken, there is of the order of one microvillar cell for every 10-20 ciliated olfactory receptor cells. However, during the biopsy procedure the site of tissue removal is limited to the superior septum for safety reasons. It is possible that, when other regions of the olfactory epithelium are examined, different relative cell populations may be observed.

A transmission electron micrograph of a longitudinal section through a microvillar cell is shown in Figure 15. It can be seen that the nucleus is located near the basal pole of the cell, and is surrounded by a cytoplasm that contains mitochondria, vesicles, and many free ribosomes. The supranuclear cytoplasm is filled with tubules of the smooth endoplasmic reticulum. Although not evident in this particular section, several basal bodies are often observed near the apical pole of the cell, which becomes quite narrow as it nears the epithelial surface. A short tuft of microvilli extend from the apical pole of the cell into the nasal cavity itself. The microvilli, evident in Figure 15, are better imaged by scanning electron microscopy in Figure 16, which shows the surface of a microvillar cell after the surrounding supporting cells have been cloven away during specimen preparation. The cell body is pear-shaped; a slender cytoplasmic projection emanates from the basal pole of the cell. The cell narrows at its apex, and a tuft of short, straight microvilli extend into the nasal cavity. It is interesting to compare the scanning electron image of these microvilli with the irregular microvilli of adjacent supporting cells and with the cilia of a nearby ciliated olfactory receptor cell.

E. Supporting Cells

Most of the biomass of the olfactory epithelium is occupied by supporting cells. These cells, which exhibit a marked morphological polarization from apex to base, surround the microvillar cells. They also encircle the cell bodies and dendrites of the ciliated olfactory receptor cells, as well as the intraepithelial portion of their axons. Figures 17-19 are low magnification images of supporting cells from the human olfactory epithelium, all oriented sideways so that the nasal cavity is toward the right side of the image, and the lamina propria toward the left. At the epithelial surface, the supporting cells send large numbers of microvilli, of irregular shape and length, into the mucus layer that lines the nasal cavity. Just beneath the microvilli, the apical pole of the cell has a dense cytoplasm filled with many mitochondria. Deeper down, a large number of electron-dense bodies of unknown composition are evident, as are many bundles of filaments that spiral as they wend their way toward the base of the cell. Although these filaments have not yet been identified, it is tempting to speculate they are bundles of actin filaments that participate in the shortening of the cell. If so, they could provide the force for the "olfactomotor response" (Graziadei 1973) in which the olfactory epithelium is believed to contract in response to certain stimuli.

The nuclei of the supporting cells, usually located in a diffuse layer superior to those of the ciliated olfactory receptor cells, tend to be long and thin; their shape mirrors that of the cell they serve. A prominent nucleolus is evident, as are distinct patches of heterochromatin. The cytoplasm surrounding the nucleus is quite electron-dense. Just beneath the level of the nucleus, however, the cytoplasm undergoes a profound

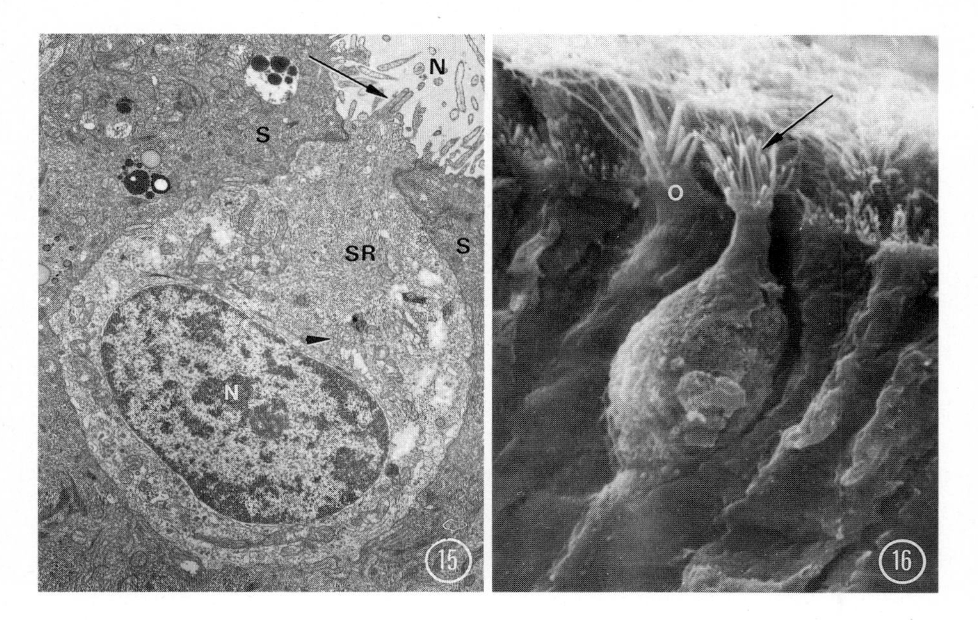

Fig 15. Transmission electron micrograph of a microvillar cell in the human olfactory epithelium. The Golgi apparatus (arrowhead) is above the nucleus (N), as is a well-developed system of the smooth endoplasmic reticulum (SR). A tuft of microvilli (arrow) project into the nasal cavity (N). Supporting cells (S) encircle the microvillar cell. 11,000X

Fig 16. Scanning electron micrograph of a microvillar cell in the human olfactory epithelium. Note tuft of microvilli (arrow) projecting into the layer of mucus (removed during specimen preparation) that covers the epithelial surface. O, olfactory vesicle. Micrograph courtesy of Dr. Edward E. Morrison (Morrison and Constanzo 1990). 7,000X

morphologic transformation, and becomes electron-lucent. This is quite evident in Figure 19, in which the basal pole of a supporting cell is caught in longitudinal section. Here, it is filled with clear, membrane-limited tubules and vesicles. This striking morphological polarization is clearly illustrated at low magnification by electron microscopy in Figure 5, where a supporting cell associated with a ciliated olfactory receptor cell has been cut along its entire length from apex to base.

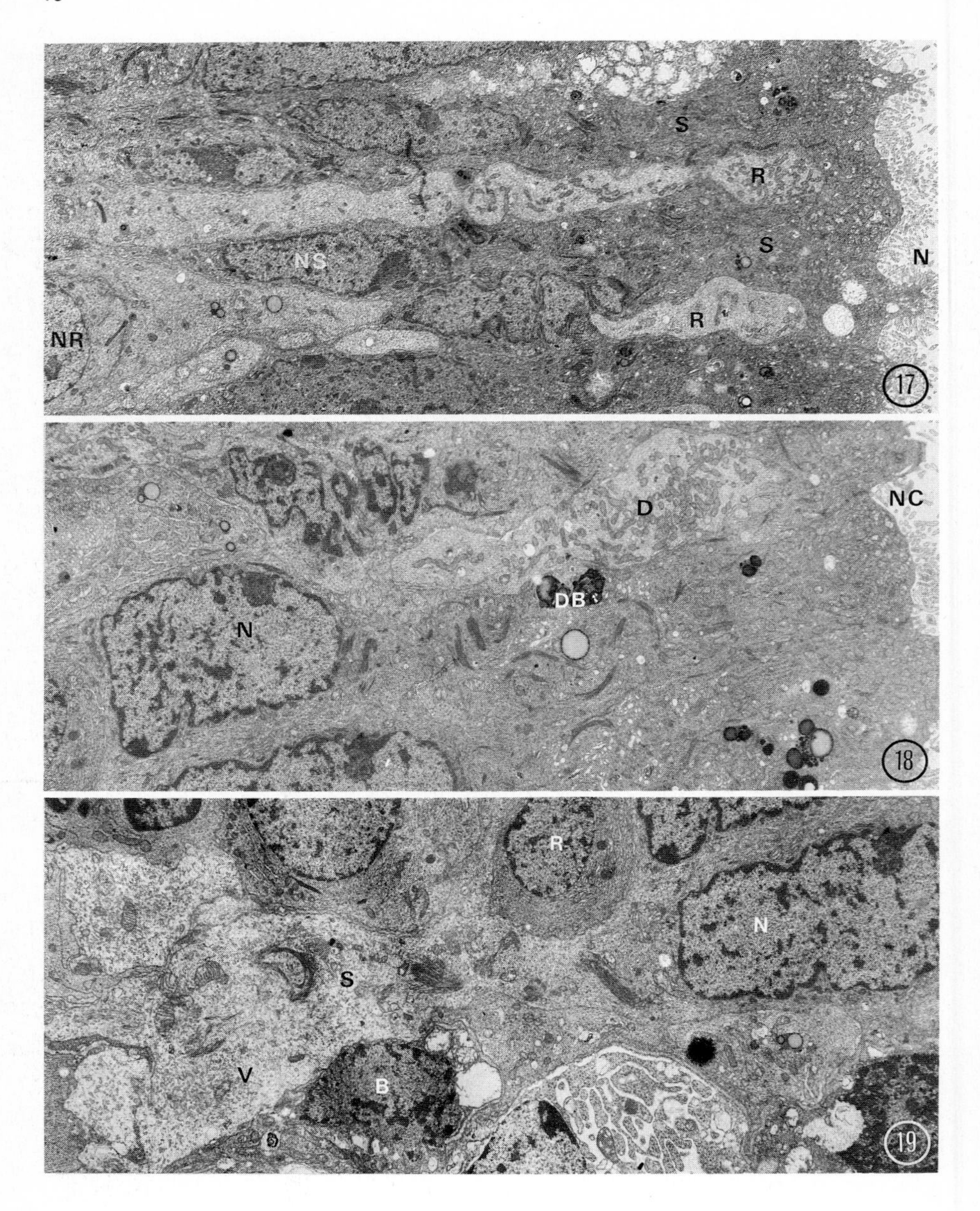
S
R
S
N
NS
R
NR
17
D
NC
DB
N
18
R
N
S
V
B
19

The precise functions of the supporting cells in the human olfactory epithelium await discovery. From their morphology, it is clear that they are not primarily adapted formucus secretion, as are the supporting cells of some amphibian olfactory epithelia which resemble goblet cells (Reese 1965). It seems likely that human supporting cells participate in secretion of materials into, and removal of other materials from the surface mucus layer (Moran et al.1982a). Recently, a class of odorant-binding proteins has been described that may assist in the shuttling of odorants to and from the receptor sites of the cell membranes of the olfactory receptor cells. It is tempting to speculate that the supporting cells are involved in the synthesis, secretion, and perhaps recycling of odorant-binding proteins. The supporting cells may also function in the uptake of odorant molecules into the olfactory epithelium. Hornung and Mozell (1981), for example, have shown that a high percentage (78%) of radio-labelled odorants are sequestered by the olfactory mucosa and remain there for 30 min after inhalation. The ultrastructure of the supporting cells suggest they are well equipped for the pinocytotic uptake and subsequent enzymatic treatment of substances, both native and inhaled, in the mucus layer that lines the nasal cavity.

F. Basal Cells

The olfactory epithelium is an example of a pseudostratified columnar epithelium. All known examples of pseudostratified epithelia contain a population of small, undifferentiated basal cells, sitting on top of the basement membrane, that serve as stem cells. These basal cells are mitotically active, and their daughter cells differentiate to become functional members of the epithelium itself (Moran and Rowley 1988). In this sense, the olfactory epithelium is a typical pseudostratified columnar epithelium. In another sense, it

---

Fig 17. Survey electron micrograph of a longitudinal section through several supporting cells (S) that surround ciliated olfactory receptor cells (R). N, nasal cavity; NS, nucleus of supporting cell; NR, nucleus of receptor cell. 4,000X

Fig 18. Longitudinal section through the apical region of several supporting cells. Note dense bodies (DB) in the cytoplasm above the nucleus (N). D, dendrite of receptor neuron; NC, nasal cavity. 6,700X

Fig 19. Longitudinal section through the basal region of several supporting cells (S). The lamina propria would be to the left and the nasal cavity to the right. Note the clear, vesicle-packed cytoplasm (V) at the base of the cell, and contrast that with the dense cytoplasm near the nucleus (N). B, basal cell; R, ciliated olfactory receptor cell. 6,700X

is quite remarkable, since in adult primates its basal cells differentiate to form nerve cells which are the ciliated bipolar olfactory receptor neurons that serve as primary sensory cells (Graziadei et al. 1980). By inference, albeit as yet unsupported by direct experimental evidence, the same process occurs in humans.

Several basal cells are depicted by electron microscopy in Figure 20, a low magnification micrograph that includes the olfactory epithelium and part of the lamina propria. These small cells, whose nuclei are very close to the basement membrane, are in their "resting" state. Their cytoplasm is limited in volume, and no part of the cell communicates with the epithelial surface. When basal cells differentiate to form ciliated olfactory receptor cells, they seem to send axons toward the brain before the apical pole of the cell grows a dendrite. This is apparent in Figure 22 of Moran et al. (1982a), in which an axon extending from a basal cell is seen to join a bundle of other thin axons of olfactory receptor cells. The fine structural data suggest the growing axon finds its way to the brain by inserting itself into an existing bundle of olfactory receptor axons and following their trajectory. How the axon is then led to make the appropriate connection with the "right" mitral cell, the second-order neuron in the olfactory bulb with which it synapses, is a fundamental problem in neurobiology that awaits resolution.

G. The Lamina Propria

The olfactory mucosa, which consists of the olfactory epithelium, basement membrane, and lamina propria, is not underlain by a true submucosa. The lamina propria is the layer of connective tissue that supports the olfactory epithelium and binds it to the underlying bone or cartilage. Illustrated in Figure 20, the lamina propria is highly cellular, and contains fibroblasts, fibrocytes, mast cells, and cells of the immune series ready to defend this sensitive region, so close to the brain, from foreign antigens presented to the nose. In addition, it is highly vascular, and supplies all the gases and nutrients needed by the olfactory epithelium, which, like all epithelia, has no microcirculation of its own. Amongst the most conspicuous elements in the lamina propria are the Bowman's glands, which are large compound tubuloalveolar secretory glands. These glands, which appear to be more serous than mucous, have ducts that penetrate the olfactory epithelium and open out onto the epithelial surface. The precise products of the Bowman's glands in humans have not been well documented to date. The ultrastructure of the secretory cells suggests

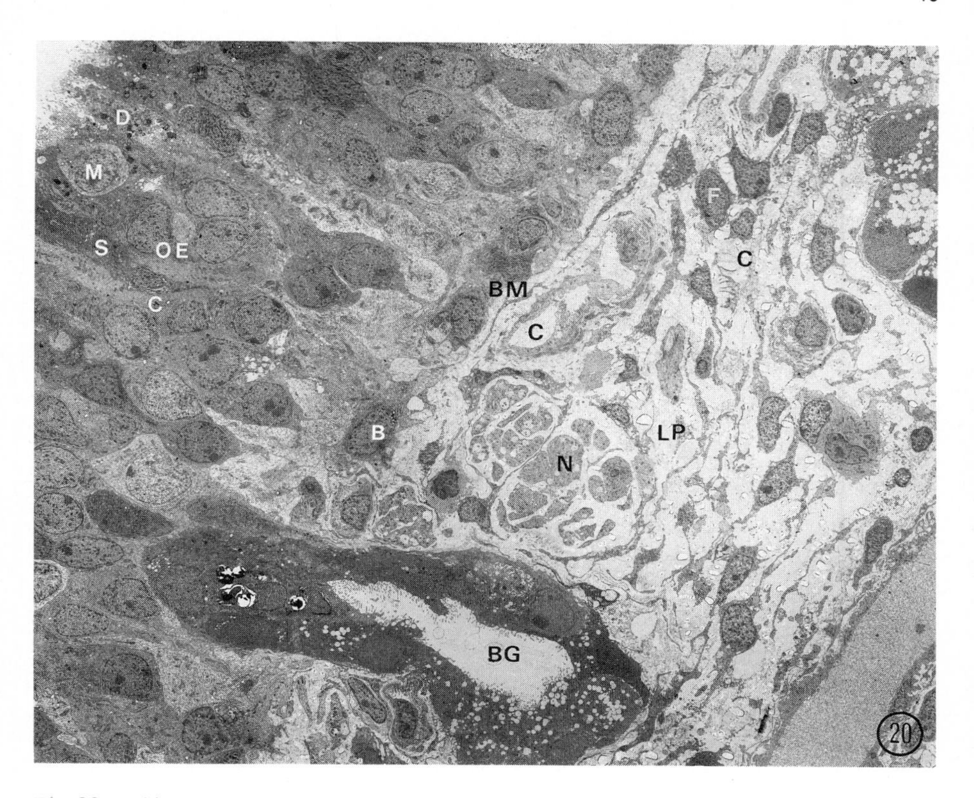

Fig 20. This extremely low magnification electron image shows the olfactory epithelium (OE), basement membrane (BM), and lamina propria (LP). A microvillar cell (M), supporting cells (S), ciliated olfactory receptor cells (C white), basal cells (B), and a degenerating cell (D) are visible in the olfactory epithelium. The loose connective tissue of the lamina propria contains fibroblasts (F), capillaries (C black), axons bundles from olfactory receptor cells (N), Bowman's glands (BG), and collagen fibrils (C). 1,400X

they are specialized for the synthesis and secretion of proteins and/or glycoproteins for export. The presence of secretory vesicles at the apical poles of the cells is reminiscent ofthe secretory cells of the exocrine pancreas and those of the parotid gland, both of which synthesize, package, and release enzymes for use outside of the cell. Like the supporting cells, the Bowman's glands seem likely candidates for the site of synthesis of proteins, such as the recently described odorant-binding proteins, found in the surface mucus. In addition, many nerves pass through the lamina propria, and most of these are axons of ciliated olfactory receptor cells. These tiny axons, often less than 0.2 μm in diameter, form conspicuous bundles such as those evident in Figure 20. These axon bundles are traveling toward the cribriform plate, through whose perforations they will pass as fila olfactoria that, taken together, form Cranial Nerve I, the olfactory nerve.

1V. COMPARATIVE ANATOMY OF MICROVILLAR CELLS FROM DIFFERENT CHEMOSENSORY NEUROEPITHELIA

As described earlier in this Chapter, the function of the microvillar cells in the human olfactory epithelium is unknown. In this section, the question; can functional inferences be drawn by comparing their location and fine structure to those of microvillar cells in other vertebrate chemosensory neuroepithelia?, is addressed. The following discussion concentrates on the comparative anatomy of four cell types (Figs 21-24): 1) microvillar cells in the human olfactory epithelium (Fig 21); 2) microvillar cells in the olfactory epithelium of the rat (Fig 22); 3) microvillar receptor cells in the vomeronasal organ of the rat (Fig 23); and 4) microvillar receptor cells in the olfactory epithelium of the Brown trout (Fig 24).

The microvillar cells in human and rat olfactory epithelia are shown side-by-side in Figures 21 and 22 respectively. They are morphologically similar in some respects, and different in others. In both species, the cells have nuclei located near the epithelial surface, and their apical poles bear short, straight microvilli that project into the surface mucus. The human microvillar cell is flask-shaped, and has a clear cytoplasm; that of the rat is more fusiform, and has an electron-dense cytoplasm. The human microvillar cell has only a few mitochondria, and its cell apex is narrow. The microvillar cell of the rat is filled with mitochondria, and its cell apex is broad and flat. In recent experiments (Rowley et al. 1989), the neurobiological tracer macromolecule horseradish peroxidase (HRP) was injected into the rat olfactory bulb. Subsequent examination of the olfactory epithelium, reacted with diaminobenzidine (DAB) to reveal the presence of HRP in the electron microscope, showed HRP to be present in both the ciliated olfactory receptor cells and the microvillar cells. These data suggest that the microvillar cells, in addition to the ciliated olfactory receptor cells, send axons to the olfactory bulbs in the brain. Whether these results can be extrapolated to human microvillar cells is an open question and awaits further investigation.

In addition to the olfactory epithelium, many mammals have a vomeronasal organ which is sensitive to odorants and rich in chemosensory bipolar neurons that send axons to the accessory olfactory bulb in the brain (Wysocki 1979). Figure 23 illustrates a vomeronasal receptor cell of the rat. Here, the tip of the dendrite of the bipolar chemosensory neuron is shown, and it is seen to be long, thin, and rich in mitochondria near the site of stimulus reception. A cluster of centrioles sits just beneath the cell surface. At the cell surface, a number of slender microvilli project out into the vomeronasal cavity. Thin portions of supporting cells surround the receptor cell dendrites. These microvillar receptor cells bear a strong resemblance to microvillar olfactory receptor cells of teleost fishes, shown by juxtaposing a micrograph of vomeronasal receptor cells of the rat (Fig 23) with that of an olfactory microvillar receptor cell of the trout (Fig 24). The trout has two known types of olfactory receptor cells, ciliated olfactory receptor cells and microvillar receptor cells (Moran, Rowley and Jafek, in press). The microvillar receptor cells, such as the one depicted in Figure 24, have long, thin dendrites, filled with mitochondria, whose tips present a cluster of microvilli to the site of stimulus reception at the epithelial surface. As in vomeronasal receptor cells, centrioles are often seen near the dendrite tip.

What does the comparative anatomy of these cells tell us? First, that human and rat microvillar cells have some ultrastructural similarities, and some differences; they may or may not be functionally equivalent. Second, the microvillar receptor cells in rat olfactory epithelium and vomeronasal organ are ultrastructurally very different. Third, the vomeronasal receptor cells of the rat and the microvillar olfactory receptor cells of the trout are strikingly similar in ultrastructure. Finally, trout olfactory receptor cells and rat vomeronasal receptor cells, which look alike, are very different in ultrastructure from the microvillar cells in human and rat olfactory epithelia. This strongly suggests that the microvillar cells evident in the olfactory epithelia of humans do not represent "displaced" receptor cells from the vomeronasal organ, which until recently, has been considered to be vestigial.

A class of cells called brush cells, originally described by Rhodin and Dalhamn (1956) in the trachea of the rat, have been observed in the respiratory epithelia of a variety of mammals (Rowley et al. 1989). Although these cells vary in appearance in different

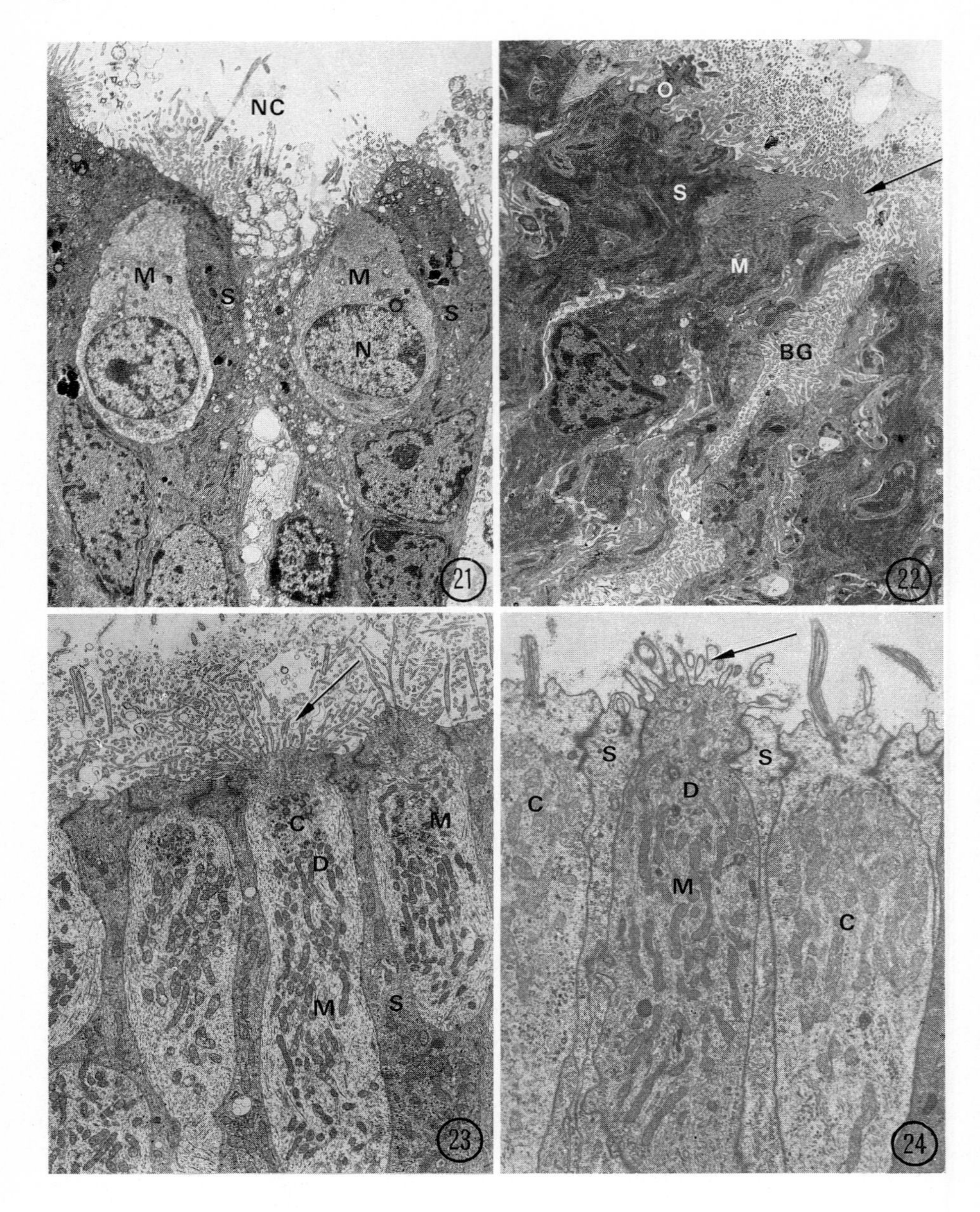
NC
M
S
M
S
N
21
O
S
M
BG
22
C
M
D
M
S
23
S
S
D
C
M
C
24

animals, they share several features in common; they are located at the epithelial surface and have an apical tuft of short straight microvilli. In the absence of functional data, it is impossible to say whether the microvillar cells in the human olfactory epithelium areequivalent to the brush cells found elsewhere. It is interesting to note, however, that human microvillar cells are very similar in fine structure to certain cells observed in the bronchioles of the monkey (Moran and Rowley 1988, Plate 11-2B). Some interesting possibilities arise when microvillar and brush cells are compared. If the microvillar cells in the rat, which seem to have axons that go to the brain, are sensory receptor cells, the human microvillar cells turn out to be sensory receptor cells, and microvillar cells in rat and human olfactory epithelia are equivalent to brush cells found elsewhere in the respiratory epithelium, then it is possible that a class of sensory cells is distributed throughout the epithelium that lines the airways of the respiratory system. Clearly the water is murky here; further research is required to determine whether or not this is the case. To confuse matters further, cells that look very much like microvillar cells have also been observed in the enteroendocrine cells of the digestive system (Junquiera et al. 1989).

Figs 21-24. These four micrographs depict different examples of cells, equipped with apical microvilli, found in chemosensory neuroepithelia of three different vertebrate animals.

Fig 21. A brace of microvillar cells (M) in the human olfactory epithelium. Note the pear-shaped cell body, clear cytoplasm, the large, round nucleus (N), and the small tuft of apical microvilli. NC, nasal cavity; S, supporting cell. 3,700X

Fig 22. Microvillar cell (M) in the olfactory epithelium of the rat. Note the flattened, microvillate cell apex (arrow), and the dense cytoplasm filled with mitochondria. BG, duct of Bowman's gland; O, olfactory vesicle; S, supporting cell. 3,200X

Fig 23. Microvillar olfactory receptor cells (M) in the chemosensory neuroepithelium of the rat vomeronasal organ. Note the apical microvilli (arrow) and numerous centrioles (C) near the tip of the long, slender dendrite (D). S, supporting cell. 3,500X

Fig 24. Microvillar olfactory receptor cells (M) in the olfactory epithelium of the Brown trout. These bipolar neurons have a long, slender dendrite (D) topped by a cluster of microvilli (arrow). C, non-sensory ciliated epithelial cell; S, supporting cell. 15,700X

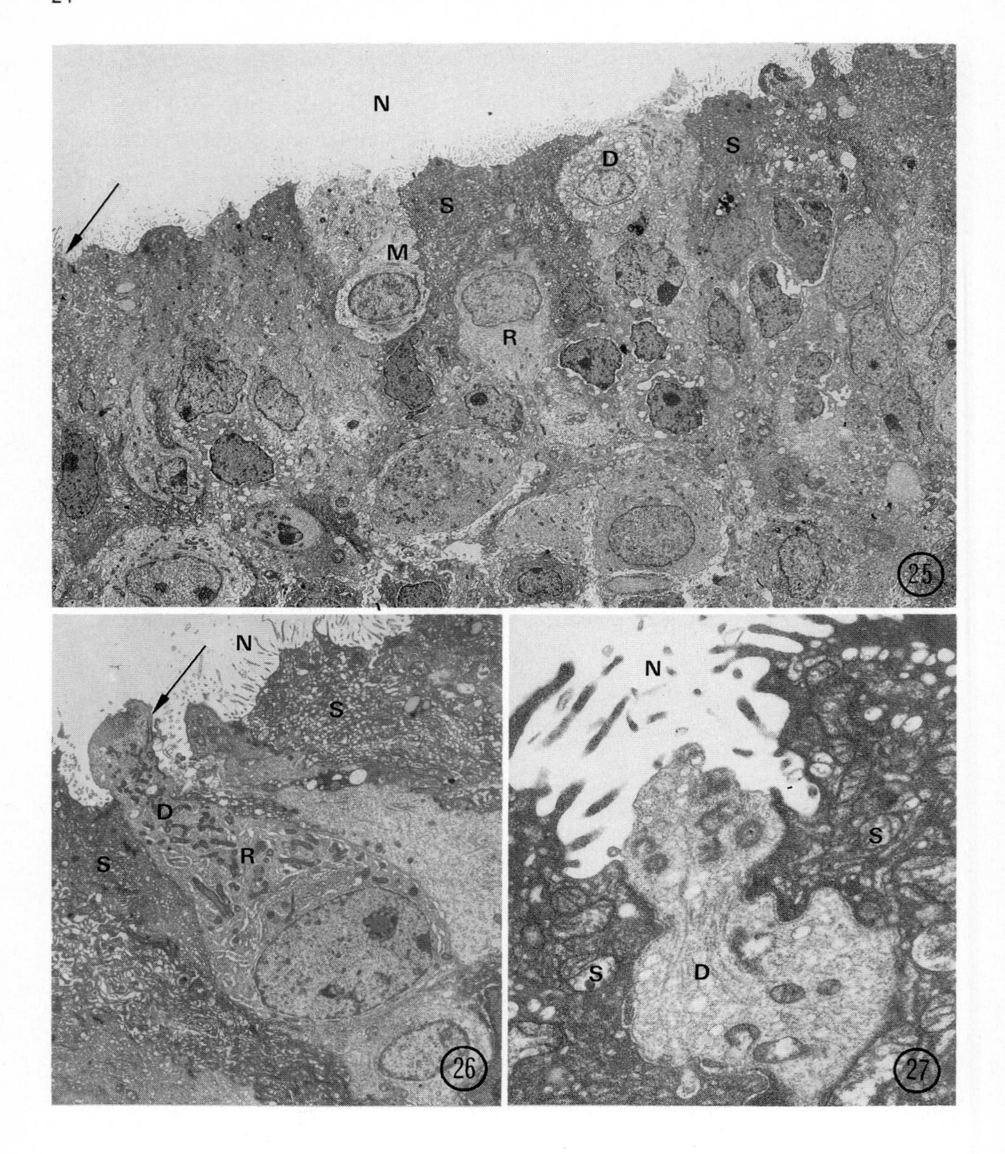
N
S
D
S
M
R
25
N
S
D
R
S
26
N
S
S
D
27

V. THE HISTOPATHOLOGY OF OLFACTORY DYSFUNCTION

A large number of patients have olfactory dysfunctions that stem from diverse causes. Since the olfactory epithelium is the only part of the olfactory system available for microscopic observation in living persons, it is important to determine if morphological alterations of the olfactory mucosa in cases of olfactory dysfunction can be detected, (see Moran, Jafek and Rowley, in press for review). Recent work with patients having post-traumatic anosmia (Moran et al. 1985, Jafek et al. 1989) and post-viral olfactory dysfunction (Jafek et al. 1990), has revealed dramatic changes in the fine structure of their olfactory epithelia that may well have functional correlates. Figures 25-27, for example, are electron micrographs of the olfactory epithelium of a 59-year old woman who had suffered complete, irrevocable loss of her sense of smell after a fall that caused head trauma. In Figure 25, a survey micrograph of the olfactory epithelium, it is obvious by inspection that the epithelium is quite disorganized. The supporting cells, olfactory receptor cells and basal cells do not display the "layering" of nuclei evident in normal tissue. The number of ciliated olfactory receptor cell bodies is dramatically decreased, and it is difficult to find any of their dendrites. Olfactory vesicles at the epithelial surface are rare, and the few that are present lack olfactory cilia. Figure 26, an electron micrograph of a lone ciliated olfactory receptor cell, shows the cell body is unusually close to the epithelial surface. The dendrite is short and thick instead of being long and thin, and its olfactory vesicle is aciliate. Figure 27 shows another dendrite tip; here the

---

Fig 25. Abnormal olfactory epithelium from a head injury patient who had experienced traumatic anosmia. The epithelium is thinner than normal, olfactory receptor cell bodies (R) are rare, and only one olfactory vesicle (arrow) is present. Normal (M) and degenerating (D) microvillar cells are present. Most of the tissue is occupied by supporting cells (S), which have an abnormally vesicular cytoplasm. N, nasal cavity. 2,650X

Fig 26. Abnormal olfactory receptor cell from the same patient depicted in Figure 25. The cell body of the receptor cell (R) is high in the epithelium, and the truncated dendrite (D) terminates in an olfactory vesicle (arrow) devoid of cilia. The adjacent supporting cells (S) present an abnormal ultrastructure as well. N, nasal cavity. 1,500X

Fig 27. This olfactory vesicle, from the same patient depicted in Figures 25-26, is typical of those observed in cases of head injury-related traumatic anosmia; it has basal bodies, but no cilia. The dendrite (D) is twisted and thickened. Note the large number of mitochondria in the apical pole of the adjacent supporting cell (S). N, nasal cavity. 18,700X

olfactory vesicle contains basal bodies, but has no cilia. What are the functional correlates of these histopathologic observations? Head trauma, long associated with olfactory dysfunction (Douek 1974), is thought to cause anosmia by sudden movement of the brain, which shears the delicate axons of the ciliated olfactory receptor cells within the thin fila olfactoria that pass through the cribriform plate (Zusho 1982). Although olfactory receptor cells are known to regenerate after axotomy (Monti-Graziadei and Graziadei 1979), it seems likely that the tiny holes in the lamina cribrosa, the cribriform plate of the ethmoid bone, scar over after the fila olfactoria are cut by a blow to the head, thus preventing new-growing axons from finding their way to the brain. The latter finding is consistent with the histopathologic observations of Chuah and Farbman (1983), who, using organ culture, found that rodent olfactory epithelium co-cultured with explants of olfactory bulb experience greater differentiation of ciliated olfactory receptor cells, more ciliogenesis of olfactory cilia, and increased levels of olfactory marker protein after synaptic contact has been established between receptor cells and the bulb. In cases of traumatic anosmia in which the sense of smell does not recover, it seems likely that basal cells attempt to differentiate into ciliated olfactory receptor cells. Their growing axons, unable to reach the brain, are stymied, and their parent cells become cases of arrested development, unable, for the most part, to elaborate dendrites and olfactory vesicles replete with functional olfactory cilia.

V1. SUMMARY

Although the ultrastructure of the human olfactory mucosa has been described, and some studies on the histopathology of olfactory dysfunction performed, much remains to be learned about the structure and function of the human peripheral olfactory system. Several important questions to be addressed are; what is the precise surface area, location and distribution of the olfactory epithelium in persons of different ages? Does the number -- and function -- of olfactory receptors change with growth, development, and aging? To what odorants are different bipolar olfactory neurons sensitive? What are the functions of the microvillar cells? Do the foramina within the cribriform plate become narrower with age, constraining the fila olfactoria as they pass to the olfactory bulbs of the brain? When the answers to these questions are in hand, we will be closer to understanding the basic biology of olfaction - one of the most primitive, important, and yet least intensely investigated of the primary human senses.

ACKNOWLEDGEMENTS

Supported by NIH Program Project Grant No. 2 PO1 NS20486 from the National Institute of Communicative Disorders and Stroke. The authors thank J. Walker for helpful discussion, and Ms. Holly Golightly for typing the manuscript.

REFERENCES

Afzelius BA, Eliasson R (1978) Flagellar mutants in man: On the heterogeneity of the immotile-cilia syndrome. J Ultrastruct Res 69: 43-52

Chuah MI, Farbman AI (1983) Olfactory bulb increases marker protein in olfactory receptor cells. J Neurosci 3: 2197-2205

Doty RL, Shaman P, Applebaum SL, Giberson R, Sikorsky L, Rosenberg L (1984a) Smell identification ability: Changes with age. Science 226: 1441-1443

Doty RL, Shaman P, Dann M (1984b) Development of the University of Pennsylvania Smell Identification Test: A standardized microencapsulated test of olfactory function. Physiol Behav 32: 489-502

Douek E (1974) The sense of smell and its abnormalities. Churchill Livingstone, Edinburgh

Gibbons IR (1967) The organization of cilia and flagella. In: Allen JM (ed) Molecular organization and biological function. Harper and Row, New York

Goss RJ (1964) Adaptive growth. Academic Press, New York

Graziadei PPC (1973) The ultrastructure of vertebrates olfactory mucosa. In: Friedmann I (ed) The ultrastructure of sensory organs. Elsevier, Oxford, p 267

Graziadei PPC, Karlin MS, Monti-Graziadei GA (1980) Neurogenesis of sensory neurons in the primate olfactory system after section of the fila olfactoria. Brain Res 186: 289-300

Hornung DE, Mozell MM (1981) Accessibility of odorant molecules to the receptors. In: Cagan RH (ed) Biochemistry of taste and olfaction. Academic Press, New York

Jafek, BW, Eller PM, Esses BA, Moran DT (1989) Post-traumatic amosmia: ultra-structural correlates. Arch Neurol 46: 300-304

Jafek BW, Hartman D, Eller PM, Johnson EW, Strahan RC, Moran DT (1990) Post-viral olfactory dysfunction. Am J Rhinol 4: 91-100.

Junquiera LC, Carniero J, Kelley RO (1989) Basic histology. Lange, Norwalk USA, p 297

Lancet D (1986) Vertebrate olfactory reception. Annu Rev Neurosci 9: 329-355

Lovell MA, Jafek BW, Moran DT, Rowley JC III (1982) Biopsy of human olfactory mucosa: an instrument and a technique. Arch Otolaryngol 108: 247-249

Monti-Graziadei GA, Graziadei PPC (1979) Neurogenesis and neuron regeneration in the olfactory system of mammals. II. Degeneration and reconstruction of the olfactory sensory neurons after axotomy. J Neurocytol 8: 197-213
Moran DT (1987) Evolutionary patterns in sensory receptors: an exercise in ultrastructural palaeontology. Ann NY Acad Sci 510: 1-8
Moran DT, Jafek BW, Rowley JC III (in press) Ultrastructural histopathology of olfactory dysfunction. J Electron Micros Tech
Moran DT, Jafek BW, Rowley JC III, Eller PM (1985) Electron microscopy of olfactory epithelia in two patients with anosmia. Arch Otolaryngol 111: 122-126
Moran DT, Rowley JC III (1987) Biological specimen preparation for correlative light and electron microscopy. In: Hayat MA (ed) Correlative microscopy. Academic Press, New York, p 1
Moran DT, Rowley JC III (1988) Visual histology. Lea and Febiger, Philadelphia
Moran DT, Rowley JC III, Jafek BW (1982b) Electron microscopy of human olfactory epithelium reveals a new cell type: the microvillar cell. Brain Res 253: 39-46
Moran DT, Rowley JC III, Jafek BW (in press) The ultrastructural neurobiology of the olfactory mucosa of the Brown trout (Salmo trutta trutta). J Electron Micros Tech
Moran DT, Rowley JC III, Jafek BW, Lovell MA (1982a) The fine structure of the olfactory mucosa in man. J Neurocytol 11: 721-746
Morrison EE, Costanzo RM (1990) Morphology of the human olfactory epithelium. J Comp Neurol 297: 1-13
Nakashima T, Kimmelman CP, Snow JB Jr (1984) Structure of human fetal and adult olfactory neuroepithelium. Arch Otolaryngol (Stockh) 110: 641-646
Reese TS (1965) Olfactory cilia in frog. J Cell Biol 25: 209-230
Rhodin J, Dalhamn T (1956) Electron microscopy of the tracheal ciliated mucosa in the rat. Z Zellforsch Mikrosk Anat 44: 345-412
Rowley JC, Moran DT, Jafek BW (1989) Peroxidase backfills suggest the mammalian olfactory epithelium contains a second morphologically distinct class of bipolar sensory neuron: the microvillar cell. Brain Res 502: 387-400
Schiffman SS (1979) Changes in taste and smell with age. In: Ordy LSM, Briggs K (eds) Sensory systems and communication in the elderly. Raven Press, New York
Shipley MT (1985) Transport of molecules from nose to brain: transneuronal anterograde and retrograde labelling in the rat olfactory system by wheat germ agglutinin horse-radish to the nasal epithelium. Brain Res Bull 15: 129-142
Wysocki CJ (1979) Neurobehavioral evidence for the involvement of the vomeronasal system in mammalian reproduction. Neurosci Behav Rev 3: 301-341
Zusho H (1982) Post-traumatic anosmia. Arch Otolaryngol 108: 90-92

2

ANATOMY OF THE HUMAN OLFACTORY BULB AND CENTRAL OLFACTORY PATHWAYS

M SHIPLEY

P REYES

1. INTRODUCTION

The olfactory system subserves important functions, including the perception of flavor and smell. Although olfaction has traditionally received limited attention in both normal and pathological conditions in humans, this sensory system is now of considerable interest to neurobiologists and medical practitioners.This interest has been brought about, in part, by neuropathological investigations that described the presence of histological lesions in olfactory-related structures in Alzheimer's disease (Ferreyra-Moyano 1989, Reyes et al. 1987, Pearson et al. 1985, Esiri and Wilcock 1984), a condition clinically characterized by progressive intellectual decline and behavioral abnormalities. Importantly, olfactory deficits have been demonstrated not only in patients with Alzheimer's disease (Doty et al. 1987), but in ones with Parkinson's disease (Doty et al. 1988) and schizophrenia (Hurwitz et al. 1988). These new findings suggest that the olfactory system is compromised in a number of neurodegenerative disorders. Unfortunately, our knowledge of the anatomy and physiology of the olfactory system in humans is limited and based largely on rodent and non-human primate studies. Since significant anatomical and functional variations occur between species, future work on human olfactory pathways are sorely needed.

The olfactory system can be divided into a peripheral and central component. The former is extracranial, represented by the olfactory epithelium within the nasal cavity, whereas the latter is intracranial and includes the olfactory bulb and various cortical and subcortical brain regions which have both direct and indirect connections with olfactory bulb neurons. Such anatomical relationships provide a direct link between the central nervous system and the organism's external environment and set up the possibility for entry of pathogens and neurotoxic chemicals into the brain (McLean et al. 1989). Since the anatomy of the peripheral olfactory system has been described in detail in the previous Chapter no further discussion of this region is given here.

II. CENTRAL OLFACTORY COMPONENTS:

A. Olfactory Bulb

Our knowledge of central olfactory anatomy derives almost entirely from experimental studies; the vast majority of these studies have been performed in laboratory animal species, primarily the rat, hamster and rabbit. Thus, the following summary can only be regarded as a provisional sketch of olfactory circuits in the human brain. Given that the major central olfactory circuits are similar across infraprimate species, the general plan of central olfactory circuits reviewed below is likely to hold for the human brain, but experimental studies in primates are badly needed to help bridge the gap between macrosmatic laboratory species and humans.

Second order olfactory neurons are located in the olfactory bulb, an oval structure that lies on the ventral surface of each frontal lobe and dorsal to the cribriform plate of the ethmoid bone. The olfactory bulb is made up of neurons, afferent and efferent fibers, astrocytes, microglia and blood vessels enveloped by a thin layer of pia-arachnoid cells (Barr and Keirnan 1983, Leopold 1986). These cellular elements are arranged in layers, namely the olfactory nerve layer, the glomerular layer, the external plexiform layer, the mitral cell layer, the internal plexiform layer, which in its deepest portion contains myelinated fibers and the granule cell layer (Fig 1). Within the core of the olfactory bulb are also found small clusters of ependymal cells which represent vestiges of the extension of the lateral ventricles into the bulb during embryonic development (Barr and Kiernan 1983).

Although these layers are quite distinct in macrosmatic species and in fetal human olfactory bulbs (Barr and Kiernan 1983), they are less well demarcated in adult humans

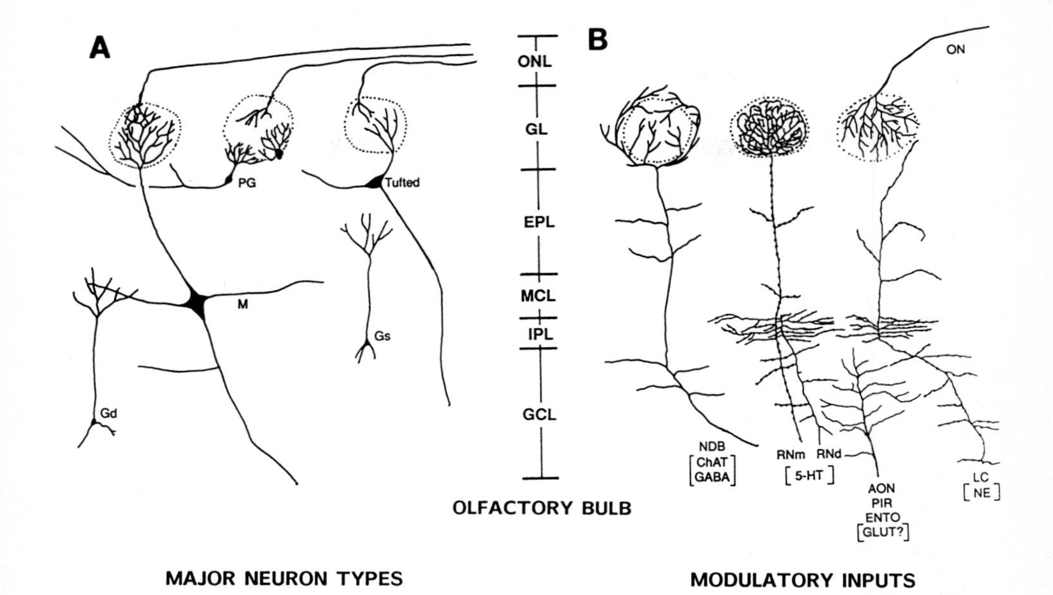

Fig 1 A. Major neuronal types of the olfactory bulb. B. Major afferent inputs to the bulb and their neurotransmitters. Abbreviations: ChAT, choline acetyltransferase; RNd & RNm, dorsal and median raphe nuclei; ENTO, entorhinal cortex; EPL, external plexiform layer; Gd and Gs, Granule cells; GCL, granule cell layer; GLUT, glutamate; GL, glomerular layer; IPL, internal plexiform layer; LC, locus coeruleus; M, mitral cell; MCL, mitral cell layer; NDB, nucleus of the diagonal band; NE, norepinephrine; ON, olfactory nerve axon; ONL, olfactory nerve layer; PIR, piriform cortex; 5-HT, serotonin.

(Leopold 1986) (Fig 2). Normal aging is accompanied by atrophy of the olfactory bulb, intrabulbar neuronal loss, and gliosis (Reyes et al. unpublished observations). Furthermore, in certain pathological conditions such as Alzheimer's disease, neuropathological alterations have been demonstrated including neurofibrillary tangles in olfactory bulb neurons and neuritic plaques in the neuropil (Esiri and Wilcock 1984).

There are two major neuronal groups in the olfactory bulb, the mitral and tufted cells, which are second-order neurons, and the interneurons, represented by the juxtaglomerular and granule cells (Fig 1A). The axons of olfactory receptor cells terminate exclusively in the glomeruli. The latter structures are formed by the terminal arbors of olfactory nerve fibers, and are surrounded by several classes of juxtaglomerular neurons, including external tufted cells, short axon cells and the periglomerular cells which receive synapses from olfactory nerve terminals. The glomeruli are also richly invested by ramifications of the apical dendrite of mitral and tufted cells, and these dendrites are heavily targeted by olfactory nerve synaptic endings (Pinching 1970, Pinching and Powell 1971a, Pinching and Powell 1971b). Therefore, each glomerulus is comprised of axon terminals of several thousand olfactory cells, the dendritic arbors of mitral, tufted and periglomerular cells and other juxtaglomerular neurons, and synaptic interactions among these elements are extensive (Pinching 1970). The majority of juxtaglomerular cells are periglomerular cells which are GABAergic (GABA: Gamma amino butyric acid), and small external tufted cells many of which are dopaminergic (DA). The GABAergic periglomerular cells, thus, are inhibitory interneurons, and the periglomerular neurons are presynaptic to mitral and tufted cells. The synaptic action of dopamine in the glomeruli is not known, although recent unpublished studies from this laboratory indicate that DA receptors are primarily located on the terminals of olfactory nerve axons and primary olfactory neurons express the mRNA for D-2 receptors. Thus, the predominant action of juxtaglomerular DA neurons is probably inhibitory to olfactory nerve terminals. In addition, axons of some juxtaglomerular neurons enter the external plexiform layer and project to nearby glomeruli where they synapse.

Deeper in the bulb, below the layer of mitral cells, is the second major class of interneurons, the granule cells (Fig 1A). These neurons make extensive dendrodendritic synapses with the lateral or secondary dendrites of mitral and tufted cells in the external plexiform layer (EPL). Axons of the mitral and tufted cells constitute the primary output pathway of the olfactory bulb (Price and Powell 1970, Haberly and Price 1977, Scott et al. 1980). Thus, interneurons of the olfactory bulb influence bulbar output via excitatory or inhibitory modulation of mitral and tufted cells. This modulation is organized into two distinct and largely separate levels.

The mitral and tufted cells have a single apical dendrite that extends from the cell body through the EPL to a single glomerulus. This apical dendrite does not branch in the EPL but ramifies extensively in the glomerulus. The mitral and tufted cells also have four to six lateral or 2° dendrites (Mori et al. 1983, Kishi et al. 1984, Orona et al. 1984) which

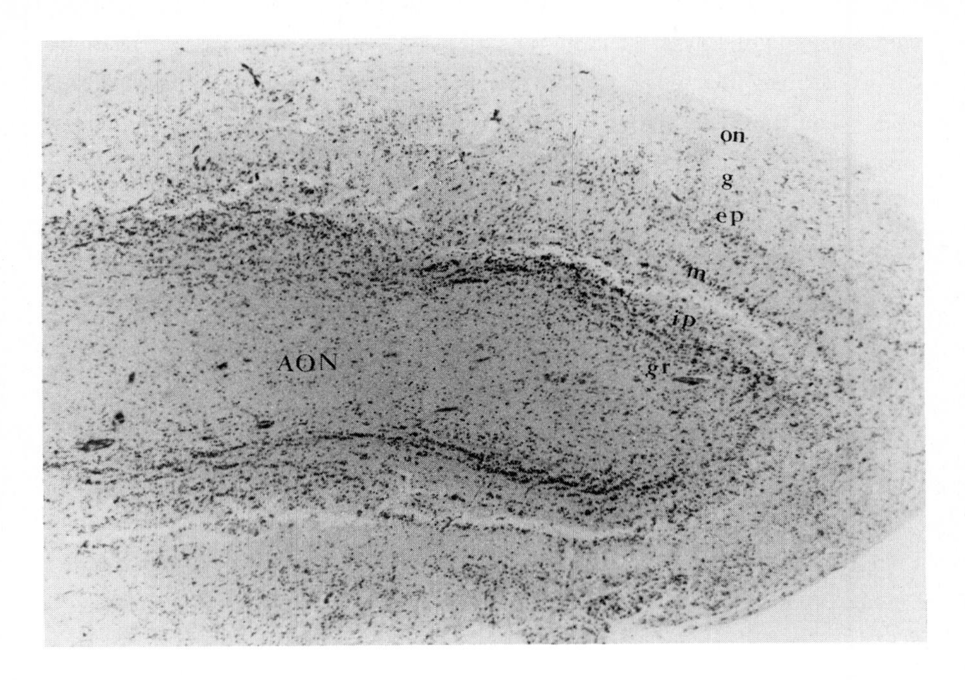

Fig 2. Low power view of human olfactory bulb (4x, cresyl violet-luxol fast blue stain). Approximate layers: on, olfactory nerve layer; g, glomerular layer; ep, external plexiform layer; m, mitral cell layer; ip, internal plexiform layer; gr, granule cell layer; AON, distal segment of anterior olfactory nucleus.

extend for considerable distances in the EPL where they may branch repeatedly. In the rodent, for example, they can extend up to one-half the circumference of the bulb. Because these two classes of dendrites ramify in two distinct layers of the bulb, they are contacted by different interneural elements and thus, may be independently regulated by different sources of neural activity. As a result, the input-output functions of the bulb requires a consideration of the influences operating on the two classes of mitral and tufted cell dendrites.

The ramified, terminal portion of the apical dendrite is synaptically contacted by olfactory nerve terminals and, via dendrodendritic synapses, with the dendrites of periglomerular and possibly other (short axon cells) juxtaglomerular neurons (Pinching 1970, Pinching and Powell 1971a, Pinching and Powell 1971b) (Fig 1A). In addition, these apical dendrites may be directly or indirectly influenced by so-called centrifugal afferents, the axons of neurons from other parts of the brain. Cholinergic axons from the basal forebrain terminate in the glomeruli (Macrides et al. 1981, Nickell and Shipley 1988a,b) as does one class of serotonergic axons from the midbrain raphe (McLean and Shipley 1987) (Fig 1B). Thus, the apical dendrites of mitral and tufted cells are influenced by olfactory nerve terminals, interneurons (juxtaglomerular cells) and centrifugal modulatory systems including cholinergic and serotonergic fibers. The predominant synaptic input, however, is from olfactory nerve terminals. Thus, the apical dendrite of mitral and tufted cells may be considered the "sensory dendrite".

By contrast, the lateral or secondary dendrites of mitral and tufted cells ramify exclusively in the EPL of the bulb. These dendrites are not influenced by olfactory nerve terminals, juxtaglomerular cells or centrifugal afferents that terminate in the glomeruli, but rather by neural elements in deeper layers of the bulb. Chief among these elements are the dendrites of the granule cells whose cell bodies are located deep to the mitral cell (MCL) and internal plexiform layers (IPL) in the granule cell layer (GCL). The apical dendrites of granule cells pass through the IPL and MCL, then ramify and contact the lateral dendrites of mitral and tufted cells to form dendrodendritic synapses (Rall et al. 1966, Price and Powell 1970). The mitral cell dendrite is presynaptic to granule cell dendrites, and granule cell dendrites are presynaptic to mitral and tufted cell dendrites. Frequently these synapses are side by side forming the classical "reciprocal synapse", with the mitral cell dendrite synapsing on the granule cell dendrite, which in turn, synapses back onto the mitral cell dendrite. Based on morphological criteria and electrophysiological studies, it appears that the mitral cell dendrite is excitatory to the granule cell dendrite, which in turn, is inhibitory to the mitral cell dendrite (Rall et al. 1966, Rall and Shepherd 1968, Price and Powell 1970, Rall 1972, Ribak et al. 1977). These dendrodendritic synapses may function to provide lateral inhibition or a temporal filter in the processing of olfactory information from the primary olfactory neuron to the output of mitral and tufted cells (Rall et al. 1966, Rall and Shepherd 1968, Rall 1972).

In addition to dendrodendritic synapses between mitral and tufted cells and granule cells, the mitral and tufted cell dendrites may also be influenced by other axons that terminate in the EPL, including axons from higher olfactory cortical areas and subcortical structures

(Davis and Macrides 1981, Luskin and Price 1983), although it is not known whether mitral cell dendrites in the EPL are contacted directly or indirectly via centrifugal synapses onto granule cell dendrites or onto other interneurons scattered in the EPL.

The predominant synaptic input to the lateral dendrites of mitral and tufted cells, however, is from the dendrites of granule cells. Thus fiber systems that synapse on granule cells are the major source of indirect modulation of the lateral dendrites of mitral and tufted cells. The reciprocal synapse between the dendrites of the latter two cell types and granule cells has been much emphasized because it is such an unusual kind of synaptic organization and because it seems to provide a mechanism for "self or lateral inhibition", i.e., by synapsing on the dendrite of a granule cell, the mitral cell causes the release of the neurotransmitter GABA back onto itself, providing a kind of self-inhibition (Rall et al. 1966, Rall and Shepherd 1968, Rall 1972, Shepherd 1972). However, not all mitral to granule or granule to mitral synapses are reciprocal (Price and Powell 1970), and this may indicate that another function of this dendrodendritic coupling may be to allow mitral and tufted cells to inhibit other mitral and tufted cells providing lateral inhibition. Which of these two modes predominates during the normal function of the olfactory bulb is not known.

Clearly, however, any input which preferentially targets the granule cells of the bulb will also potently modulate the lateral dendrites of the mitral and tufted cells. This modulation should be inhibitory to mitral and tufted cells if the granule cell is induced to release its transmitter, GABA. The granule cells are heavily targeted by synaptic inputs from ipsi- and contralateral olfactory cortical structures (Davis and Macrides 1981, Luskin and Price 1983). Physiological studies indicate that synapses on granule cells from these olfactory regions are excitatory, causing the granule cells to release GABA, thus inhibiting mitral cells (Kerr and Hagbarth 1955, von Baumgarten et al. 1962, Yamamoto et al. 1963, Mori and Takagi 1978). In addition to these "associational inputs", the granule cells appear to be heavily, and selectively targeted by norepinephrine (NE) containing terminals from the locus coeruleus (LC) (Shipley et al. 1985, McLean et al. 1989 (Fig 1B). Cholinergic (Macrides et al. 1981, Nickell and Shipley 1988a,b) and serotonergic fibers (McLean and Shipley 1987) also terminate in the EPL and GCL and may directly or indirectly modulate the lateral dendrites of mitral and tufted cells (Fig 1B). Thus, diverse centrifugal afferents terminate upon granule cells and presumably influence mitral/tufted cell activity via granule to mitral or tufted cell synapses. Granule cell regulation of currents in mitral and tufted cell dendrites may also be facilitated by tight coupling among granule cells as recent studies have shown that granule cell somata are coupled by gap junctions (Reyher et al. in press). Thus, centrifugal inputs to the granule cell layer may evoke coordinated activation of gap junction-coupled clusters of granule cells, which could "amplify" granule cell inhibition of mitral and tufted cell lateral dendrites.

It is clear from the foregoing discussion that different inputs to the bulb selectively modulate the apical or the lateral dendrites of mitral and tufted cells. The apical dendrite is influenced by olfactory nerve terminals, periglomerular cells and centrifugal inputs from the raphe (serotonin) and basal forebrain (cholinergic), whereas the lateral dendrites of mitral and tufted cells are modulated primarily by the dendrites of granule cells. Major centrifugal inputs from ipsi- and contralateral olfactory cortical areas and the subcortical modulatory projection from the locus coeruleus primarily influence granule cells, and hence exert their major influence on the lateral dendrites of mitral and tufted cells.

The functional circuitry of the olfactory bulb, may be usefully considered from the perspective of mitral and tufted cells. These neurons receive direct sensory input from primary olfactory neurons and "relay" an output to higher olfactory structures. Mitral and tufted cells are modulated not only by sensory signals arriving via the olfactory nerve, but also by centrifugal fibers acting via two classes of interneurons, the juxtaglomerular and granule cells, and possibly by direct centrifugal inputs and/or centrifugal inputs to interneurons in the EPL. Thus, functional inputs to mitral and tufted cells are segregated so as to selectively modulate either the apical or lateral dendrites of these principal neurons. Inputs acting on the apical dendrites would presumably have the most direct influence on the "sensory throughput" of mitral and tufted cells. If GABA-containing periglomerular cells are activated by primary olfactory nerve terminals, by other periglomerular cells, or by 5-hydroxytryptamine (5-HT) or acetylcholine (ACh)-containing fibers (Fig 1B), then "sensory signals" from the apical or "sensory dendrite" will have less impact on mitral and tufted cell soma, and accordingly less impact on the output targets of the olfactory bulb. By contrast, inputs that selectively target granule cells would tend to increase or decrease the amount of GABA released on the lateral dendrites of mitral and tufted cells. Since the majority of ipsi- and contralateral centrifugal inputs appear to "excite" granule cells, they will tend to increase inhibition on the lateral dendrites. Thus, whether mitral or tufted cells discharge and convey signals to higher olfactory structures, depends on the moment-to-moment influences on their two classes of dendrites, and these influences are anatomically independent.

A critical issue for understanding the nature of sensory processing in the olfactory bulb, therefore, is the degree of coupling between the apical and lateral dendrites of mitral and tufted cells. Since the lateral dendrites of these cells may extend for several hundred microns, it is important to know if inhibition caused by granule cells on the distal portions of mitral and tufted cell lateral dendrites appreciably modifies spike generation at the axon hillock. The foregoing discussion has implicitly assumed that granule cells interact exclusively with the lateral dendrites of mitral and tufted cells in the EPL. However, the apical dendrite passes through the EPL en route to the glomeruli. Although the apical dendrite does not branch in the EPL, it could be contacted by synapses from granule cell dendrites or by synapses from other interneurons in the EPL, or by synapses of centrifugal fibers in the EPL. This is an important issue because if the apical dendrite is immune from synaptic influences in the EPL, its "message" to the soma/spike generation site may be considered purely a function of sensory input and juxtaglomerular modulation. Recent studies in the rat olfactory bulb indicate that the apical dendrite receives few, if any, synapses in the EPL; and its synaptic inputs are almost entirely confined to the branches of the dendrite that ramify in the glomerular layer. Preliminary estimates indicate that the lateral dendrites receive 150-200 times as many synapses per unit of dendrite surface in the EPL as does the pre-glomerular shaft of the apical dendrite (Shipley and Zahm, in press). Further, the shaft of the apical dendrite appears to be ensheathed along much of its entire length by filets of glial processes giving the appearance of a myelin-free, glial insulation. There is evidence that the apical dendrites of some mitral cells are myelinated in primates (Pinching 1970). While these observations need to be confirmed in other species including humans, it is likely, given the generally conserved cytological organization of the olfactory bulb across vertebrate species, that this striking degree of dendritic specialization is a general feature of olfactory bulb organization. Thus it appears that the apical and lateral dendrites of mitral and tufted cells are, at least in terms of their synaptic inputs, largely independent. This anatomical independence may have important functional implications.

III. ANTERIOR OLFACTORY NUCLEUS

The olfactory bulb is connected to the base of the temporal lobe by a stalk of tissue called the olfactory peduncle. In most infra-primate species the peduncle consists of a population of neurons, termed the anterior olfactory nucleus (AON), and two major tracts of fibers, the lateral olfactory tract (LOT) and the rostral limb of the anterior commissure (Fig 3). Though historically referred to as a "nucleus", the AON is now considered a cortical structure with several subdivisions distinguished on the basis of cellular architecture and connectional patterns (de Olmos et al. 1978, Haberly and Price 1978b). Neither the architecture nor the connections of the primate AON has been well characterized.

A substantial number of AON neurons are pyramidal cells whose apical dendrite extends towards the pial surfaces of the AON (Haberly and Price 1978b) where they are synaptically contacted by mitral and tufted cell axons. The core of the AON is white matter, analogous to the white matter of the cerebral cortex (Haberly and Price 1978a,b). The axons of AON neurons collect in this white matter to project either rostrally to the ipsilateral bulb or caudally where they join the anterior commissure to cross the midline and terminate in the contralateral AON and/or bulb. In addition, some AON axons project caudally to terminate in the superficial layer of the ipsilateral primary olfactory cortex, and a few AON axons, after crossing the midline, terminate in the rostral parts of the contralateral primary olfactory cortex.

The projections of the AON to the olfactory cortex terminate in a laminar arrangement complimentary to the terminal projection of mitral and tufted cells from the bulb. The axons of the latter cells in the lateral olfactory tract (LOT) synapse in the superficial half of layer I (layer Ia) of the AON and continue on to terminate in layer Ia of the primary olfactory cortex (Price 1973, de Olmos et al. 1978). Axons of AON neurons also project to the primary olfactory cortex where they terminate immediately subjacent to the LOT

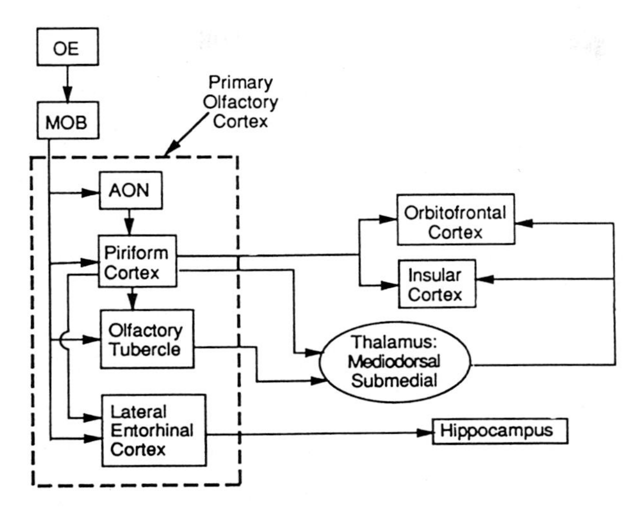

Fig 3. Major efferent connections of the olfactory system. Regions within the thick dashed lines comprise the primary olfactory cortex. Cortical regions are indicated with rectangles and subcortical regions with ovals. Abbreviations: AON, anterior olfactory nucleus; MOB, main olfactory bulb; OE, olfactory epithelium.

fibers in the deep half of layer I (layer Ib) (Price 1973, Haberly and Price 1978a,b, Davis and Macrides 1981).

In primates there are fewer AON neurons surrounding the white matter that connects the bulb to the base of the temporal lobe. This white matter, which in gross anatomical terms is referred to as the LOT, is probably a combination of the LOT and the rostral limb of the anterior commissure. In the primate brain, the AON appears to have been subjected to a longitudinal stretch with the result that part of the AON lies in the olfactory bulb itself, deep to the granule cell layer. The AON is then interrupted by a stretch of the aforementioned white matter, and reappears at the junction of the LOT and the base of the temporal lobe. There have been few tract tracing studies of the primate olfactory pathways, so the present account of the olfactory peduncle and organization of the AON, LOT and the anterior commissure is based largely on analogy with rodents and may be modified by future experimental studies in primates.

The major connections of the AON are with the ipsi- and contralateral bulbs, the contralateral AON, and the ipsilateral primary olfactory cortex (Figs 3 and 4). Two major sources of afferents to the AON arise in i) the ipsilateral bulb, and ii) the contralateral AON (de Olmos et al. 1978, Haberly and Price 1978a,b, Davis and Macrides 1981). In addition, the AON receives afferents from more caudal primary olfactory cortical structures and from a variety of cortical and subcortical structures associated with the limbic system (Haberly and Price 1978b, Davis and Macrides 1981). These afferents have been characterized in greatest detail in the rat by Haberly and Price (1978a,b), and are exceedingly complex. Different subdivisions of the AON receive selective or preferentially heavy inputs from the anterior and posterior parts of the piriform cortex, the lateral entorhinal cortex, subiculum, olfactory tubercle and tenia tecta, as well as associational projections from other subdivisions of the ipsi- and contralateral AON. Current understanding of these complex patterns of connections has been summarized by Haberly and Price (1978b). In addition to these major cortical inputs, the AON is also innervated by subcortical modulatory systems that include cholinergic inputs from the basal forebrain, serotonergic inputs from the raphe nuclei, and noradrenergic inputs from the locus coeruleus. These modulatory inputs to the AON, however, have not been systematically characterized in any species. Other possible inputs to the AON from the cortical amygdaloid complex, certain hypothalamic nuclei and the medial septal complex have been reported in various species, but remain to be characterized with more rigor.

The major connections of the AON are with the ipsi- and contralateral bulb, the contralateral AON and the ipsilateral piriform cortex. These connections are discussed in more detail below because they comprise important linkages between intrabulbar and interbulbar association circuits. The pars externa system, one subdivision of the AON, will be discussed first, because this component of the AON seems to be a major linkage between a set of circuits that have a degree of topographical organization that is unusual in the olfactory system.

One of the most puzzling but consistent observations about the anatomical organization of the olfactory system, is the lack of point-to-point spatial topography in the pathways from the epithelium to the bulb and from the bulb to other olfactory cortical structures. This lack of topographical organization is in marked contrast to other sensory systems for

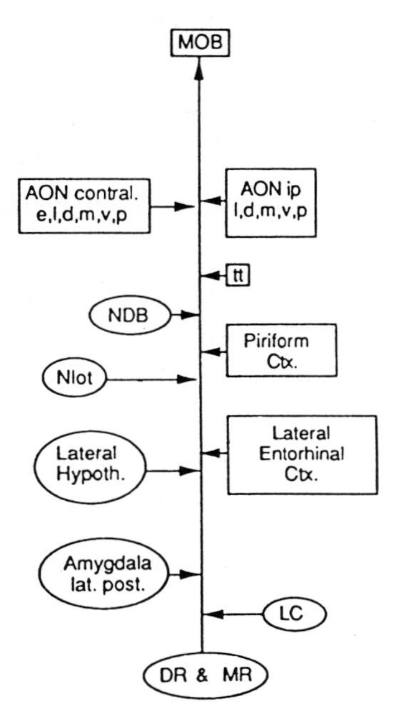

Fig 4. Centrifugal inputs to the olfactory bulb. Abbreviations: AON contral e,l,d,m,v,p, contralateral anterior olfactory nucleus-pars externa, lateralis, dorsalis, medialis, ventralis, posterior: AON ip, ipsilateral AON; DR, dorsal raphe nucleus; LC, locus coeruleus; MR, median raphe nucleus; NDB, nucleus of the diagonal band; Nlot, nucleus of the lateral olfactory tract; tt, tenia tecta.

which there is typically a precise point-to-point mapping of the receptor surface across several synaptically linked stages of the sensory pathway. This "diffuse" organization has been presumed to mean that the neural coding of "odor space" does not depend in any straightforward way on an orderly spatial arrangement of olfactory receptor neurons in the epithelium, or upon their orderly mapping into the glomerular layer of the olfactory bulb. This may be a premature conclusion as the mapping of the epithelium to the bulb has been suggested by several studies to consist of both diffuse and focal components, and the focal components may represent some degree of topography (Adamek et al. 1986, Land 1973, Land and Shepherd 1974, Stewart 1985). The resolving power of current anatomical tracing methods, therefore, may obscure a more subtle but, nonetheless, precisely organized spatial mapping of primary olfactory neurons into the bulb.

Recent anatomical studies have also demonstrated a previously unsuspected degree of point-to-point topography for some inter- and intrabulbar circuits. Macrides and co-workers have shown that a subpopulation of external tufted cells send axons to the opposite side of the same bulb (Schoenfeld et al. 1985). This intrabulbar association system is discretely organized, with neighboring tufted cells projecting to adjacent focal sites on the opposite side of the same bulb. The organization of this intrabulbar association pathway is reminiscent of the kind of point-to-point topography in other sensory pathways. The functional significance of this topographically organized intrabulbar system is unclear, but it might represent a neural substrate that could further maintain the "focal-diffuse" organization reported by others in the primary olfactory projection from the epithelium to the glomerular layer (Adamek et al. 1986, Land 1973, Land and Shepherd 1974, Stewart 1985). For example, several studies have reported that focal lesions or discrete application of an anterograde tract tracer in the epithelium labels olfactory nerve terminals in glomeruli over a broad expanse of the olfactory bulb. However, within this diffuse terminal field there are discrete foci of much denser terminal labelling. If these dense terminal foci represent a preferential targeting of a few glomeruli by a discrete population of receptor cells, then the point-to-point organization of the intrabulbar association pathway (Schoenfeld et al. 1985) could function to preserve the topography of the focal projection from the epithelium to the bulb across one additional synaptic linkage to the opposite side of the bulb.

The focal organization of this intrabulbar association pathway may be further preserved in at least one of the output pathways from the bulb originating in the AON *viz* the pars externa system of the anterior olfactory nucleus (AONpE). The AONpE is one of the architectonic subdivisions of the AON. Macrides and co-workers have shown that mitral and tufted cells giving rise to the intrabulbar association pathway, project to the AONpE (Schoenfeld and Macrides 1984, Macrides et al. 1985). Axons of AONpE neurons project caudally, enter the anterior commissure, cross the midline and terminate in the contralateral olfactory bulb. The AONpE system terminates in the contralateral bulb in a modified point-to-point fashion; pars externa neurons terminate with circumferential, but not longitudinal specificity, that is, the pars externa projection terminates all along the longitudinal (rostrocaudal) axis of the bulb, but is restricted to a limited band along the circumference of the bulb (Schoenfeld and Macrides 1984). Taken together, the topography in the combined intra- and interbulbar pathway is such that tufted cells associated with a focal part of the glomerular layer in one olfactory bulb project via the intrabulbar and interbulbar association pathways to a discrete region in the contralateral bulb (Macrides et al. 1985). However, the pars externa system terminates in a longitudinal (rostro-to-caudal) strip in the contralateral bulb. This strip appears to have the same circumferential specificity as the intrabulbar association pathway but lacks its longitudinal i.e., rostro-caudal specificity. There is also evidence that the projection from the epithelium to the bulb has a higher degree of circumferential than longitudinal specificity (Astic and Saucier 1986, Macrides et al. personal communication). Thus, the circumferential specificity seen in the intrabulbar association pathway may relate specifically to circumferential specificity in the epithelium to bulb projection.

Taken together, these recent anatomical findings suggest that there may be epithelium to bulb, and intrabulbar and interbulbar pathways that are anatomically organized so as to preserve a plan of circumferential specificity in both the initial stages of the olfactory pathway and its connection to the contralateral bulb. The functional significance of this circumferential organization remains to be determined. In addition, it is not known whether the circumferential mapping pattern is preserved at higher levels of the olfactory system. Nonetheless, it is tempting to wonder if there are additional schemes of point-to-point or point-to-strip anatomical networks in the olfactory system. If, for example, there were other networks capable of addressing "slabs" of bulb perpendicular to the longitudinal axis of the bulb, these networks would be orthogonal to the circumferential networks and could have the effect of creating grid-like organization in the olfactory bulb. Further research is needed, therefore, to investigate the functional significance of the topographical mapping patterns in intra- and interbulbar circuits and if these or similar organizational features are reflected at other levels of the olfactory pathway.

The AONpE system is only one component of the interbulbar association system. The AON has several other subdivisions, each containing populations of neurons that interconnect the olfactory bulbs on the two sides of the brain. These other AON subdivisions contain many more neurons than the pars externa subdivision and their connectional organization differs considerably from that of the latter system.

In the rat approximately 60% of the non-pE AON neurons that project back to the ipsilateral bulb also project to the contralateral bulb (Alheid et al. 1984) via the anterior commissure. The non-pE AON projections to the bulb do not appear to have any point to point or point to strip organization, but do have some degree of radial organization. Axons of these AON neurons terminate in a laminar-like fashion at characteristic depths in the ipsi- and contralateral bulbs (Luskin and Price 1983). Most AON neurons project to the granule cell layer of the two bulbs, but different AON divisions preferentially terminate in the superficial or deep part of the granule cell layer. This superficial-to-deep dimension represents the radial axis of the bulb and this axis is orthogonal to both the circumferential and longitudinal axes. Thus, the intrabulbar association pathway and the pars externa component of the interbulbar pathway have a degree of longitudinal and circumferential organization, and the bulk of the AON interbulbar pathway and the centrifugal projections from the AON back to the ipsilateral bulb have a degree of radial but no suggestion of circumferential or longitudinal organization.

This radial or laminar organization of the non-pE AON system may relate to a proposed plan of a radial or laminar organization between mitral or tufted cells and granule cells. Scott and co-workers have used microinjections of horseradish peroxidase to analyze the dendritic organization of mitral and granule cells (Orona et al. 1983, Orona et al. 1984), and their studies suggest that the lateral dendrites of different mitral cells preferentially arborize at a particular depth of the EPL. The dendrites of some mitral cells arborize in the deeper parts of the EPL just above the mitral cell layer, while the lateral dendrites of other mitral cells preferentially arborize at more superficial locations in the EPL. Orona et al.(1983) have shown that the apical dendrites of granule cells located in the superficial part of the GCL may arborize in the superficial part of the EPL, while the apical dendrites of granule cells located deeper in the GCL preferentially arborize in the deeper parts of the EPL. Since the apical dendrites of granule cells have significant synaptic interactions with the lateral dendrites of mitral or tufted cells, it is possible that superficial granule cells preferentially synapse with mitral cells whose lateral dendrites arborize in the superficial EPL, and deeper lying granule cells preferentially interact with mitral cells whose apical dendrites arborize in the deeper parts of the EPL. Interactions between mitral or tufted cells and granule cells, therefore, may have a significant degree of radial organization. Since different divisions of the non-pE AON system terminate at different depths of the granule cell layer, there is a potential anatomical substrate to link, and possibly coordinate mitral and granule cells having similar radial organization in the two olfactory bulbs. This radial organization may also differentially influence the outputs of the bulb as Scott and co-workers have shown that superficially located tufted cells tend to project preferentially to rostral parts of the olfactory cortex, while deeper tufted and most mitral cells project preferentially to more caudal parts of the olfactory cortex (Scott et al. 1980, Scott 1981, Schneider and Scott 1983, Scott and Harrison 1987). Since the non-pE AON system does not appear to have any circumferential or longitudinal organization, the relation between the radial organization of the mitral-granule cells and the non-pE AON system on the one hand, and the circumferential and longitudinal organization of tufted cells, the intrabulbar association pathway and the circumferential organization of the pars externa system, on the other hand, is presently unclear.

However, notwithstanding the frequent statement that the olfactory system lacks topographical organization, it should not be inferred that this system is devoid of anatomical organization. As this review indicates, the connections of different populations of bulbar and centrifugal neurons preferentially express different degrees of circumferential, longitudinal or radial organization. An interesting problem for future research in olfactory neuroanatomy, therefore, is to determine if there are additional dimensions of organization and how the organization of different sets of olfactory networks relate to each other.

Our understanding of the functional significance of the AON is still rudimentary. Clearly, the major interbulbar connections of the AON implicate this structure in the interhemispheric processing of olfactory information. There is evidence that binasal mechanisms may function in the spatial localization of odors (Bennett 1968) and the AON system would be suspected to play a significant role in such mechanisms. There is also evidence from animal studies that the AON plays a key role in the interhemispheric transfer of olfactory memories (Kucharski and Hall 1987).

IV. OLFACTORY TUBERCLE

The olfactory tubercle in rodents, rabbits and other macrosmatic mammals is a prominent bulge on the base of the hemisphere just caudal to the olfactory peduncle. In such species, axons of mitral and tufted cells (Heimer 1968, Price 1973, de Olmos et al. 1978) terminate in the superficial layer of the tubercle as in the AON and primary olfactory cortex. The tubercle has a superficial plexiform layer like the AON and primary olfactory cortex (POC), but the deeper cellular architecture of the tubercle is intermediate between a cortical and a striatal structure (Heimer et al. 1986). Immediately deep to the plexiform layer is a layer of neurons with apical dendrites that extend into the plexiform layer. Neurons deep to this so-called cortical layer, however, are not like layer III pyramids of the POC but rather are polymorphic, and their dendrites do not appear to preferentially extend into the plexiform layer like those of the pyramidal cells of layer III in the POC. These polymorphic neurons are more akin to neurons of the striatum, indeed, Heimer's extensive neuroanatomical analysis of the tubercle and adjacent basal telencephalic grey matter has led to the concept of the "ventral striatum" (Heimer et al. 1986). This concept is beyond the scope of this Chapter but its definition is based on parallel patterns of cytoarchitecture, transmitters and connectivity with the dorsal neostriatum. Heimer and co-workers concluded that the ventral striatum is anatomically similar to the more familiar dorsal striatum (including the caudo-putamen), except that whereas the extrinsic and subcortical connections of the dorsal striatum are with the neocortex and associated parts of the intralaminal thalamic nuclei, those of the ventral striatum are connected with cortical structures of the limbic system and the mediodorsal thalamus (Fig 3). Consistent with the striatal motif of the tubercle, the part of this structure that receives olfactory bulb afferents is much reduced in humans and other microsomatic primates despite the continued absolute expansion of the tubercle (Heimer et al. 1977). Viewed in this context the prominent olfactory input to the tubercle of macrosmatic species may reflect an expanded role for olfactory influence in such species on ventral striatal functions which are thought to be more related to emotions than those of the dorsal striatum. The tubercle also differs from the POC in that it does not send a reciprocal projection back to the bulb. This generalization needs to be somewhat qualified as the projection from the magnocellular basal forebrain, the nucleus of the diagonal band (NDB), which is the sole source of cholinergic innervation of the bulb (Macrides et al. 1981, Nickell and Shipley 1988a), may be considered a kind of centrifugal return from the ventral striatum. However, at present there is no evidence that the olfactory bulb projections to the plexiform layer of the tubercle have direct or indirect anatomical linkages to influence the NDB neurons. Thus, the tubercle differs from the rest of the POC in that it does not contain a population of cortical neurons that reciprocate the projection from the bulb.

V. PRIMARY OLFACTORY CORTEX

Beginning at the caudal limits of the AON, the cortex of the basal temporal lobe begins to expand in the caudal direction forming a pear shaped structure, the piriform lobe. The rostral part of this structure contains the piriform cortex (Fig 3), which is the major cortical component of the POC. Further caudally, the medial part of the piriform cortex overlies the amygdaloid complex and this part of the piriform cortex is referred to as the periamygdaloid cortex. Further caudally and medially the piriform-periamygdaloid cortex gives way gradually to the lateral entorhinal cortex. The architecture of the lateral entorhinal cortex changes and gives way to the medial entorhinal cortex. Caudomedially the entorhinal cortex gives way to the parasubiculum, presubiculum and subiculum; the latter blends with field CA1 of Ammon's horn of the hippocampus. Thus, there is a continuous expanse of gradually changing cortical architecture leading from the olfactory bulb, the AON, piriform, periamygdaloid, entorhinal and subicular cortices leading directly into the CA fields of the hippocampus. This orderly anatomical arrangement was one reason why early anatomists referred to the olfactory cortex, parahippocampus and hippocampus as the "rhinencephalon", believing that the entire expanse of cortex constituted the "smell brain". This view was tempered by subsequent tract tracing experiments showing that the projections of the olfactory bulb directly innervate only the AON, piriform, periamygdaloid and lateral entorhinal cortices. Thus, the parahippocampal region (entorhinal cortex, para- and presubiculum and subiculum) and the hippocampus proper came to be included in the "limbic system". Notwithstanding this considerable loss of cortical real estate, the olfactory system can still claim to be the sensory system with the most direct access to the hippocampus, because there are direct projections of the olfactory bulb to the lateral entorhinal cortex, and the latter structure is the major source of afferent input to the hippocampus (Fig 3). Moreover, the piriform cortex also has direct connections with the entorhinal cortex and, in addition, the lateral entorhinal cortex sends reciprocal projections back to the olfactory bulb (Fig 4).

The organization of the terminal projections from the bulb to the primary olfactory cortex appears to be constant throughout the latter structure. The rostral parts of the POC receive terminals from both tufted and mitral cells (Haberly and Price 1977, Schneider and Scott 1983, Scott et al. 1980, Scott 1981). The caudal parts of the piriform, the periamygdaloid and lateral entorhinal cortex receive inputs primarily or exclusively from mitral cells (Scott et al. 1980, Scott 1981). In all parts of the POC, however, the olfactory bulb projection terminates in the superficial half of layer I (Heimer 1968, Price 1973) which is designated as layer Ia. Thus, beginning with the intrabulbar association pathway, which primarily or exclusively involves tufted cells, through the AON, olfactory tubercule and anterior parts of the piriform cortex which involves both mitral and tufted cells and, finally, the caudal piriform, periamygdaloid and lateral entorhinal cortices which involve primarily or exclusively mitral cells, there is a systematic plan of radial or laminar organization. The more superficial tufted cells project to more proximal targets and the deeper tufted and mitral cells project to more distal output targets of the bulb. With the sole exception of the tufted cell-mediated intrabulbar association pathway discussed earlier, there is little indication of point-to-point topography within this complex output network. Thus, anterograde tracer injections limited to only a small population of mitral cells in the bulb, label axons and terminals throughout the entire expanse of the POC, and discrete injections of retrograde tracer in all parts of the POC label mitral cells scattered throughout the mitral cell layer of the bulb (Haberly and Price 1977, Scott et al. 1980, de Olmos et al. 1978). Thus, there is no indication that discrete sites in the bulb preferentially communicate with discrete sites in the POC. This tends to reinforce the idea that point-to-point topography does not play an obvious role in central representation of "odor space" and has led several workers in the field to suggest that processing of olfactory information in olfactory cortex can best be understood by considering the olfactory cortex as a content addressable, distributed neural network (Haberly 1985, Haberly and Bower 1989, Ambros-Ingerson et al. 1990). In this view, the functioning of a network is inherent in the organization of its microcircuits, the patterns of connection between input and output of the circuits do not matter so much as the ability of the circuitry to form associative linkages with no (obviously) discernible spatial patterns of anatomical connectivity.

This view derives, in part, from the failure of anatomical studies to reveal any intuitively obvious plan of topography in the projections of the bulb to the POC, or in the associative connections between different parts of the POC, or any indication that different parts of the POC are specialized to differentially target higher order structures. Despite the failure of tract tracing studies to reveal any point-to-point topography in the projection of mitral and tufted cells, it is perhaps prudent to refrain from closing the book on this issue. While there have been spectacular advances in anatomical tract tracing methods in the last two decades, current methods still have their limitations. Thus, most studies have not specifically addressed the question of how the synaptic terminals of different mitral cells are distributed in the POC. One study, using intracellular fills of individual mitral cells, showed that while their axons project over a considerable expanse of the POC, these axons appear to distribute terminal synaptic arbors in a patchy-like fashion (Ojima et al. 1984). Since it is the synaptic organization that dictates the functional specificity of a projection, any definitive conclusions about the presence or absence of topography in the output projection of the bulb should be withheld until the organization of the terminal patterns of mitral and tufted axons is better understood. Moreover, the functional consequence of the radial organization of bulb outputs i.e., superficial tufted cells projecting to more proximal parts of the POC, and deeper tufted and mitral cells projecting to more caudal parts of the POC, has received little attention.

VI. CONNECTIONS OF THE PRIMARY OLFACTORY CORTEX

The connections of the POC can be discussed as four classes: 1) *intrinsic* or *local* - short connections between neurons in different layers of the POC; 2) *associative* - connections with different parts of the POC; 3) *extrinsic* - connections with other structures, and 4) *modulatry inputs* - afferents that terminate in the POC as part of a broader innervation of other cortical and subcortical neural systems.

A. Intrinsic or Local Connections of the Primary Olfactory Cortex

The POC has two principal layers of pyramidal neurons, layers II and III, which comprise several morphological classes and also several classes of non-pyramidal neurons (Haberly 1983, 1985). There are extensive translaminar or local connections among POC neurons (Haberly 1985). Layer II neurons give off axon collaterals to deeper layer III pyramidal cells, and there are local inhibitory interneurons in layers I and II that are contacted by olfactory bulb terminals and by local collaterals of pyramidal cells (Haberly 1985). Deeper pyramidal cells also give rise to extensive local collaterals that may synapse with local interneurons or with more superficial pyramidal cells (Guo-Fang and Haberly 1989). Thus, there appear to be extensive translaminar connections between superficial and deeper layers. In addition, there are several classes of GABAergic and neuropeptide-containing neurons in the POC, and although the connection of these neurons is not known, many of them have the appearance of local interneurons.

B. Association Connections of the Primary Olfactory Cortex

Cortico-cortical projections within the POC are extensive and exhibit some degree of laminar and regional organization. Axons from pyramidal cells of layer IIb are primarily directed at more caudal sites in the POC, while cells in layer III project predominantly to rostral parts of the POC. Commissural fibers to the contralateral POC arise from layer IIb pyramidal cells of the anterior parts of the POC (Haberly and Price 1978a), and as noted, projections back to the bulb arise from layers IIb and III. Thus, neurons in layer IIb project both caudally to terminate as association fibers, rostrally to terminate in the bulb and contralaterally to the rostral POC. It is not known whether any layer IIb cells project both caudally and rostrally.

The ipsilateral and commissural association projections of the POC terminate in a highly laminar fashion in layer Ib (Price 1973, Haberly and Price 1978a), immediately below the zone that contains the afferents from the olfactory bulb, and a lighter projection terminates in layer III. Projections from the POC back to the AON also terminate in layer Ib, below the bulb recipient zone (Haberly and Price 1978a).

C. Extrinsic Connections of the Primary Olfactory Cortex

The major extrinsic connections of the POC are its reciprocal connections with the olfactory bulb and the AON and its efferent projections to various non-olfactory cortical-subcortical targets (Figs 3 and 4). Mitral and some tufted cells project from the bulb to form the lateral olfactory tract; axons exit this tract at various points along its length to ramify and terminate in the superficial layer of the POC. These axons terminate in the superficial half of layer I (Ia) among the dendrites of superficial (layer II) and deeper (layer III) pyramidal cells (Heimer 1968, Price 1973). Layer II pyramidal cells appear to comprise two groups, layers IIa and IIb. Layer IIb and some layer III cells project back to the bulb (Haberly and Price 1978a). These projections back to the bulb terminate preferentially in the granule cell layer (Haberly and Price 1978a) and are thought to be excitatory to the granule cells. Thus, the POC has reciprocal projections back to the bulb forming a kind of feedback inhibition, exciting granule cells and inhibiting mitral cells, the neurons which are afferent to the POC. Thus, the POC and AON are both the major recipients of the output of the bulb and are the major sources of ipsilateral centrifugal inputs to the bulb (Figs 3 and 4). Some of the physiological features of this massive reciprocal circuitry have been characterized, but its functional significance in terms of olfactory information processing remains unclear. The connections of the POC with non-olfactory structures are discussed below (see VII Beyond the Primary Olfactory Cortex).

D. Modulatory Inputs to the Primary Olfactory Cortex

The POC also receives subcortical modulatory inputs from the locus coeruleus (norepinephrine), midbrain raphe nuclei (serotonergic), and the nucleus of the diagonal band (cholinergic, and possibly GABAergic). The terminal organization of these projections to the POC have not been characterized in detail and their functional actions are largely unknown.

VII. BEYOND THE PRIMARY OLFACTORY CORTEX

Two classes of POC outputs have already been discussed: 1) the reciprocal projection back to the olfactory bulb, and 2) the association projections between the rostral and caudal piriform cortex, and piriform cortex projections to the entorhinal cortex. A third class of outputs is treated separately because it represents projections of the POC to brain regions not generally included in the olfactory system *per se*, although their receipt of inputs from the POC obviously implicates these POC targets in olfactory function. The outputs of the piriform cortex have been studied most extensively by Price and co-workers in the rat and more recently in the cat and primate (Price 1985, Russchen et al. 1987).

The extrinsic outputs of the piriform cortex are both to cortical and subcortical structures. Projections from the POC to the lateral entorhinal cortex and subsequently to the hippocampus have already been described above as associative connections of the POC. In addition, the piriform cortex projects to closely adjacent cortical fields such as the perirhinal, orbital and insular cortex. The major subcortical projections of the piriform cortex are to the thalamus, hypothalamus and ventral striatum.

VIII. NEOCORTICAL PROJECTIONS

In the mouse and rat the projection of the olfactory bulb to the POC extends dorsally beyond the cytoarchitectural limits of the POC into the ventral parts of the agranular insular and perirhinal cortices (Shipley and Geinisman 1984). This ventral part of the insular cortex in rodents contains the cortical representation of ascending pathways arising in the nucleus of the solitary tract (NTS) in the medulla (Shipley 1982). The NTS is the initial subcortical relay for gustatory and visceral sensory input to the central nervous system. The representation of the NTS in insular cortex appears to comprise both a primary sensory cortical map for gustation and visceral sensations, and also via descending corticofugal projections a route whereby cortex can modify visceral-autonomic and possibly gustatory function (Shipley 1982). Thus, the direct olfactory bulb projection into ventral insular cortex in rodents has been suggested as one relatively direct route for the integration of olfactory and gustatory information in the neural representation of flavor and the integration of olfactory and autonomic information (Shipley and Geinisman 1984). The existence of direct projections from the olfactory bulb to homologous insular cortical areas in the primate have not been established with anatomical methods. In rodents there are also direct projections from the POC to the dorsally adjacent insular cortical fields involved in gustation and visceral sensation (Price 1985, Shipley, unpublished observations), and there are projections from the mediodorsal thalamus to the insular cortex and to medial cortical fields (Fig 3) that also preferentially project to hypothalamic and brainstem regions involved in autonomic function (Price 1985). Thus, there are direct POC projections to gustatory-autonomic cortical areas, and though less well characterized, potential indirect circuits from the POC to the mediodorsal-submedial thalamus and thence to the lateral and medial neocortical areas involved in gustatory and autonomic function. The precise homologies among the lateral and medial cortical fields linking olfactory with gustatory-autonomic systems and the corresponding fields in the primate brain remain to be resolved. However, electrophysiological studies indicate that neurons in potentially homologous cortical areas in primates respond to odors with a higher degree of selectivity than neurons in either the olfactory bulb or the POC (Takagi 1986, Yarita et al. 1980). Thus, it may be that cortico-cortical and cortico-thalamo-cortical circuits from the POC to neighboring cortical areas play a role in flavor perception and in linking olfactory stimuli to the hypothalamic-autonomic axis.

IX. SUBCORTICAL PROJECTIONS

A. Hypothalamus

The heaviest and most direct projections to the hypothalamus derive from neurons in the deepest layers of the piriform cortex and the anterior olfactory nucleus. These projections terminate most heavily in the lateral hypothalamic area (Price 1985). Some polymorphic neurons of the olfactory tubercle and part of the anterior olfactory nucleus (Price 1985), also project to the hypothalamus. Olfactory-recipient parts of the cortical and medial amygdaloid nuclei also project to medial and anterior parts of the hypothalamus.

B. Thalamus

In infra-primate species, anatomical experiments have demonstrated a strong projection from the POC to the magnocellular, medial part of the mediodorsal thalamic nucleus and the submedial nucleus (nucleus gelatinosa) (Price and Slotnick 1983) (Fig 3). Retrograde tracing studies show that these olfactory cortico-thalamic projections arise from neurons in the deepest layer of the piriform, periamygdala and entorhinal cortex and the polymorphic cell layer of the olfactory tubercle (Price and Slotnick 1983). These projections have apparently not yet been established with anatomical methods in primates, but neurophysiological studies in primates indicate that neurons in the magnocellular medial part of the mediodorsal and in the submedial thalamic nuclei are responsive to olfactory bulb stimulation and odors (Benjamin and Jackson 1974, Russchen et al. 1987, Yarita et al. 1980). Thus a strong output from all parts of the POC to the mediodorsal and submedial thalamic nuclei appears to be a fundamental feature of olfactory circuitry and represents a potentially important route for the dissemination of olfactory information to other cortical and subcortical areas.

X. FUNCTIONAL ORGANIZATION OF THE OLFACTORY SYSTEM

Many features of the intrinsic anatomical organization and classical electrophysiology of the POC have been studied in some detail. Unfortunately, this body of literature gives little indication of the sensory processing characteristics of the POC. Situated as the primary cortical projection target of the output of the bulb, it has been tempting to think that the POC is involved in some kind of hierarchical or higher order stimulus feature extraction. This expectation is based upon analogy with the organization of the other major sensory systems where it has been possible to infer how neurons, at successive levels from the periphery through subcortical relays to the primary sensory cortex, transform inputs to extract different features of the sensory signal. In the case of the olfactory system this prevailing "sensory systems paradigm" has provided little insight into the neural operations that lead to the perception of odor qualities or the features of odors that modify behaviors. It is reasonable, therefore, to wonder why the paradigm that has worked so well in understanding relationships between sensory stimuli and the neural mechanisms that encode stimulus features in other sensory systems has fallen so short in the olfactory system. There are many possible answers to this puzzle but none seem to be uniquely compelling. One clear problem is that we have a relatively poor understanding of olfactory stimuli and the relationship between olfactory stimuli and sensory transduction in olfactory receptor neurons. In the case of vision, audition and somatic sensations, there is a fairly intuitive, Newtonian-like appreciation of the relevant dimensions of light, sound and pressure that is not immediately obvious for the relationship between a molecule and its odor. Complex patterns of light can be thought of in terms of contrast, hue, reflection and absorption of light, but such dimensions do not leap readily to mind for odors. There is again an almost intuitively obvious grasp of the utility of a place code in other sensory systems. Light patterns falling on different parts of the retina correspond to different external distributions of contrast. An itch on the tip of the nose, or a socially inaccessible place on the trunk, defines a spatial locus that invites a directed behavioral response. The position of the stimulus is used to guide the motor system to the correct location; foveation or scratching have no obvious equivalents in the olfactory system. It is difficult to overemphasize how critical the simple place coding principle has been to the analysis of neural mechanisms in other sensory systems. The poverty of our intuitive sense of the odor world, combined with the lack of any obvious meaning to a place-code relationship between the olfactory receptor sheet and subsequent levels of the olfactory system, have greatly impeded our attempts to understand olfactory function by analogy with other sensory systems. Notwithstanding, there have been repeated efforts to understand olfactory anatomy and physiology by comparison with other sensory systems. Almost every decade someone likens the olfactory bulb to the retina. The analogy between the alternating layers of neuron types with intervening plexuses of synaptic integration make it irresistibly tempting to try to understand the circuitry of the bulb in terms of the retina. This analogy is easily maintained so long as one ignores equally compelling and fundamental differences between the retina and the bulb. Chief among these is the existence of massive centrifugal inputs to the olfactory bulb and the total absence of centrifugal inputs to the retina in mammals. In some non-mammalian vertebrates there is a midbrain nucleus that projects to the retina but this feature is lacking in mammals. The retina transduces light, processes the neural responses via networks of intrinsic neurons and then conducts the output to the geniculo-striate or collicular systems. In the olfactory bulb the outputs of the bulb are paralleled at every step by massive feedback pathways that modulate populations of bulbar interneurons which in turn directly regulate the output neurons. In this sense, the bulb is not so much a relay to higher olfactory structures, but is more like the first stage in a circuit that feeds back upon itself at every stage of synaptic transfer. From this perspective, attempts to draw analogies between the olfactory bulb and the retina may be frustrated because the functional status of bulb neurons at any moment reflects not only activity in sensory afferents and interneurons, but also the activity in feedback pathways from the anterior olfactory nucleus and the POC. Given that these latter two structures are the predominant output targets of the bulb, olfactory bulb neurons must be continuously modulated by return signals that represent transformations of its own output. The retina does not work this way. The output of ganglion cells is modulated by intra-retinal processing and not by operations performed by the output targets of the retina.

XI. SUMMARY

The olfactory system remains one of the most poorly understood of all the sensory systems. Part of this is due to the historical intractability of the sensory transduction process, ignorance about the specificity of transduction mechanisms for different odorants, lack of insight about the sensory "code", and the seeming lack of point-to-point specificity and a "place code" in the initial stages of olfactory circuitry. However, it is perhaps worth considering that as one moves from the initial levels of thalamocortical circuitry in the better understood visual and somatosensory pathways to "higher" levels of these systems, there seems to be a progressive loss of the familiar principles of topographical point-to-point remapping, and both the nature of the "sensory code" and what precisely is being encoded is less obvious than at lower levels. Perhaps the olfactory sense does not employ these familiar principles from even its earliest stages of processing because it is not an inherently two-dimensional sense. Thus, the olfactory system may employ principles of neural processing that have more in common with those used at higher levels in other sensory systems.

REFERENCES

Adamek GD, Nickell WT, Shipley MT (1986) Evidence for diffuse and focal projections from the olfactory epithelium to the bulb. Chem Sens 11: 575

Alheid GF, Carlsen J, de Olmos J, Heimer L (1984) Quantitative determination of collateral anterior olfactory nucleus projections using a fluorescent tracer with an algebraic solution to the problem of double retrograde labelling. Brain Res 292: 17-22

Ambros-Ingerson J, Granger R, Lynch G (1990) Stimulation of paleocortex performs hierarchial clustering. Science 247: 1344-1347

Astic L, Saucier D (1986) Anatomical mapping of the neuroepithelial projection to the olfactory bulb in the rat. Brain Res Bull 16: 445-454

Barr ML, Kiernan JA (1983) In: The human nervous system, an anatomical viewpoint. Harper and Row, New York

Benjamin RM, Jackson JC (1974) Unit discharges in the mediodorsal nucleus of the squirrel monkey evoked by electrical stimulation of the olfactory bulb. Brain Res 75: 181-192

Bennett MH (1968) The role of the anterior limb of the anterior commissure in olfaction. Physiol Behav 3: 507-515

Davis BJ, Macrides F (1981) The organization of centrifugal projections from the anterior olfactory nucleus, ventral hippocampal rudiment, and piriform cortex to the main olfactory bulb in the hamster: An autoradiographic study. J Comp Neurol 203: 475-493

de Olmos J, Hardy H, Heimer L (1978) The afferent connections of the main and accessory olfactory bulb formations in the rat: An experimental HRP study. J Comp Neurol 181: 213-244

Doty RL, Deems D, Stellar S (1988) Olfactory dysfunction in Parkinson's disease: A general deficit unrelated to neurologic signs, disease stage, or disease duration. Neurology 38: 1237-1244

Doty RL, Reyes PF, Gregor T (1987) Presence of both odor identification and detection deficits in Alzheimer's disease. Brain Res Bull 18: 597-600

Esiri MM, Wilcock GK (1984) The olfactory bulbs in Alzheimer's disease. J Neural Neurosurg Psychiatry 47: 52-60

Ferreyra-Moyano H (1989) The olfactory system and Alzheimer's disease. Int J Neurosci 49: 157-197

Guo-Fang T, Haberly LB (1989) Deep neurons in piriform cortex. I. Morphology and synaptically evoked responses including a unique high-amplitude paired shock facilitation. J Neurophysiol 62: 369-385

Haberly LB (1983) Structure of the piriform cortex of the opossum. I. Description of neuron types with Golgi methods. J Comp Neurol 213: 163-187

Haberly LB (1985) Neuronal circuitry in olfactory cortex: Anatomy and functional implications. Chem Sens 10: 219-238

Haberly LB, Bower JM (1989) Olfactory cortex: Model circuit for study of associative memory? TINS 12: 258-264

Haberly LB, Price JL (1977) The axonal projection patterns of the mitral and tufted cells of the olfactory bulb in the rat. Brain Res 129: 152-157

Haberly LB, Price JL (1978a) Associational and commissural fiber systems of the olfactory cortex of the rat. I. Systems arising in the piriform cortex and adjacent areas. J Comp Neurol 178: 711-740

Haberly LB, Price JL (1978b) Association and commissural fiber systems of the olfactory cortex in the rat. II. Systems originating in the olfactory peduncle. J Comp Neurol 178: 781-808

Heimer L (1968) Synaptic distribution of centropetal and centrifugal nerve fibers in the olfactory system of the rat. An experimental anatomical study. J Anat 103: 413-432
Heimer L, Alheid GF, Zaborsky L (1986) The basal ganglia. In: Paxinos G (ed) The rat nervous system, vol 1. Academic Press, Orlando, Florida, p 37
Heimer L, Van Hoesen GW, Rosene DL (1977) The olfactory pathways and the anterior perforated substance in the primate brain. Int J Neurol 12: 42-52
Hurwitz T, Kopala L, Clark C (1988) Olfactory deficits in schizophrenia. Biol Psychiatr 23: 123-128
Kerr DIB, Hagbarth KE (1955) An investigation of the olfactory centrifugal system. J Neurophysiol 18: 362-374
Kishi K, Mori K, Ojima H (1984) Distribution of local axon collaterals of mitrals, displaced mitrals, and tufted cells in the rabbit olfactory bulb. J Comp Neurol 225: 511-526
Kucharski D, Hall WG (1987) New routes to early memories. Science 238: 786-788
Land LJ (1973) Localized projection of olfactory nerves to rabbit olfactory bulb. Brain Res 63: 153-166
Land LJ, Shepherd GM (1974) Autoradiographic analysis of olfactory receptor projections in the rabbit. Brain Res 70: 506-510
Leopold DA (1986) Physiology of olfaction. In: Cummings CW, Frederickson JM, Harker LA, Kraus CJ, Schuller DE (eds) Otolaryngology - Head and Neck Surgery. CV Mosby Co., St Louis
Luskin MB, Price JL (1983) The topographic organization of associational fibers of the olfactory system in the rat, including centrifugal fibers to the olfactory bulb. J Comp Neurol 216: 264-291
Macrides F, Davis BJ, Youngs WM, Nadi NS, Margolis FL (1981) Cholinergic and catecholaminergic afferents to the olfactory bulb in the hamster: A neuroanatomical, biochemical and histochemical investigation. J Comp Neurol 203: 495-514
Macrides F, Schoenfeld TA, Marchand JE, Clancy AN (1985) Evidence for morphologically, neurochemically and functionally heterogeneous classes of mitral and tufted cells in the olfactory bulb. Chem Sens 10: 175-202
McLean JH, Shipley MT (1987) Serotonergic afferents to the rat olfactory bulb: I. Origins and laminar specificity of serotonergic inputs in adult rat. J Neurosci 7: 3016-3028
McLean JH, Shipley MT, Nickell WT, Aston-Jones G, Reyher CKH (1989) Chemoanatomical organization of the noradrenergic input from locus coeruleus to the olfactory bulb of the adult rat. J Comp Neurol 285: 339-349
Mori K, Kishi K, Ojima H (1983) Distribution of dendrites of mitral, displaced mitral, tufted, and granule cells in the isolated turtle olfactory bulb. J Neurosci 219: 339-355
Mori K, Takagi SF (1978) Activation and inhibition of olfactory bulb neurones by anterior commissure volleys in the rabbit. J Physiol (Lond) 279: 589-604
Nickell WT, Shipley MT (1988a) Two anatomically specific classes of candidate cholinoceptive neurons in the rat olfactory bulb. J Neurosci 8: 4482-4491
Nickell WT, Shipley MT (1988b) Neurophysiology of magnocellular forebrain inputs to the olfactory bulb in the rat: Frequency potentiation of field potentials and inhibition of output neurons. J Neurosci 8: 4492-4502
Ojima H, Mori K, Kishi K (1984) The trajectory of mitral cell axons in the rabbit olfactory cortex revealed by intracellular HRP injection. J Comp Neurol 230: 77-87

Orona E, Ranier EC, Scott JW (1984) Dendritic and axonal organization of mitral and tufted cells in the rat olfactory bulb. J Comp Neurol 226: 346-356
Orona E, Scott JW, Ranier EC (1983) Different granule cell populations innervate superficial and deep regions of the external plexiform layer in rat olfactory bulb. J Comp Neurol 217: 227-237
Pearson RC, Esiri MM, Hoins RW, Wilcock GK, Powell TPS(1985) Anatomical correlates of the distribution of the pathological changes in the neocortex in Alzheimer's disease. Proc Natl Acad Sci USA 82: 4531-4535
Pinching AJ (1970) Synaptic connections in the glomerular layer of the olfactory bulb. J Physiol (Lond) 210: 14P-15P
Pinching AJ, Powell TPS (1971a) The neuropil of the glomeruli of the olfactory bulb. J Cell Sci 9: 347-377
Pinching AJ, Powell TPS (1971b) The neuropil of the periglomerular region of the olfactory bulb. J Cell Sci 9: 379-409
Price JL (1973) An autoradiographic study of complimentary laminar patterns of termination of afferent fibers to the olfactory cortex. J Anat 150: 87-108
Price JL (1985) Beyond the primary olfactory cortex: Olfactory-related areas in the neocortex, thalamus and hypothalamus. Chem Sens 10: 239-258
Price JL, Powell TPS (1970) The synaptology of granule cells of the olfactory bulb. J Cell Sci 7: 125-155
Price JL, Slotnick BM (1983) Dual olfactory representation in the rat thalamus: An anatomical and electrophysiological study. J Comp Neurol 215: 63-77
Rall W (1972) Dendritic neuron theory and dendrodendritic synapses in a simple cortical system. In: Schmitt FO (ed) The neurosciences: Second study program. Rockefeller University Press, New York, p 552
Rall W, Shepherd GM (1968) Theoretical reconstruction of field potentials and dendrodendritic synaptic interactions in the olfactory bulb. J Neurophysiol 31: 884-915
Rall W, Shepherd GM, Reese TS, Brightman MW (1966) Dendrodendritic synaptic pathway for synaptic interactions in the olfactory bulb. Exp Neurol 14: 44-56
Reyes PF, Golden G, Fagel P, Fariettlo RG, Katz L, Carner B (1987) The prepyriform cortex in dementia of the Alzheimer type. Arch Neurol 44: 644-645
Reyher CKH, Larsen WJ, Skowron-Lomneth C, Baumgarten HG, Shipley MT (in press) Olfactory bulb granule cell aggregates: Morphological evidence for pericaryal electrotonic coupling via gap junctions. J. Neurosci.
Ribak CE, Vaughn JE, Saito K, Barber R, Roberts E (1977) Glutamate decarboxylase localization in neurons of the olfactory bulb. Brain Res 126: 1-18
Russchen FT, Amaral DG, Price JL (1987) The afferent input to the magnocellular division of the mediodorsal thalamic nucleus in the monkey macaca fascicularis. J Comp Neurol 256: 175-210
Schneider SP, Scott JW (1983) Orthodromic response properties of rat olfactory bulb mitral and tufted cells correlate with their projection patterns. J Neurophysiol 50: 358-378
Schoenfeld TA, Macrides F (1984) Topographic organization of connections between the main olfactory bulb and pars externa of the anterior olfactory nucleus in the hamster. J Comp Neurol 227: 121-135
Schoenfeld TA, Marchand JE, Macrides F (1985) Topographic organization of tufted cell axonal projections in the hamster main olfactory bulb: An intrabulbar associational system. J Comp Neurol 235: 503-518

Scott JW (1981) Electrophysiological identification of mitral and tufted cells and distributions of their axons in the olfactory system of rat. J Neurophysiol 46: 918-931
Scott JW, Harrison TA (1987) The olfactory bulb: Anatomy and physiology. In: Finger TE, Silver WL (eds) Neurobiology of taste and smell. John Wiley and Sons, New York, p 151
Scott JW, McBride RL, Schneider SP (1980) The organization of projections from the olfactory bulb to the piriform cortex and olfactory tubercle in the rat. J Comp Neurol 194: 519-534
Shepherd GM (1972) Synaptic organization of the mammalian olfactory bulb. Physiol Rev 52: 864-917
Shipley MT (1982) Insular cortex projection to the nucleus solitarius and brain stem visceromotor regions in the mouse. Brain Res Bull 8: 139-148
Shipley MT, Geinismann YG (1984) Anatomical evidence for convergence of olfactory, gustatory and visceral pathways in mouse cerebral cortex. Brain Res Bull 12: 221-226
Shipley MT, Halloran FR, de la Torre J (1985) Surprisingly rich, orderly projection from locus coeruleus to the olfactory bulb in the rat. Brain Res 329: 294-299
Shipley MT, Zahm DS (in press) Differential synaptic processing on apical versus lateral mitral/tufted cell dendrites. Chem. Sens
Stewart WB (1985) Labelling of olfactory glomeruli following horseradish peroxidase lavage of the nasal cavity. Brain Res 347: 200-203
Takagi SF (1986) Studies on the olfactory system of the old world monkey. Prog Neurobiol 27: 195-249
von Baumgarten R, Green JD, Mancia M (1962) Slow waves in the olfactory bulb and their relation to unitary discharges. Electroenceph Clin Neurophysiol 14: 621-634
Yamamoto C, Yamamoto T, Iwama K (1963) The inhibitory systems in the olfactory bulb studied by intracellular recording. J Neurophysiol 26: 403-415
Yarita H, Iino M, Tanabe T, Kogure S, Takagi SF (1980) A transthalamic olfactory pathway to the orbitofrontal cortex in the monkey. J Neurophysiol 43: 69-85

3

PHYSIOLOGY OF OLFACTORY RECEPTION AND TRANSDUCTION: GENERAL PRINCIPLES

THOMAS V. GETCHELL

MARILYN L. GETCHELL

I. INTRODUCTION

The purpose of this Chapter is to provide an overview of the proposed mechanisms by which olfactory receptor neurons detect odorants and transduce this information into electrical activity. The cellular and membrane substrates that are identified with these processes are also described in sufficient detail to characterize the sequence. It is germane to recognize, except where noted, that the results and conclusions discussed in this Chapter were obtained experimentally from a variety of non-human animal species. The Chapter concludes with a summary of provisionally identified first principles of sensory transduction and coding. The text provides representative citations to assist the reader's entrance into the primary literature.

II. CELLULAR ORGANIZATION OF THE OLFACTORY MUCOSA

A. Sensory Mucosa: Location and Neural Organization

The olfactory mucosa, which subserves the sense of smell, is located in the dorsal recess of the nasal chamber (Ecker 1857 as reviewed in Zippel 1988, Moran et al. 1982, Jafek 1983, Nakashima et al. 1984. Fig 1, A and B). In humans, it is estimated to have a total bilateral surface area of 2 cm^2 (Moran et al. 1982). It partially lines the dorsal recess proper, the medial nasal septum and the superior nasal turbinates.

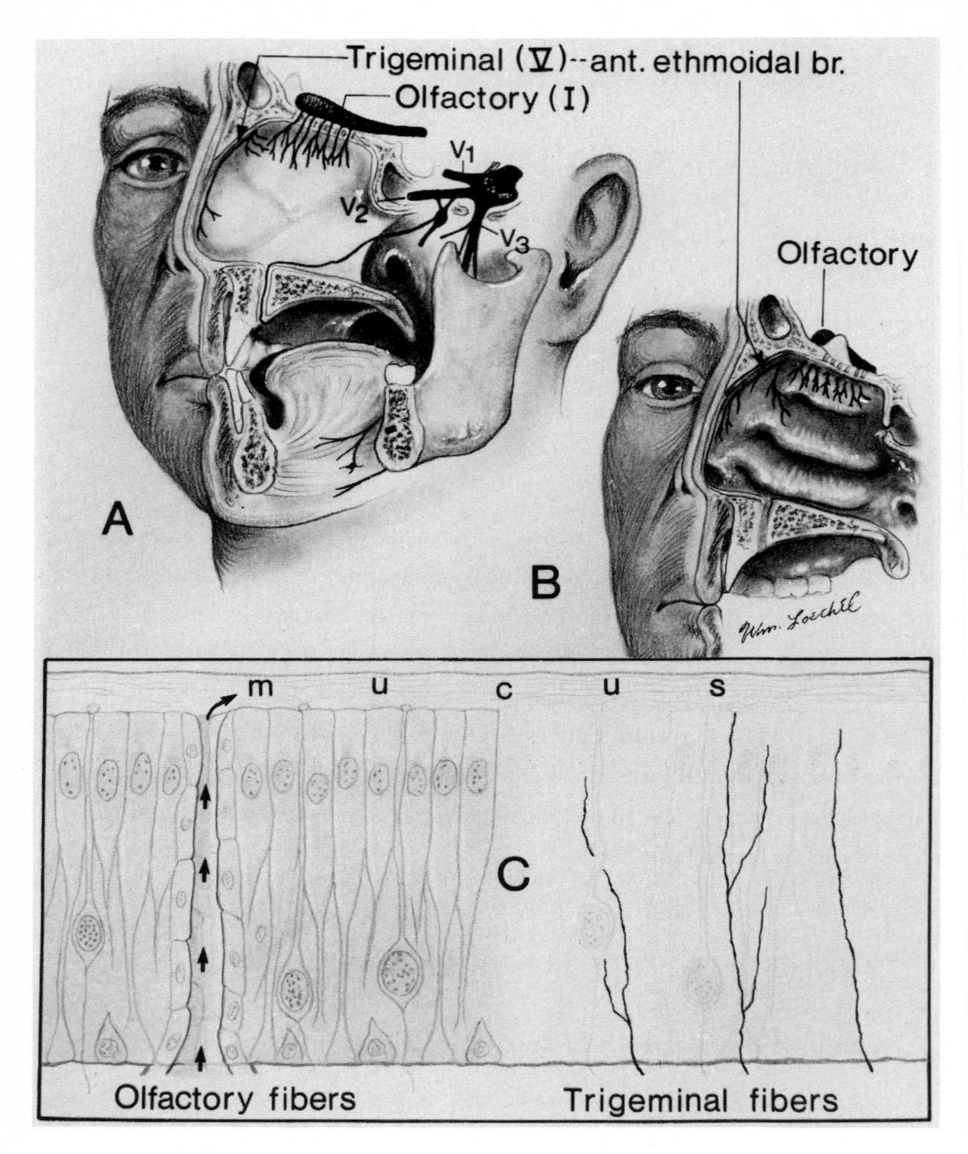

Fig 1. Location and innervation of the human olfactory mucosa. The distribution of olfactory (I) and trigeminal (V) nerve fibers is shown in a subject's right (A) and left (B) nasal chambers. In C, the cellular organization of the olfactory epithelium is shown to the left, and the intraepithelial distribution of trigeminal fibers is shown to the right.

Upon macroscopic examination, the mucosa is more highly pigmented in some non-human species than the surrounding epithelium, appearing yellowish-brown in color. In humans, however, it is difficult to discern the olfactory epithelium from the surrounding respiratory epithelium on the basis of color alone. The olfactory mucosa is a highly vascularized tissue that is covered by a slowly moving layer of mucus derived primarily from the subepithelial olfactory (Bowman's) glands. The mucosa receives two types of innervation: intrinsic and extrinsic. The intrinsic innervation is characterized by the presence of olfactory receptor neurons whose axons coalesce to form the fila olfactoria of the olfactory nerve (cranial nerve I). The diffuse fila penetrate foramina in the cribriform plate, and subsequently their axons make synaptic contacts with second-order neurons in the olfactory bulb. The function of olfactory receptor neurons is to detect odorants, to transduce information related to the concentration and duration of stimulation, and to transmit this information in the form of action potentials to the brain. The extrinsic innervation is characterized by the presence of fibers of the trigeminal nerve (cranial nerve V) and of the autonomic nervous system. Sensory afferent fibers project into the sensory epithelium proper and efferent autonomic fibers innervate the blood vessels and the olfactory glands in the lamina propria. The function of the trigeminal fibers is to participate in secreto-sensory neuromodulation of receptor activity and secretomotor reflexes.

B. Olfactory Epithelium: Cell Types and Innervation

There are three morphologically defined cell types in the olfactory epithelium: olfactory receptor neurons, sustentacular cells and basal cells (Fig. 1C, left). Moran et al. (1982) calculated that there are approximately 6×10^6 ciliated olfactory receptor neurons located only in the 2 cm^2 of tissue identified as olfactory mucosa. They also reported the presence of another subclass of cells within the human olfactory mucosa, called the microvillar cell, provisionally identified as a type of olfactory receptor cell (see also Chapter 1 this book). These flask-shaped cells appear to have a unique immunocytochemical reactivity in human olfactory mucosa (Yamagishi et al. 1989). The ciliated receptor cell is a bipolar neuron whose soma-containing nucleus lies at intermediate epithelial depths. A single dendrite projects to the surface of the epithelium where it terminates in several cilia that extend into the overlying olfactory mucus. A single nonmyelinated axon projects to the olfactory bulb without axon collaterals or branching.

The morphological specialization of the neuron has led to the spatial segregation and localization of membrane function (Getchell 1986). Odorant receptor molecules and ligand-gated channels are localized in the ciliary region; voltage-gated channels are localized in the axon hillock and initial segment, where action potentials are initiated, as well as in the axon.

The sustentacular cells, which are modified ependymal cells (Rafols and Getchell 1983), extend from the epithelial surface to the basal lamina. They have five morphologically distinct regions that appear to have specialized functions (Zielinski et al. 1988). For example, the apical microvillar region is characterized by the presence of furosemide-sensitive cotransport ion channels (Persaud et al. 1987), exocytotic secretory vesicles and endosome-like transport vesicles. The basilar expansion, which abuts the basement membrane, is characterized by the presence of ouabain-sensitive (Na^{+}/K^{+})-ATPase transport channels and endosome-like transport vesicles. Both receptor neurons and sustentacular cells have been thoroughly characterized using biochemical and immunocytochemical techniques (for review, see Getchell et al. 1988).

The basal cells serve as progenitor cells for olfactory receptor neurons. The receptor neurons have a life span of less than several months (Hinds and McNelly 1981, Farbman 1986); thus, the olfactory epithelium is a site of continual neurogenesis. In adult humans, large areas of olfactory neuroepithelium are often disorganized and replaced by respiratory epithelium (Nakashima et al. 1984).

The extrinsic innervation of the epithelium consists primarily of substance P-like immunoreactive fibers that project from the trigeminal nerve into the olfactory epithelium (for reviews, see Getchell et al. 1988, Finger et al. 1988, Getchell et al. 1989, and Fig 1C, right). The fibers terminate near the surface of the epithelium and lie in close apposition to the apico-lateral membrane of the sustentacular cells (Zielinski et al. 1989). The density of these fibers has not been determined for humans, but there are estimated to be approximately 2400 substance P-like immunoreactive fibers/mm^2 in amphibian olfactory epithelium. Cytological and electrophysiological evidence indicates that these fibers regulate the secretory activity of the sustentacular cells that, in turn, may regulate the access of odorants to the receptor neurons (Getchell and Getchell, in press).

C. Lamina Propria: Principal Elements and Extrinsic Innervation

Four principal elements are found in the lamina propria: Bowman's glands, whose ducts project to the epithelial surface (Fig 1C, left); large bundles of nonmyelinated axons of the olfactory receptor neurons; small bundles of myelinated axons from the trigeminal nerve; and blood vessels. Bowman's glands produce and secrete the mucus that covers the olfactory epithelium; in human Bowman's glands, serous acini predominate over mucous ones (Nakashima et al. 1984, Lucheroni et al. 1986). The average diameter of human olfactory axons was found to be 0.25 μm; numerous axons are ensheathed by a single Schwann cell (Koling et al. 1986). Fibers of the trigeminal nerve and postganglionic autonomic fibers that travel with it innervate the acinar cells of Bowman's glands and blood vessels. The adrenergic, cholinergic and peptidergic properties of these fibers and their terminals have been described as they relate to the regulation of secretion and vasomotor tone (for review, see Getchell et al. 1988, Getchell et al. 1989, Zielinski et al. 1989). Both physiological and psychophysical interactions between the trigeminal and olfactory systems have been described (Cain 1976, 1981, Silver 1987); the secretomotor reflexes activated by peptidergic secretion from the trigeminal nerve as a result of chemical stimulation of the olfactory mucosa provides an additional mechanism by which the access of odorants is regulated through changes in the composition of the olfactory mucus.

III. SENSORY TRANSDUCTION

A. Stimulus Access: Perireceptor Events

Olfactory mucus is composed of a motile mesh-like network in which the glycoprotein molecules form the gel lattice with water, containing dissolved proteins, enzymes, polyaminoglycans and electrolytes, filling the interstices. Access and clearance of odorants to and from binding sites located on the olfactory cilia are regulated by a complex interplay of physical, molecular and cellular factors (for reviews, see Getchell et al. 1984, Getchell and Getchell, in press). There are at least five physical factors that regulate their access and diffusion times in olfactory mucus. They are; the odorant's partition coefficient and molecular size, the viscosity of the mucus, the tortuosity factor due to the presence of cilia and the length of the diffusion path, i.e., the depth of the mucus layer. The initial partitioning of commonly used odorants at the air-mucus interface favors the water phase over that of the air phase by a factor of 10 for hydrophobic odorants such as the musk ω-pentadecalactone, to 100,000 for hydrophilic odorants such as pyrroline and trimethylamine. The two major variables in the Stokes-Einstein equation, which is used to calculate the diffusion coefficient of molecules, are the odorant's molecular size as approximated by its molecular weight and the viscosity of mucus. As the molecular weight of odorants increases, for example from 130 for amyl acetate to about 350 for certain musks, the diffusion coefficient increases, from about 0.6 x 10^{-6} cm^2/sec to about 0.4 x 10^{-6} cm^2/sec, respectively. This slows the diffusion time by a factor of 5. The viscosity of olfactory mucus is estimated to range from that of water (0.01 poise) to that of tracheal mucus (200 poise). The effect of increased mucus viscosity, assuming that it is a homogeneous medium, would be to retard the diffusion of odorants to receptor sites. The minimum length of the diffusion path, estimated to range from 5 to 35 μm, may be increased substantially for hydrophilic compounds by the presence of the ciliary matrix. When one takes into account these five factors, the estimated diffusion times of odorants to receptor sites located on the cilia range from about 0.3 to 1.5 s, which is in agreement with the onset latencies of excitatory discharges recorded from olfactory receptor neurons.

Although the partition coefficient of odorants favors the water phase over that of the air phase, the olfactory mucus represents a formidable watery barrier for most odorants that have hydrophobic properties. Odorant-binding proteins (OBPs) in nasal/olfactory mucus may serve as a molecular transport mechanism by which hydrophobic odorants are carried from the air-mucus interface through the olfactory mucus to receptor sites located on olfactory cilia (Snyder et al. 1989, Getchell and Getchell, in press). OBP is a homodimer with each subunit having a molecular mass of about 19 kDa; its site of synthesis has been localized by *in situ* hybridization to the lateral nasal gland, and its release is under adrenergic regulation. It constitutes about 1% of the soluble nasal protein and can bind most odorants with affinities that range from 1 to 60 μm. The amino acid sequence of OBP has been determined, and a cDNA clone has established that OBP belongs to a family of proteins whose primary function is to transport lipophilic molecules. Thus, the function of OBP in nasal/olfactory mucus may be to facilitate the transport of hydrophobic odorants from the mucus surface to olfactory receptor sites.

B. Stimulus Reception-Transduction

1. Characterization of odorant receptors

The interaction of odorants with olfactory receptor neurons initiates a sequence of membrane events associated with molecular recognition and sensory transduction (e.g., Bruch et al. 1988, Dodd 1988, Lancet et al. 1988, Snyder et al. 1988, Brand et al. 1989). Three mechanisms for olfactory transduction have been proposed that involve odorant activation of; 1) transmembrane receptors coupled to G-proteins (guanosine triphosphate proteins), adenylate cyclase and cyclic adenosine monophosphate (cAMP)-gated channels, 2) the phosphatidylinositol pathway, and 3) ion channels directly. The biochemical and molecular biological evidence indicates that activation of an odorant receptor initiates a series of molecular events involving second messengers that lead to transmembrane conductance changes through membrane channel proteins. The combined electrophysiological, microscopic and biochemical evidence supports the conclusion that the molecular components of sensory transduction lie in the ciliary membrane of olfactory receptor neurons. For example, the amplitude of the summated receptor potential systematically decreases or increases with a coincident time course that parallels the loss or recovery, respectively, of the ciliary matrix following its ablation by olfactory nerve section, bulbectomy or topical application of chemicals such as zinc sulfate. Also, odorant-evoked conductance changes are recorded from patch clamped receptor neurons only when odorants are delivered to the ciliary membrane and not to other regions such as the dendrite or soma. In addition, ciliary membranes isolated for biochemical studies showed a higher binding affinity for odorants than control tissues and proved to be a rich source of odorant-activated G-proteins and adenylate cyclase, both of which are identified as intermediates in sensory transduction. These observations do not rule out the possibility that certain molecular intermediates in the transduction sequence, such as odorant receptors, G-proteins, adenylate cyclase and odorant-gated channels cannot be identified in other tissues (e.g., Lerner et al. 1988, Kashiwayanagi and Kurihara 1984, 1985), but rather that the alignment of the molecular sequence, its localization in the ciliary membrane, and its coupling to the action potential generating mechanism of a primary sensory neuron establishes a process unique to olfactory transduction.

Ten to thirty cilia, which range in length from about 100 to 150 μm, project from each receptor knob into the overlying mucus to form a dense mucociliary complex (Jafek 1983). The size, distribution and density of ciliary intramembranous particles (IMPs) have been determined and provide evidence for the density of odorant receptors (for review see Getchell 1986). There are about 10^6 IMPs/cilium with a representative surface area of about 160 μm^2. Assuming that the IMPs represent a molecular component in the transduction sequence, such as odorant receptors or receptor transducer complexes, the data suggest a density of about 800 to 2500 odorant receptors/μm^2 of ciliary membrane.

2. Second messenger systems

The laboratories of Lancet (e.g., Pace and Lancet 1987, Lancet et al. 1988) and Fesenko (e.g., Fesenko et al. 1985, 1988), have provided evidence to identify and characterize polypeptides from olfactory cilia that bind odorants specifically. These studies confirm and extend the earlier work by Cagan and coworkers (e.g., Cagan and Zeiger 1978, Cagan and Kare 1981, Rhein and Cagan 1983) who established the ligand-binding characteristics of olfactory tissue. Two sets of transmembrane glycoproteins, identified as gp 95 and gp 56 by Lancet et al. (1988) and gp 88 and gp 55 by Fesenko et al. (1988) have been isolated from olfactory cilia and fulfill many of the criteria established by Lancet (1986) to characterize odorant receptors. A G-protein, localized on the inner face of the ciliary membrane, links the odorant receptor molecules with the second messenger enzyme adenylate cyclase. Since odorant stimulation of olfactory tissue activates adenylate cyclase, the G-protein is identified as a G_s type. Molecular biological studies further indicate that this G protein, G_{olf}, is characteristic of olfactory tissue (Jones and Reed 1989). Odorant-stimulated, G protein-mediated activation of adenylate cyclase activates the enzyme cascade that opens many ion channels, leading to transmembrane conductance changes and subsequent depolarization of the receptor neuron (Lancet 1986, Anholt 1988).

Functional evidence for the role of G-proteins and second messengers in sensory transduction was obtained from electrophysiological studies of the isolated olfactory mucosa and receptor neurons. Persaud et al. (1988) used the voltage-clamped Ussing chamber preparation of the olfactory mucosa to study odorant stimulated transmucosal current. Analogues of cAMP, such as bromo-cAMP or forskolin applied to the ciliated side, caused an enhanced odorant-stimulated current, whereas bromo-cGMP (guanosine monophosphate) decreased it. G-protein analogues, such as GTP-gamma-S and GTP-beta-S, caused enhancement and inhibition, respectively, of odorant-evoked currents. Nakamura and Gold (1987) developed an isolated olfactory neuron preparation in which they patch clamped individual olfactory cilia. They demonstrated that cAMP and cGMP conductances are localized on olfactory cilia. This functional evidence, taken together with the biochemical evidence on cyclic nucleotides and adenylate cyclase (Lowe et al. 1989), demonstrates the presence of cyclic nucleotide-gated conductances localized in olfactory cilia that are associated with odorant transduction.

3. Membrane conductances

The basic electrical properties of the isolated and non-stimulated olfactory mucosa and certain of its cell types have been determined. The mucosa has an average open-circuit potential of -3.2 $\pm$ 0.43 mV with the ciliated side electronegative, a transmucosal resistance of 67 $\pm$ 26.5 ohm cm^2 and a standing short circuit current of 53.0 $\pm$ 14.5 uA/cm^2 (Persaud et al. 1987). The resting membrane potential and input resistance of olfactory receptor neurons *in situ* are 57.4 $\pm$ mV and 259 $\pm$ 157 megaohms respectively (Hedlund et al. 1987). In contrast, the same properties of sustentacular cells *in situ* are: -95 $\pm$ 13 mV and 9 $\pm$ 20 megaohms respectively (Masukawa et al. 1985). The basic electrical properties of the olfactory mucosa and individual cells change as a function of developmental stage, odorant stimulation and pharmacological manipulation.

Membrane currents are being identified and characterized in isolated olfactory neurons using patch clamp techniques (e.g., Trotier 1986, Firestein and Werblin 1987, Maue and Dione 1987). Several currents have been studied, including three voltage-dependent ionic currents that include a sustained inward Ca^{++} current, a transient inward Na^+ current and an outward K^+ current. The gating kinetics of the K^+ currents have been characterized, and the Na^+ current appears to be TTX-insensitive. Firestein and Werblin (1989) developed a tissue slice technique from which the receptor current could be recorded in response to controlled and monitored odorant pulses. The odorant-induced inward current is reversibly blocked by amiloride, suggesting that the initial influx of receptor current is carried, at least partially, by Na^{+} (Persaud et al. 1987, Frings and Lindemann 1988). Odorant evoked inward currents have also been identified in Xenopus oocytes into which mRNA isolated from olfactory mucosa has been microinjected (Getchell 1988, Getchell and Margolis 1990).

IV. ACTION POTENTIAL INITIATION AND TRANSMISSION

The initial influx of receptor current (Labarca and Bacigalupo 1988) leads to a depolarization of the chemoreceptive membrane. The magnitude and duration of the receptor potential are a direct function of the odorant concentration and duration. The site of action potential initiation has been assigned to the axon hillock initial segment using electrophysiological (Getchell 1973) and microscopic (Rafols and Getchell 1983, Small and Pfenninger 1984) techniques. As noted by Hedlund et al. (1987), the threshold for impulse initiation is very low, in the range of 30 to 100 pA for injected current. This indicates that there is tight electrical coupling between the site of odorant transduction located in the ciliary region and the site of impulse initiation at the initial segment. The nonmyelinated axons of olfactory neurons have a small diameter, about 0.2 μm. Consequently, they have a slow conduction velocity, about 0.2 m/sec, which means that it would require about 30 ms for an action potential to travel from the site of impulse initiation to the first synapse in the olfactory bulb about 6 mm away. These observations also indicate a specialization of the neuronal membrane, with specific regions that possess chemoreceptive, passive and active membrane properties. The sequence of events initiated by odorant binding to olfactory ciliary membranes and culminating in action potentials in olfactory axons is summarized in Figure 2.

When the responsiveness of olfactory receptor neurons is examined with a test battery of odorants, approximately 50 to 75% of the neurons respond to one or more odorants (Getchell 1986). Of those neurons responding, a single neuron may respond to one or several odorants in the test battery. These observations suggest a differential distribution of odorant receptors across a population of olfactory receptor neurons. The odorant selectivity displayed by a receptor neuron does not appear to fall along precisely constructed classifications based on molecular structure, chemical functional groups or odor quality.

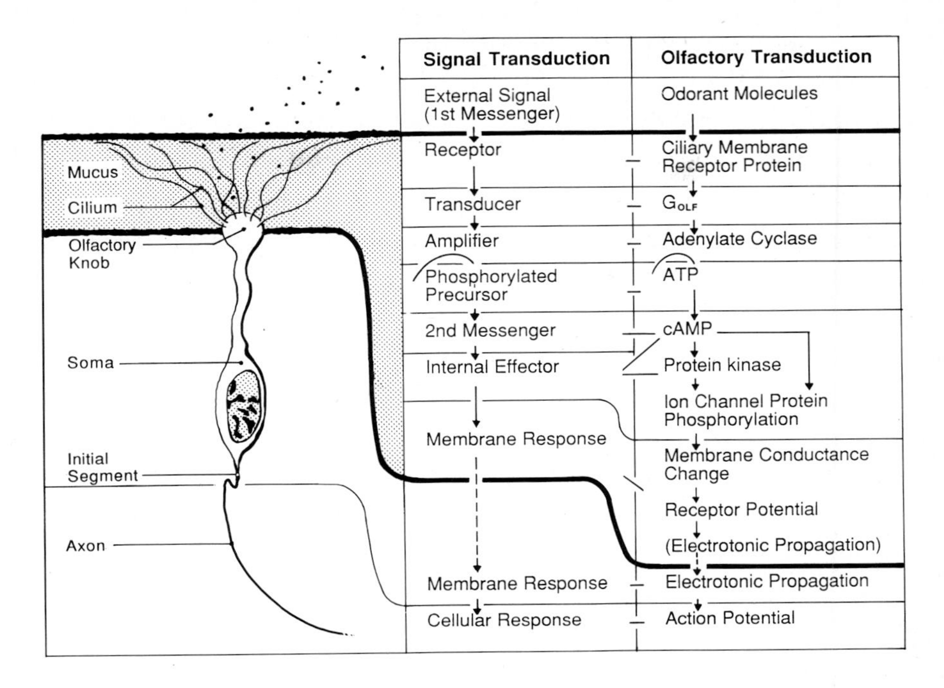

Fig 2. Sequence of events leading to transmission of olfactory information to the central nervous system. On the left, a schematic drawing of an olfactory receptor neuron showing its cilia embedded in a layer of mucus, its dendrite and soma in the olfactory epithelium, and its axon (truncated) in the lamina propria. The center column outlines the general sequence of events in signal transduction, and the right column identifies the counterparts of those events in olfactory receptor neurons. The alignment of the columns and the horizontal lines crossing all three columns suggest the localization of transduction processes in the olfactory receptor neurons. (Concept modified from Berridge 1985). Abbreviations: G_{OLF}, G-protein unique to olfactory receptor neurons; ATP, adenosine triphosphate; cAMP, cyclic adenosine monophosphate.

The threshold for odorant stimulation of olfactory receptor neurons determined in electrophysiological studies ranges from about 10^{-5} to 10^{-9} M. The reported thresholds for a given odorant varies among different receptor neurons and may diverge greatly from those reported in psychophysical and behavioral studies. The dynamic range of most receptor neurons is narrow. The concentration-excitatory response functions are typically steep and show threshold-to-saturation curves that range from less than one to about three log units (Getchell and Shepherd 1978, Firestein and Werblin 1989).

Olfactory sensations in humans are classified as rapidly adapting, which means that a perceptual loss of sensation accompanies a long duration sniff or repetitive sniffs of odors at short intervals. Earlier electrophysiological studies of summed and unit recordings from the olfactory mucosa (Ottoson 1956, Duchamp et al. 1974, Getchell and Shepherd 1978) showed that olfactory receptor neurons have slowly adapting properties with long duration odorant pulses and show response fatigue with iterative stimulation (Getchell and Shepherd 1978, Getchell 1986). These properties are presumably related to the desensitization of the receptor and ancillary perireceptor events. The perceptual adaptation and fatigue to odor stimulation presumably results primarily from the activation of powerful inhibitory surround-type mechanisms by short axon cells in the olfactory bulb (Getchell and Shepherd 1975), and, to a lesser extent, by receptor desensitization.

V. SUMMARY AND CONCLUSIONS: EMERGING FIRST PRINCIPLES

First principles of sensory transduction and coding are emerging from multidisciplinary studies of olfactory receptor neurons. They include:

1. The partitioning and transport of odorants to receptor sites are influenced strongly by the physical and chemical properties of olfactory mucus and the presence of transport proteins.
2. The membrane specializations that subserve the primary functions of odorant detection and transduction, and impulse initiation and transmission, are spatially localized in the ciliary and axonal membranes, respectively.
3. Candidates for odorant receptor sites are transmembrane glycoproteins that bind odorants with high affinities.
4. Odorant-receptor interactions stimulate G-protein-mediated activation of adenylate cyclase that leads to the opening of ion channels.
5. Amiloride reversibly blocks the receptor current, suggesting that Na^+ influx partially accounts for transduction currents.
6. Olfactory receptor neurons show varying degrees of odorant selectivity.
7. The magnitude and duration of the receptor current are a direct function of the odorant concentration and duration.

8. The site of odorant transduction is tightly coupled with the site of impulse initiation.
9. Action potentials initiated at the initial segment are transmitted actively to the olfactory bulb.
10. The onset latency, frequency and duration of excitatory discharges are a direct function of odorant concentration and stimulus duration.
11. Olfactory receptor neurons are slowly adapting receptors that exhibit the property of response fatigue.

REFERENCES

Anholt RRH (1988) Functional reconstitution of the olfactory membrane: Incorporation of the olfactory adenylate cyclase in liposomes. Biochemistry 27: 6464-6468

Berridge MJ (1985) The molecular basis of communication within the cell. Sci Am 253: 142-152

Brand JG, Teeter JH, Cagan RH, Kare MR (eds) (1989) Chemical senses, vol 1. Receptor events and transduction in taste and olfaction. Marcel Dekker, New York

Bruch RC, Kalinoski DL, Kare MR (1988) Biochemistry of vertebrate olfaction. Ann Rev Nutr 8: 21-42

Cagan RH, Kare MR (eds) (1981) Biochemistry of taste and olfaction. Academic Press, New York

Cagan RH, Zeiger WN (1978) Biochemical studies of olfaction: Binding specificity of radioactively labeled stimuli to an isolated olfactory preparation from rainbow trout (Salmo gairdneri). Proc Natl Acad Sci USA 75: 4679-4683

Cain WS (1976) Olfaction and the common chemical sense: some psychophysical contrasts. Sensory Processes 1: 57-67

Cain WS (1981) Olfaction and the common chemical sense. Similarities, differences, and interactions. In: Moskowitz HR, Warren CB (eds) ACS Symposium Series, No 148. Odor quality and chemical structure. American Chemical Society, New York

Dodd GH (1988) The molecular dimension in perfumery. In: Van Toller S, Dodd GH (eds) Perfumery. The psychology and biology of fragrance. Chapman and Hall, London, p 19

Duchamp A, Revial MF, Holley A, MacLeod P (1974) Odor discrimination by frog olfactory receptors. Chem Sens Flavor 1: 213-233

Farbman AI (1986) Prenatal development of mammalian olfactory receptor cells. Chem Sens 11: 3-18

Fesenko EE, Novoselov VI, Bystrova MF (1988) Properties of odour-binding glycoproteins from rat olfactory epithelium. Biochim Biophys Acta 937: 369-378

Fesenko EE, Novoselov VI, Novikov JV (1985) Molecular mechanisms of olfactory reception. VI. Kinetic characteristics of camphor interaction with binding sites of rat olfactory epithelium. Biochim Biophys Acta 839: 268-275

Finger TE, Getchell ML, Getchell TV, Kinnamon JC (1988) Affector and effector functions of peptidergic innervation of the nasal cavity. In: Green BG, Mason JR, Kare MR (eds) Chemical irritation in the nose and mouth. Marcel Dekker, New York, p 1

Firestein S, Werblin FS (1987) Gated currents in isolated olfactory receptor neurons of the larval tiger salamander. Proc Natl Acad Sci USA 84: 6292-6296

Firestein S, Werblin F (1989) Odor-induced membrane currents in vertebrate-olfactory receptor neurons. Science 244: 79-82

Frings S, Lindemann B (1988) Odorant response of isolated olfactory receptor cells is blocked by amiloride. J Membr Biol 105: 233-243

Getchell TV (1973) Analysis of unitary spikes recorded extracellularly from frog olfactory receptor cells and axons. J Physiol 234: 533-551

Getchell TV (1986) Functional properties of vertebrate olfactory receptor neurons. Physiol Rev 66: 772-817

Getchell TV (1988) Induction of odorant-evoked current transients in Xenopus oocytes injected with mRNA isolated from the olfactory mucosa of Rana pipiens. Neurosci Lett 91: 217-221

Getchell ML, Bouvet J-F, Finger TE, Holley A, Getchell TV (1989) Peptidergic regulation of secretory activity in amphibian olfactory mucosa: immunohistochemistry, neural stimulation, and pharmacology. Cell Tiss Res 256: 381-389

Getchell TV, Getchell ML (in press) Regulatory factors in vertebrate olfactory mucosa. Chem Sens

Getchell TV, Margolis FL (1990) The Xenopus oocyte as an in vitro translation and expression system for chemosensory-specific gene products. In: Schild D (ed) NATO ASI Series, Series H: Cell Biology, vol 39. Springer Verlag, Berlin, p 87

Getchell TV, Margolis FL, Getchell ML (1984) Perireceptor and receptor events in vertebrate olfaction. Prog Neurobiol 23: 317-345

Getchell TV, Shepherd GM (1975) Short-axon cells in the olfactory bulb: dendrodendritic synaptic interactions. J Physiol 251: 523-54

Getchell TV, Shepherd GM (1978) Responses of olfactory receptor cells to step pulses of odour at different concentrations in the salamander. J Physiol 282: 521-5408

Getchell ML, Zielinski B, Getchell TV (1988) Odorant and autonomic regulation of secretion in the olfactory mucosa. In: Margolis FL, Getchell TV (eds) Molecular neurobiology of the olfactory system. Plenum Press, New York, p 71

Hedlund B, Masukawa LM, Shepherd GM (1987) Excitable properties of olfactory receptor neurons. J Neurosci 7: 2338-2343

Hinds JW, McNelly NA (1981) Aging in the rat olfactory system: Correlation of changes in the olfactory epithelium and olfactory bulb. J Comp Neurol 203: 441-453

Jafek BW (1983) Ultrastructure of human nasal mucosa. Laryngoscope 93: 1576-1599

Jones DT, Reed RR (1989) G_{olf}: An olfactory neuron specific-G protein involved in odorant signal transduction. Science 244: 790-795

Kashiwayanagi M, Kurihara K (1984) Neuroblastoma cell as model for olfactory cell: mechanism of depolarization in response to various odorants. Brain Res 293: 251-258

Kashiwayanagi M, Kurihara K (1985) Evidence for non-receptor odor discrimination using neuroblastoma cells as a model for olfactory receptor cells. Brain Res 359: 97-103

Koling A, Rask-Andersen H, Deuschl H (1986) A freeze-fracture study of receptor axons and Schwann cells in the human olfactory mucosa. Acta Otolaryngol 102: 494-499

Labarca P, Bacigalupo J (1988) Ion channels from chemosensory olfactory neurons. J Bioenerg Biomembr 20: 551-569

Lancet D (1986) Vertebrate olfactory reception. Annu Rev Neurosci 9: 329-55

Lancet D, Lazard D, Heldman J, Khen M, Nef P (1988) Molecular transduction in smell and taste. Cold Spring Harbor Symp Quant Biol 53: 343-348

Lerner MR, Reagan J, Gyorgyi T, Roby A (1988) Olfaction by melanophores: What does it mean? Proc Natl Acad Sci USA 85: 261-264

Lowe G, Nakamura T, Gold GH (1989) Adenylate cyclase mediates olfactory transduction for a wide variety of odorants. Proc Natl Acad Sci USA 86: 5641-5645

Lucheroni A, Maurizi M, Spreca A, Palmerini CA, Binazzi M (1986) Some aspects of the secretory activity of the human olfactory glands. Rhinology 24: 57-60

Masukawa LM, Hedlund B, Shepherd GM (1985) Changes in the electrical properties of olfactory epithelial cells in the tiger salamander after olfactory nerve transection. J Neurosci 5: 136-141

Maue RA, Dionne VE (1987) Patch-clamp studies of isolated mouse olfactory receptor neurons. J Gen Physiol 90: 95-125

Moran DT, Rowley JC III, Jafek BW, Lovell MA (1982) The fine structure of the olfactory mucosa in man. J Neurocytol 11: 721-746

Nakamura T, Gold GH (1987) A cyclic nucleotide-gated conductance in olfactory receptor cilia. Nature 325: 442-444

Nakashima T, Kimmelman CP, Snow JB Jr (1984) Structure of human fetal and adult olfactory neuroepithelium. Arch Otolaryngol 110: 641-646

Ottoson D (1956) Analysis of the electrical activity of the olfactory epithelium. Acta Physiol Scand 35 Suppl 122: 1-83

Pace U, Lancet D (1987) Molecular mechanisms of vertebrate olfaction: Implications for pheromone biochemistry. In: Prestwick GD, Blomquist GJ (eds) Pheromone biochemistry. Academic Press, New York, p 529

Persaud KC, DeSimone JA, Getchell ML, Heck GL, Getchell TV (1987) Ion transport across the frog olfactory mucosa: the basal and odorant-stimulated states. Biochim Biophys Acta 902: 65-79

Persaud KC, Heck GL, DeSimone SK, Getchell TV, DeSimone JA (1988) Ion transport across the frog olfactory mucosa: the action of cyclic nucleotides on the basal and odorant-stimulated states. Biochim Biophys Acta 944: 49-62

Rafols JA, Getchell TV (1983) Morphological relations between the receptor neurons, sustentacular cells and Schwann cells in the olfactory mucosa of the salamander. Anat Rec 206: 87-101

Rhein LD, Cagan RH (1983) Biochemical studies of olfaction: binding specificity of odorants to a cilia preparation from rainbow trout olfactory rosettes. J Neurochem 41: 569-577

Silver WL (1987) The common chemical sense. In: Finger TE, Silver WL (eds) Neurobiology of taste and smell. John Wiley and Sons, New York, p 65

Small RK, Pfenninger KH (1984) Components of the plasma membrane of growing axons. I. Size and distribution of intramembrane particles. J Cell Biol 98: 1422-1433

Snyder SH, Sklar PB, Hwang PM, Pevsner J (1989) Molecular mechanisms of olfaction. TINS 12: 35-38

Snyder SH, Sklar PB, Pevsner J (1988) Molecular mechanisms of olfaction. J Biol Chem 263: 13971-13974

Trotier D (1986) A patch-clamp analysis of membrane currents in salamander olfactory receptor cells. Pflugers Arch 407: 589-595

Yamagishi M, Hasegawa S, Takahashi S, Nakano Y, Iwanaga T (1989) Immunohistochemical analysis of the olfactory mucosa by use of antibodies to brain proteins and cytokeratin. Ann Otol Rhinol Laryngol 98: 384-388

Zielinski BS, Getchell ML, Getchell TV (1988) Ultrastructural characteristics of sustentacular cells in control and odorant- treated olfactory mucosae of the salamander. Anat Rec 221: 769-779

Zielinski BS, Getchell ML, Wenokur RL, Getchell TV (1989) Ultrastructural localization and identification of adrenergic and cholinergic nerve terminals in the olfactory mucosa. Anat Rec 225: 232-245

Zippel HP (1988) Die Entdeckung der Riechrezeptoren. Spektrum Wissenschaft, Oktober: 120-132

4

MOLECULAR STRUCTURE AND SMELL

M CHASTRETTE

D ZAKARYA

I. INTRODUCTION

Olfactory perception results from a cascade of events beginning with the arrival of airborne odorant molecules at the periphery of the olfactory system, and ending in physiological and psychological effects, defining a response to these stimuli.

In human olfaction the explicit responses are generally verbal, including detection, recognition, discrimination and estimation of intensity. Most known Structure-Odor Relationships (SOR) are based on correlations between chemical structures and odor descriptions of a collection of molecules. In this paper SOR will be presented for a few odors only, to illustrate the techniques used and the current level of knowledge in this field.

II. STRUCTURE-ODOR RELATIONSHIPS

The structure of molecules can play a role in two important steps in the olfactory process; the arrival of the molecules in the vicinity of receptors and receptor-molecule interaction.

Odoriferous substances have first to reach the nasal area,and dissolve into and diffuse through the mucus layer to reach the receptors at the surface of the cilia of the olfactory cell. Substances with very low vapor pressure have very low air concentrations and commonly but not always have high olfactory thresholds. Molecules present in air have to dissolve in the mucus layer, diffuse in it and then leave or bind to molecules within the mucus. This limits the minimum and maximum solubilities of an odorant. Methane for instance which has a high vapor pressure but a very low solubility, is odorless, but becomes odorous when breathed by divers at 13 absolute atmospheres (Laffort and Gortan 1987). In a plot of vapor pressure against solubility, the combined influences of these variables determine the limits of an area outside of which substances are odorless. A limiting curve resembling a parabola was found by Boelens (1983), who plotted the octanol/water partition coefficient P (as log P) against boiling point for several chemical families. Recent work in our laboratory using experimental and calculated partition coefficients and vapor pressures, produced a more satisfying closed curve (Tiyal 1987). Other molecular properties have been used to explain quantitative and even qualitative odor discrimination (Dravnieks and Laffort 1972, Chastrette 1981).

It is widely accepted though not absolutely proven that odorant molecules interact with receptors located at the surface of the olfactory cilia. As a result of these interactions, ion channels are opened, ions are exchanged, and finally a train of spike potentials is generated at the other pole of the cell. A number of hypotheses have been put forward to determine which part of the membrane is actually involved in the interaction. Davies (1953), has suggested interaction with membrane phospholipids. Another hypothesis, more generally accepted, is that odorants interact with proteins embedded in the membrane. In fact, techniques such as freeze-fracture microscopy show that the ciliary membrane contains embedded protein islets. It has been shown (Mason and Clark 1984) that membrane proteins can function as olfactory receptors in mammals. Such interactions provide an explanation for the known ability of humans to discriminate between the odors of enantiomers, which implies diastereotopic associations with chiral binding sites (Ohloff et al. 1980). The receptor protein hypothesis has been discussed by Price (1984) and Lerner et al (1988). Many authors are in favor of a reversible physical contact between receptors and molecules.

Davies (1953, 1965) suggested that odorant molecules might cause a temporary puncture of the olfactory cell membrane initiating a nerve impulse. This theory which received some support from studies on olfactory thresholds has been maintained in recent publications (Kurihara 1986).

Interaction mechanisms between molecules and receptors (Dyson 1928, Wright 1954), based on energy transfer through space at very short range, now seem questionable since they do not explain olfactory differences between enantiomers. Moreover they are not in agreement with mechanisms generally accepted for other types of chemoreception.

Another approach was that of Amoore (1952,1964), who proposed that receptors are sensitive to the shape, size and/or electronic status of the odorant molecules and postulated the existence of seven "primary odors", which correspond to highly specific receptor sites. On the basis of the number of specific anosmias reported, which might correspond to protein receptors (Pelosi 1989), the number of primary odors was later estimated at about thirty. In this approach, importance was attached to the complementarity between shapes of molecules and receptor sites, and to dispersion interactions. For instance, molecules with a musk odor fit into an oval site of 11.5 x 9 Å with a depth of 4 Å. However, frog olfactory cells have been found to be generalists i.e., to respond to many chemicals (Holley et al. 1974, Blank and Mozell 1976), and a study with humans by Schiffman (1974) has shown that changes in a molecule which do not appreciably alter its size and shape can induce important changes in odor quality.

Beets's theory (1957, 1964), on the other hand, differs from that of Amoore on several important points. Beets did not accept the concept of primary odors, and proposed instead the concept of informational modality i.e., the "common informational element in the sensory messages received from all similar compounds in our environment" (Beets 1982).

Another important difference between the two theories was Beets's statistical approach. He considered that stimulation by a pure odorant i.e., a collection of molecules structurally, configurationally and chirally homogeneous but differing by their conformations and orientations, generates a pattern of information containing several modalities at various levels of intensity. Two molecular features are important; the shape of the molecule and the nature and disposition of its functional group(s), which determine the direction of the molecular dipole vector and consequently the orientation of the molecule. The term, functional group, may denote a polar group such as carbonyl or a non-polar group such as methylene involved in the interaction (Beets 1982).

Chastrette and Zakarya (1988a) proposed a theory based on hydrogen bonding and dispersion interactions. Beets's model of interaction with generalist receptors, based on both the shape of the molecule and its functional group(s) was retained, but emphasis was put on interactions through hydrogen bonding rather than on the orientation of dipoles. Each odorant may interact with a given number of types of receptors. For some odorants, however, the main character of the odor may result from a predominant interaction with a single type of receptor. In such cases it should be relatively easy to establish structure-odor relationships. To select odors of this type an original method based on similarities between descriptors was used.

A databank including descriptions of the odors of 2467 pure substances was derived (Chastrette et al. 1986a, 1988b) from a book by Arctander (1969). For a single compound the number of descriptors varies from one to eight, with a mean of 2.8. Two descriptors co-occurring in the same set of compounds would appear identical, while two descriptors which never co-occur would be very dissimilar. Similarities among descriptors were estimated using Ochiai's (1957) similarity coefficients. The olfactory space defined by the compounds and their descriptors appears as a continuum with no hierarchical structure. However, among 74 descriptors used by Arctander, musk, amber, anisic and sandalwood appear isolated, suggesting that the corresponding odors present a very small number of modalities and offer good opportunities for studying SOR.

It seemed convenient to consider for each odor the largest possible collection of pure substances, including commercial compounds with low olfactory thresholds. As molecular shape was an important feature, rigid molecules were carefully selected and their odors checked by perfumers to avoid any doubt of their authenticity.

In the second part of this paper we describe the SOR for musk, sandalwood and bell pepper odors and the associated interaction models. The knowledge gained from extensive work on ambergris odorants (Ohloff 1982,1986) has led to the postulation of the "triaxial rule of odor sensation". This rule and the interesting SOR proposed for ambergris by Ohloff (1986) will not be described here.

III. STRUCTURE-ODOR RELATIONSHIPS FOR MUSKY ODORANTS

In an important review on musky odorants, Wood (1968) divided these compounds into several classes:

Nitromusks include benzene, indan and tetralin derivatives. Among benzene derivatives four important musks are found: Musk Xylol 1, Musk Ketone 2, Musk Tibetene 3 and Musk ambrette 4.

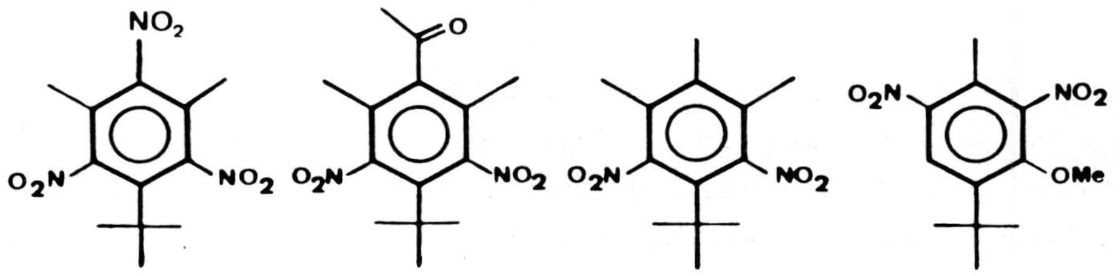

NO_2 O_2N NO_2 O O_2N NO_2 O_2N NO_2 O_2N NO_2 OMe

1   2   3   4

Carbonyl benzene derivatives such as 5 and 6 possess a carbonyl group (CHO or COR) and two t-butyl groups.

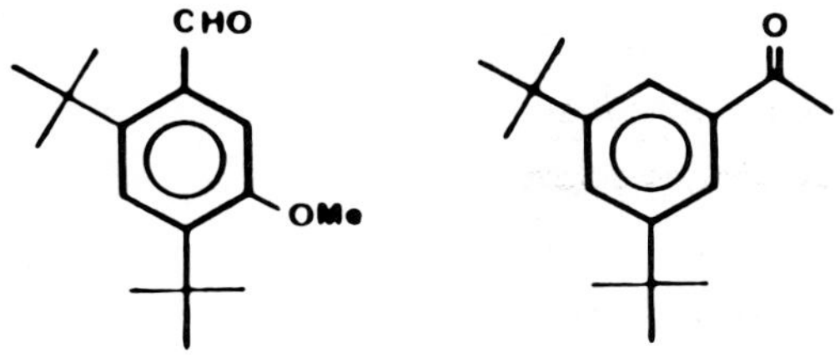

CHO OMe   O

5   6

Compounds 7 to 11 are examples of important indan and tetralin derivatives.

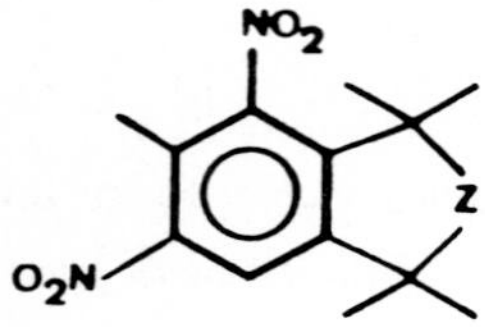

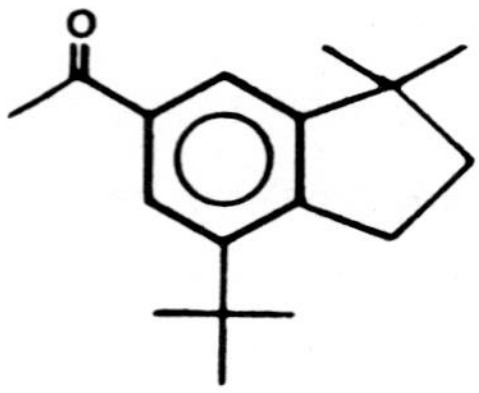

NO_2 O_2N Z

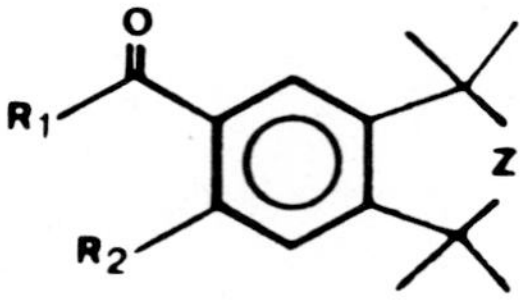

7 Z = CH_2

8 Z = CH_2-CH_2

O R_1 R_2 Z

9 Z = CH_2

10 Z = CH_2-CH_2

O

11

Tricyclic musks are represented by galaxolide 12, 13, and 14.

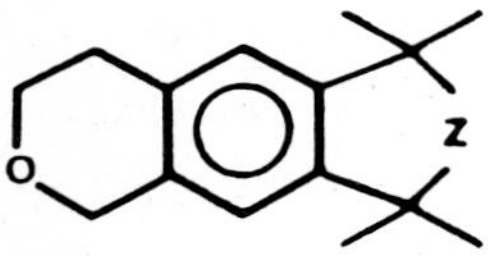

O Z

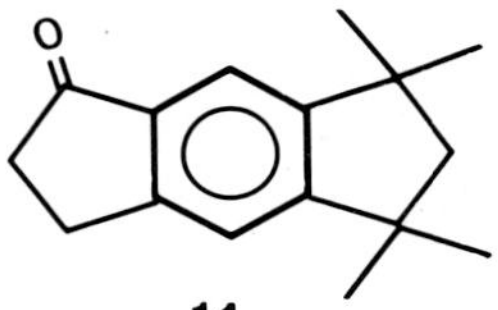

12 Z = CH-CH_3

13 Z = CH_2-CH_2

O

14

Macrocyclic musks may present a variety of heteroatoms and polar groups. Muscone 15 and exaltolide 16 are important compounds in this family.

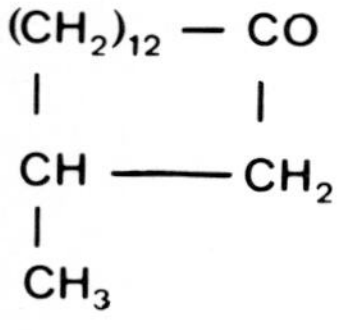

$(CH_2)_{12}$ — CO

CH —— CH_2

CH_3

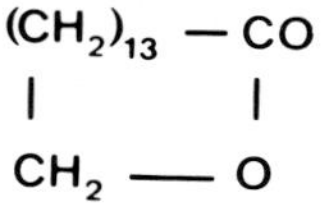

15

$(CH_2)_{13}$ — CO

CH_2 —— O

16

Several relationships between structure and musk odor of chemical compounds have been proposed (Wright 1967, Wright and Burgess 1969, Amoore 1964, Wood 1982). Beets (1957,1978) defined several patterns for musks including ortho or pseudo ortho e.g., 1, 3, and 5, meta or pseudo meta e.g., 4, and 8, and their analogs, and macrocyclic musks. Beets proposed that, a nitro group in a sterically unhindered position may act as a functional group analogous to the acetyl group, while a nitro group ortho to a t-alkyl group may be equivalent to a t-butyl group.

Jurs and colleagues (Ham and Jurs 1985, Narvaez et al. 1986) investigated the relationships between molecular structure and musk odor using pattern recognition techniques. Several separate studies were conducted, with 100% correct classification of musks and non-musks in each case. The descriptors which were retained in the optimized discriminants were closely related to molecular shape, volatility, and electronic effects.

Chastrette et al. (1986b), Chastrette and Zakarya (1988a) studied a set of 244 nitrobenzenes, acetylindans and tetralins, including 104 musks, and determined two sets of structural characteristics called muskophore patterns (P_1 and P_2) for nitrobenzenes and one (P_3) for indans and tetralins (Fig 1) which are necessary for the presence of the musk odor. These patterns are very similar to those suggested by Beets except that they include explicit conditions for the absence of unfavourable substituents such as Y in P_1 and P_2.

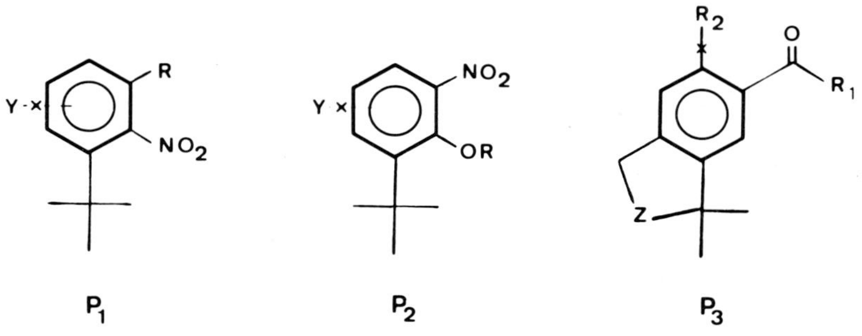

R= Me,Et R_1= H,Me,Et R_2= OR,COOH,CH_2OR
Y= NH_2,OH,COOH Z= CH_2 or CH_2-CH_2

Fig 1. Muskophore patterns.

To design an interaction model explaining how odorants interact with receptor molecules, the patterns were combined with hypotheses on the possible interactions (hydrogen bonding, hydrophobic interactions) due to functional groups in the odorant molecule.

It was not possible to find a unique site able to accommodate these three patterns when only their geometrical and dipolar characteristics were taken into account, in accordance with results from Schiffman and Dackis (1976). We nevertheless assumed that these patterns may interact with a common site and tested the possibility of interactions through hydrogen bonding. The problem of finding a unique site was reformulated as: Is it possible to find acceptable conformations of patterns P_1 to P_3 in which hydrogen atoms, hydrogen bonded to the functional group, would occupy the same position relative to the aromatic ring and to the tertiary butyl group?

Information concerning hydrogen bonding to functional groups was found for carbonyl groups only (Murray-Rust and Glusker 1984, Vedani and Dunitz 1985). As nitro groups possess geometrical characteristics that are very similar to those of carbonyl groups we assumed that the geometrical constraints for bonding were the same for both groups. Once this point is accepted, the similarity between patterns P_2 and P_3 becomes obvious and we only have to compare P_1 and P_2. If P_1 and P_2 are to interact with the same site through a hydrogen bond, the associated hydrogen-bonded atoms H_1 and H_2 should lie in the same position relative to the aromatic ring and to the t-butyl group (Fig 2).

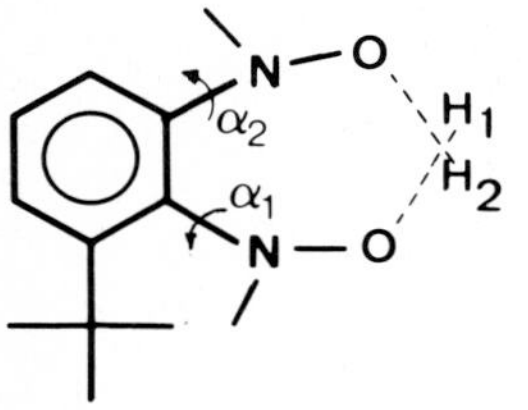

Fig 2. Compatibility of P_1 and P_2.

By designating α_1 and α_2 to be the angles between the ring and the nitro group in P_1 and P_2 respectively, attempts were made to find a pair of α_1 and α_2 values such that the atoms H_1 and H_2 coincide when the aromatic rings are superimposed. It was possible, using the Alchemy II software program to obtain a minimum distance $d(H_1\text{-}H_2)$ of about 0.04 Å for α_1 = -52, α_2 = 51°. The distance between the associated hydrogen atoms and the quaternary carbon atom is about 5Å.

The high values of the α angles mean that the nitro group has lost a large part of its conjugation with the aromatic ring. Confirmation of these values was sought in several ways. An estimation of the lower energy conformations of molecules possessing patterns P_1 and P_2 led to calculated α values in agreement with α_1 and α_2. A second confirmation comes from crystallographic data on compounds structurally similar to P_1 and P_2 (Havere et al. 1982, Bryden 1972).

This interaction model implies that there exists an optimum for α angles depending on the size of substituents. This was shown to be true in two series of indan and tetralin derivatives where the stronger musks are found for $R_1=R_2=$ methyl in 9, and $R_1=H$, $R_2=$ methyl or ethyl in 10.

Thus the interaction model consists of a set of structural elements, geometrical characteristics and physicochemical interactions. The t-butyl group interacts with the hydrophobic part of the site through dispersion forces. The functional groups interact with the hydrophilic part of the site through hydrogen bonds, within the geometrical constraints of hydrogen bonding. The sensory and chemical information indicate that odor intensity depends on the quality of fit between molecules and receptor sites, but other molecular properties such as vapor pressure and solubility in the mucus play an important role.

The hydrogen bonding interaction model correctly predicts the musk odor of various compounds such as isochromans, indans and tetralins, and benzene and coumarin derivatives. Macrocyclic musks set a more difficult problem as they have highly variable conformations and no t-butyl group. The molecular model of muscone in its lowest energy conformation and P_2 were superimposed, using Alchemy II. It was found that when the carbonyl groups are superimposed, the methylene groups corresponding to carbons numbered 5, 6 and 7 are superimposed on the t-butyl group and are probably those involved in the hydrophobic interaction with the site (Fig 3).

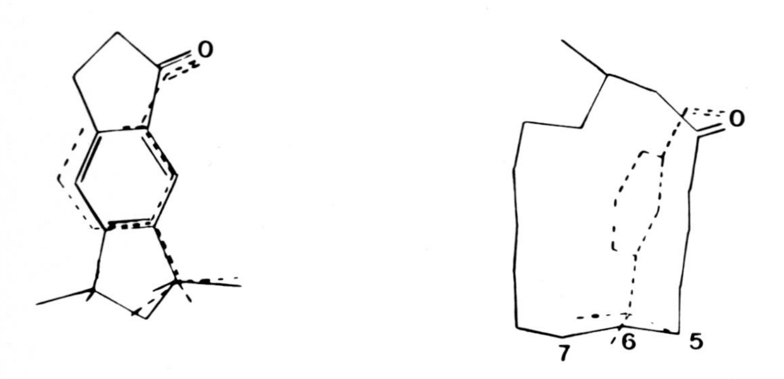

Fig 3. Superposition of P_2 on 14 and muscone 15.

To test the limits of the interaction model further, the constraints on the hydrophobic part of molecules were also examined. The optimum size of the hydrophobic group, which corresponds to a t-butyl or t-amyl group, was more precisely determined using results from Wrobel and Wannagat (1982) on a series of analogs of musk xylol 1 in which the carbon atom of the t-butyl group was replaced by a heteroatom. The derivative containing a silicium atom is a stronger musk than 1, while the germanium derivative is weaker.

IV. STRUCTURE-ODOR RELATIONSHIPS FOR SANDALWOOD ODORANTS

Two important reviews have been published by Naipawer (1987) and Brunke and Klein (1982), in which sandalwood odorants were classified into the following groups:
Norbornyl derivatives are similar to α and β-santalol 17 and 18. The side chain can be replaced by a cyclohexyl radical as in Sandela 19.

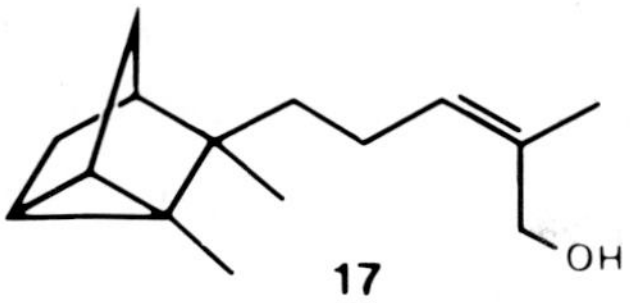

17 OH

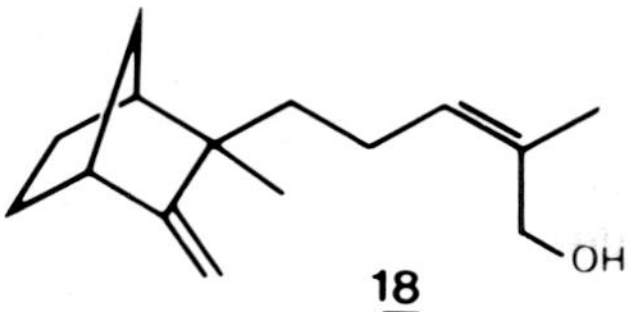

18 OH

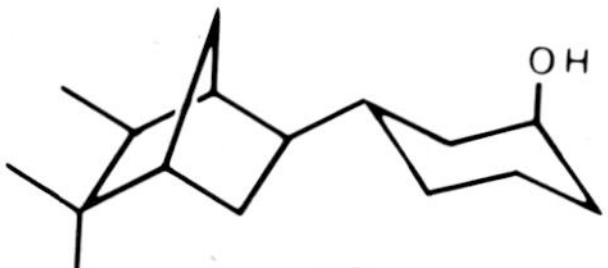

19 OH

Campholenyl derivatives present a campholenic ring bearing a group R which can be an alkyl, cyclohexyl, or cyclopentyl group. Several campholenyl derivatives such as 20 are important commercial products.

Cyclohexene derivatives are somewhat similar to campholenyl compounds. Cyclohexane derivatives represented by Sandrox 21 and Decalin derivatives such as 22, possess a rigid chemical structure. The most important Acyclic compound is Osyrol 23.

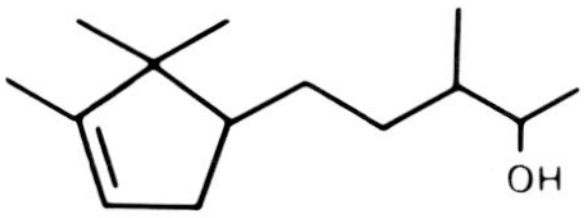

20 OH

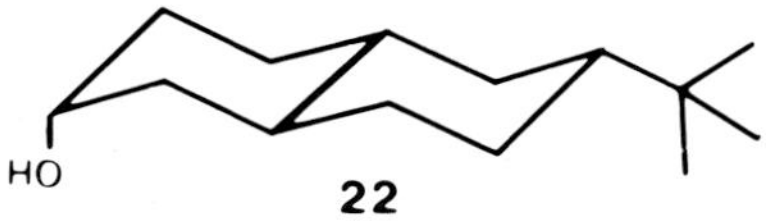

OH 21

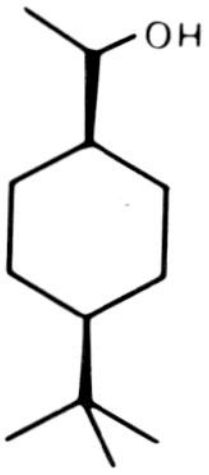

OH OMe 23

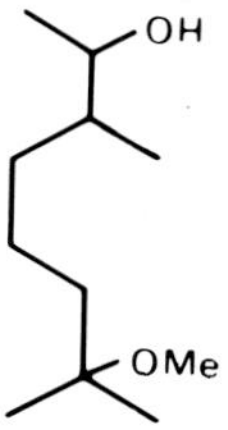

HO 22

Empirical rules given by Naipawer (1987) and Brunke and Klein (1982) state that sandalwood odor requires structural features such as a hydroxyl group, a bulky part, a double bond, and a substituent on the carbinol or on the α carbon atom. A statistical evaluation of these rules (Chastrette et al. unpublished data) for a set of 139 compounds (57 sandalwood and 82 non-sandalwood) confirmed that a bulky part (t-amyl group or equivalent) and a hydroxyl group are necessary to produce sandalwood fragrance.

These structural elements are parts of santalophore patterns S_1 and S_2 which were obtained by superimposing models of molecules having a strong sandalwood odor and possessing a relatively rigid structure. The distance between the hydroxyl oxygen atom and the quaternary carbon atom is approximately 7.1 Å in S_1 and 6.5 Å in S_2 (Fig 4). Thus S_1 and S_2 cannot be fused in a unique santalophore pattern as they would be if only one type of receptor existed for sandalwood odor.

Considering the small difference between the distances measured in S_1 and S_2, it was assumed that molecules possessing these patterns may interact with different parts of the same receptor, which is larger than any of them. This led to the concept of a superpattern defined by the envelope of santalophore patterns. An interaction model was then obtained by combination of the superpattern S structure with hypotheses according to which the bulky part is responsible for dispersion interactions and the hydroxyl group is responsible for hydrogen bonding interactions (Fig 4).

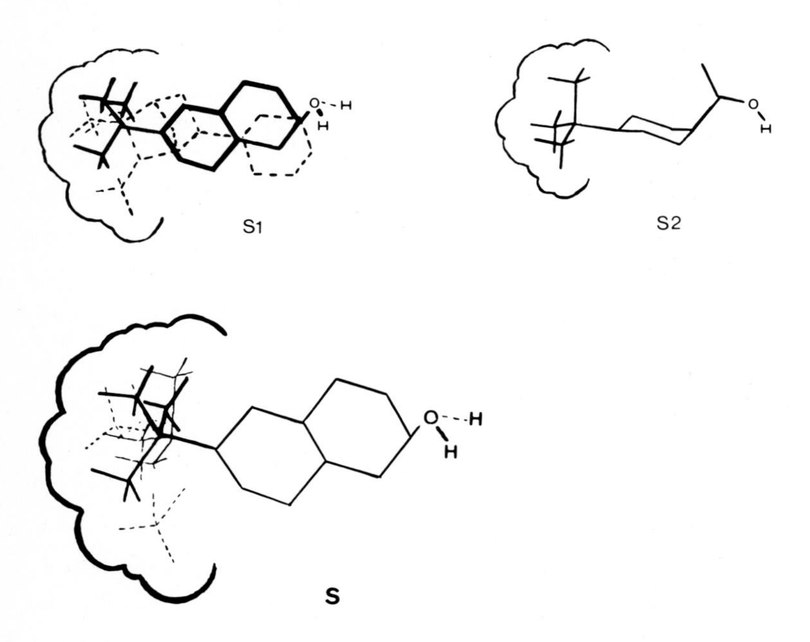

Fig 4. Santalophore patterns S_1, S_2 and superpattern S.

Using a model based on superpattern S, and a set of pertinent distances and angles, 100% correct predictions were made for active and inactive compounds in a test set of 17 relatively rigid molecules. This model proved useful in the design of potential sandalwood odorants and in predicting intensities of sandalwood odors for several pairs of enantiomeric compounds.

V. STRUCTURE-ODOR RELATIONSHIPS FOR BELL PEPPER ODORANTS

Masuda and Mihara (1988) determined the structural features essential for bell pepper odor and proposed an interaction model as shown in Figure 5. An alkyl group, preferably n-pentyl, at the 3-position interacts with a hydrophobic part of the site. The nitrogen atoms of the ring and the oxygen atom in the methoxy group participate as hydrogen-bond acceptors. However, low olfactory thresholds were found (Pelosi et al. 1983) for substituted thiazoles and other pyrazines (Fig 5) and Pelosi (1989) concludes that "only the nitrogen atom farthest from the oxygen is important in determining the odor of bell pepper".

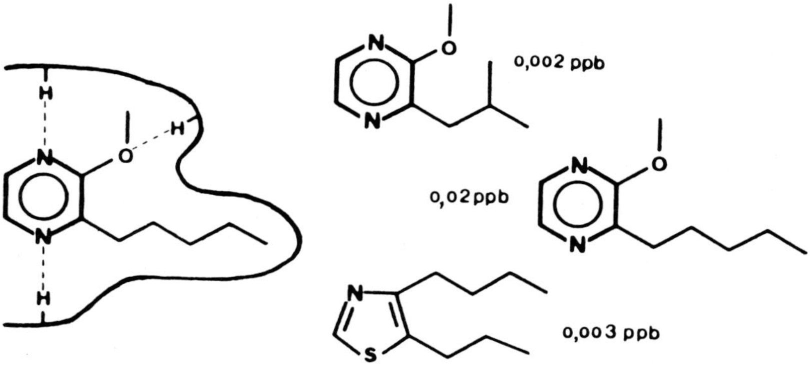

Fig 5. Interaction model for bell pepper odorants and olfactory thresholds (ppb in water). After Masuda and Mihara (1988) and Pelosi (1989).

VI. CONCLUSION

Molecule-receptor interaction models have now been established for a few specific odors, namely musk, sandalwood, and bell pepper, and have proved useful for designing new odorants. However, despite these recent developments in our knowledge of the interactions between odor molecules and receptor sites, much work remains to be done for a satisfying understanding of the first steps of the olfactory process.

REFERENCES

Amoore JE (1952) The stereochemical specificities of human olfactory receptors. Perf Ess Oil Rec 43: 321-330

Amoore JE (1964) Current status of the steric theory of odor. Ann NY Acad Sci 11: 457-476

Arctander S (1969) Perfume and flavor chemicals. Montclair, New Jersey

Beets MGJ (1957) Structure and odour.In: Molecular structure and organoleptic quality. Soc of Chem Ind, London

Beets MGJ (1964) A molecular approach to olfaction. In: Ariens EJ (ed) Molecular pharmacology, vol II. Academic Press, New York

Beets MGJ (1978) Structure-activity relationships in human chemoreception. Applied Science Publishers, London

Beets MGJ (1982) Odor and stimulant structure. In: Theimer ET (ed) Fragrance chemistry. The science of the sense of smell. Academic Press, New York

Blank DL, Mozell MM (1976) Olfactory receptor response characteristics: a factor analysis. Brain Res Bull 1: 185-192

Boelens H (1983) Structure activity relationships in chemo-reception by human olfaction. TIPS 5: 421-426

Brunke EJ, Klein E (1982) Chemistry of sandalwood fragrance. In: Theimer ET (ed) Fragrance chemistry. The science of the sense of smell. Academic Press, New York

Bryden JH (1972) The crystal structure of 2,4,6 trinitro m-xylene (TNX). Acta Cryst 28: 1395-98

Chastrette M (1981) An approach to a classification of odours using physicochemical parameters. Chem Sens 6: 157-163

Chastrette M, Elmouaffek A, Sauvegrain P (1988b) A multi-dimensional statistical study of similarities between 74 notes used in perfumery. Chem Sens 6: 295-306

Chastrette M, Elmouaffek A, Zakarya D (1986a) Etude statistique multidimensionnelle des similarités entre 24 notes utilisées en parfumerie. C R Acad Sci t303 serII: 1209-1214

Chastrette M, Zakarya D, Elmouaffek A (1986b) Relations structure-odeur dans la famille des muscs benzeniques nitrés Eur J Med Chem 21: 505-510

Chastrette M, Zakarya D (1988a) Sur le rôle la liaison hydrogène dans l'interaction entre les récepteurs olfactifs et les molecules à odeur de musc. C R Acad Sci t307 ser 2: 1185-1188

Davies JT (1953) L'odeur et la morphologie des molécules. Industr Parf 8: 74-79

Davies JT (1965) A theory of the quality of odours. J Theor Biol 8: 1

Dravnieks A, Laffort P (1972) Physicochemical basis of quantitative and qualitative odor discrimination in humans.In: Schneider D (ed) Olfaction and taste, vol IV. MBH, Stuttgart, p 142

Dyson GM (1928) Some aspects of the vibrational theory of odour. Perf Ess Oil Rec 19: 456-459

Ham CL, Jurs PC (1985) Structure-activity studies of musk odorants using pattern recognition:monocyclic nitro-benzenes. Chem Sens 10: 491-505

Havere WV, Lenstra ATH, Geise HJ (1982) 4-Terbutyl 3,5-dinitro-anisole. Acta Cryst 38: 3119-20

Holley A, Duchamp A, Revial M, Juge A, Macleod P (1974) Qualitative and quantitative discrimination in the frog olfactory receptors: Analysis from electrophysiological data. Ann NY Acad Sci 237: 102-114

Kurihara K (1986) Transduction mechanism in chemoreception. Comp Biochem Physiol A. Comp Physiol 85: 1-22

Laffort P, Gortan C (1987) Olfactory properties of some gases in hyperbaric atmosphere. Chem Sens 12: 139-142

Lerner MR, Reagan J, Gyorgyi T, Roby A(1988) Olfaction by melanophores: what does it mean? Proc Nat Acad Sci USA 85: 261-264

Mason JR, Clark L (1984) Selective deficits in the sense of smell caused by chemical modifications of the olfactory epithelium. Science 226: 1092-1094 and ref cited

Masuda H, Mihara S (1988) Olfactive properties of alkylpyrazines and 3-substituted 2-alkylpyrazines. J Agric Food Chem 36: 584-587

Murray-Rust P, Glusker JJ (1984) Directional hydrogen bonding to sp^2 and sp^3 hybridized oxygen atoms and its relevance to ligand-macromolecule interactions. J Am Chem Soc 106: 1018-25

Naipawer RE (1987) Recent advances in synthetic sandalwood odorants. Annu Congr Am Chem Soc, Washington

Narvaez JN, Lavine BK, Jurs PC (1986) Structure-activity studies of musk odorants using pattern recognition:bicyclo and tricyclobenzenoids. Chem Sens 11: 145-156

Ochiai A (1957) Zoogeographic studies on the soleoid fishes found in Japan and its neighbouring regions. Bull Jap Sci Fish 22: 526-530

Ohloff G (1982) The fragrance of ambergris. In: Theimer ET (ed) Fragrance chemistry. The science of the sense of smell. Academic Press, New York

Ohloff G (1986) Chemistry of odor stimuli. Experientia 42: 271-279

Ohloff G, Vial C, Wolf HR, Job K, Jegon E, Polonsky J, Lederer E (1980) Stereochemistry-odor relationships in enantiomeric ambergris fragances. Helv Chim Acta 63: 1932-1976 and ref cited

Pelosi P (1989 Designing odors. Quintessenza 5: 36-46

Pelosi P, Pasqualetto P, Lorenzi R (1983) Synthesis and olfactory properties of some thiazoles with bell pepper like odor. J Agric Food Chem 31: 482-484

Price S (1984) Mechanism of stimulation of olfactory neurons:an essay. Chem Sens 8: 341-354

Schiffman SS (1974) Physicochemical correlates of olfactory quality. Science 185: 112-117

Schiffman SS, Dackis C (1976) Multidimensional scaling of musks. Physiol Behav 17: 823-829

Tiyal F (1987) Diplome d'Etudes Approfondies. Université Lyon I

Vedani A, Dunitz JD (1985) Lone pair directionality in hydrogen bond potential functions for molecular mechanics calculations: the inhibition of human carbonic anhydrase II by sulfonamides. J Am Chem Soc 107: 653-58

Wood TF (1968) Chemistry of the aromatic musks. The Givaudanian

Wood TF (1982) Chemistry of synthetic musks. In: Theimer ET (ed) Fragrance chemistry: The science of the sense of smell. Academic Press, New York

Wright RH (1954) Odour and molecular vibration. I Quantum and thermodynamic considerations. J Appl Chem 4: 611-615

Wright RH (1967) The musk odour. Perf Ess Oil Rec 58: 648-650

Wright RH, Burgess RE (1969) Musk odour and far-infrared vibration frequencies. Nature 224: 1033-1035

Wrobel D, Wannagat U (1982) Sila substituted perfumes, sila derivatives of some musk scents. J Organomet Chem 228: 203-10.

PART 2

MEASUREMENT OF OLFACTORY RESPONSES

5

PSYCHOPHYSICAL MEASUREMENT OF ODOR PERCEPTION IN HUMANS

RICHARD L DOTY

1. INTRODUCTION

The olfactory system is a remarkable receiving and integrating system capable of detecting and discriminating among thousands of odorants, often at concentrations too low to be sensed by analytical instruments such as the gas chromatograph. The resultant odor sensations derive not only from single chemicals, but from mixtures which contain, in some instances, hundreds of elements. In the latter case complex summation and integration occurs which most commonly results in a unitary perception of a smell.

It is within this framework of complexity that our attempts at quantification of olfactory sensation must be made. Psychophysics, a field traditionally concerned with mathematical relationships between changes in physical stimuli and resulting psychological sensations, has been hard pressed to deal with such complexity and has provided only limited insight into the physiological mechanisms underlying qualitative odor sensations. This is due, in part, to the fact that no simple physical metric analogous to wavelength for color or frequency for pitch has been found for the sense of smell, and to the observation that mixtures of odorants do not lend themselves to predictable psychological and physiological outcomes in the same manner as do mixtures of lights or sounds. Although mathematical models utilizing physico-chemical variables have been developed in an effort to "explain" the psychological dimensions of odor quality and intensity (Laffort et al. 1974), such models have minimal explanatory value, even when regression coefficients of respectable magnitude are obtained. One limitation of this approach is that high correlations are present among many physicochemical parameters and, therefore, seemingly infinite sets of elements can be substituted into the regression equations. This situation is made even more complex by the observations that 1) enantiomers of the same compound can smell quite differently (Pike et al. 1988); 2) compounds with different chemical structures can produce similar, if not identical, odors (Beets 1971); 3) repeated testing or exposure to odorants can result in increased sensitivity to them (Wysocki et al. 1989), and 4) some odor qualities can be influenced by stimulus concentration (Gross-Isseroff and Lancet 1988).

Despite having provided little insight into basic mechanisms of odor quality perception, psychophysical measurement has led to a better understanding of several general aspects of human olfactory function. For example, the perceived intensity of a number of odorants correlates well with both the degree to which they induce adenylate cyclase enzyme activity in the frog olfactory ciliary preparation and the magnitude of the frog electro-olfactogram (Doty et al. 1990). These observations, along with the finding that pseudohypoparathyroid (PHP) patients deficient in the guanosine triphosphate protein known as G_s protein evidence an impaired ability to smell, whereas PHP patients with normal G_s protein activity do not, suggests that G proteins and the cAMP second messenger system may be involved in human odor perception (Ikeda et al. 1988, Weinstock et al. 1986). In addition, quantitative psychophysical measures have served to demonstrate the degree to which olfactory function is influenced by gender, age, tobacco smoking, environmental toxins and various disease processes. For these and other reasons, olfactory testing is now routinely performed in the food, beverage and perfume industries, as well as in a number of medical fields, including neurology, otorhinolaryngology, epidemiology and public health (see Chapters 14 and 15 this book).

In the present Chapter the basic psychophysical methods used to measure olfactory function are described. In addition, some of the factors known to influence measures determined by such methods are reviewed. The reader is referred elsewhere for more technical accounts of psychological and psychophysical measurement, as well as descriptions of psychophysical procedures, such as those based upon Thurstone's (1927) law of comparative judgment, which have rarely or never been applied to the olfactory domain (Boring 1942, Gescheider 1988, Guilford 1954, Marks 1974, Torgerson 1958).

II. ODOR QUALITY CLASSIFICATION

Attempts to quantify olfactory sensations predate the successful systematic procedural developments outlined by Fechner in his 1860 publication of Elemente der Psychophysik. Paramount was the early focus on the grouping of odors into a seemingly infinite number of categories and the use of such categories to distinguish and describe a variety of environmental situations, including miasmas which accompanied epidemics and the widespread distribution of human waste (Corbin 1986). Although, in retrospect, such classification systems provided little insight into the human olfactory process (other than to suggest complexity), they became the backbone for a number of 19th and 20th Century odor categorization schemes.

The most important of the systematic classification systems to emerge from this era was the 7-category system developed by Linneaus (1756) (for reviews see Boring 1933, Harper et al. 1968). Linneaus' system served as the basis for most subsequent odor classification schemes, including the elaborate one devised by the great Dutch scientist, Hans Zwaardemaker. In effect, Zwaardemaker (1895, 1925) sought to improve on Linneaus' system by subdividing several of Linneaus' classes and by adding the ethereal and ephreumatic classes of Lorry (1784-1785) and von Haller (1763) respectively, resulting in a 9-category system (Table 1).

A number of efforts to formally explain the seemingly infinite array of odor qualities on the basis of combinations of a finite and smaller set of odors or odor dimensions followed this early period. These included 1) Henning's (1916) smell prism which sought to represent, in three-dimensional space, all possible smell sensations in a manner analogous to the way that hues and degrees of saturation are depicted on a color wheel or solid; 2) Crocker and Henderson's (1927) system in which odors were related numerically (in steps ranging from 1 to 8) to four presumed underlying psychological odor dimensions (fragrant, acid, burnt and caprylic), and 3) Amoore's primary odor system (Amoore 1962a,b, 1970), in which all odor qualities were presumed to result from some combination of seven primary odor classes (ethereal, camphoraceous, musky, floral, minty, pungent and putrid). The latter system sought to derive empirical support for the seven classes from structure-activity associations; namely, correlations between odor quality and measures of molecular stereochemistry. Amoore's system is currently undergoing modification (Amoore, personal communication) to accept a larger number of primaries than seven, in accord with empirical studies which do not support an inclusive 7-category odor typology (Coxon et al. 1978, Schiffman 1974).

TABLE 1 : Zwaardemaker's (1925) complete 9-category odor classification system.

1. Etherial: e.g., acetone, chloroform, ethyl ether, ethyl acetate

2. Aromatic
   Subclasses:
   a. Camphorous e.g., camphor, eucalyptol, pinene
   b. Spicy e.g., eugenol
   c. Anisic e.g., anisole, thymol, menthol
   d. Citric e.g., citral, geraniol
   e. Amygdalate e.g., benzaldehyde, nitrobenzene

   Others e.g., laurel, resins, lemon, rose, cinnamon, lavender, mint, marjoram.

3. Balsamic
   Subclasses:
   a: Flower perfumes e.g., jasmine, orange blossom
   b: Lily e.g., ionone, violet root
   c: Vanilla

4. Amber-musk

5. Alliceous
   Subclasses:
   a. Garlic e.g., acetylene, H_2S, ethyl sulfide, mercaptan
   b. Cacodyl e.g., trimethylamine
   c. Halogen e.g., bromine, iodine

6. Empyreumatic e.g., roasted coffee, toasted bread, tobacco smoke, tar, benzol, phenyl xylol, toluol, cresol, guiacol, naphthalene, aniline

7. Hircine e.g., caproic acid, other fatty acids, cheese, sweat, bilberry, cat's urine; perhaps also vaginal and sperm odor, chestnut and barberry

8. Repulsive suffocating e.g., odors of the solanaceae and of coriander, some orchids, some bugs; narcotic odors

9. Nauseous e.g., rotten meat, indole, skatole, carrion flower

Despite the fact that odor classification systems have not been of much use in elucidating basic mechanisms of the olfactory system, they have found a place in applied settings. For example, the wine, beer, and drinking water industries have established standard descriptors for odors in an effort to maintain quality and to eliminate contaminants (Meilgard et al. 1982, Nobel et al. 1987). Mallevialle and Suffet (1987, p 102) succinctly state the value of odor classification to the water industry as follows: "Classification of odors that may be detected in drinking water is desirable and, in fact, necessary. Classification simplifies the description of odors, unifies the terminology used to describe odors, indicates possible sources of odors and possible relations between other odors, and, finally, may indicate the best method to remove a particular odor from water."

III. MEASUREMENT OF ABSOLUTE OLFACTORY SENSITIVITY

Two general approaches to the measurement of absolute olfactory sensitivity have been developed. The first is based upon the idea, espoused by Herbart (1824-1825) and developed operationally by Fechner in 1860, that some critical amount of a mental event must be reached in order to be consciously perceived. This "classical threshold" approach underlies the most widely used procedures for measuring olfactory sensitivity. The second approach, based upon Tanner and Swets' (1954) application of statistical decision theory to the human detection situation, emphasizes the importance of subject expectancies and payoffs in influencing performance, and implies that the traditional threshold concept is misleading. The general model derived from this application, the theory of signal detection, is useful in establishing the relative roles of sensory and non-sensory factors in a given detection task.

Regardless of whether the focus is on threshold or signal detection measures, valid olfactory sensitivity measurement requires a reliable means for presenting different concentrations of odorants to the subject. Procedures used for this purpose include the draw tube olfactometer of Zwaardemaker (1927) (an ingenious device which has not received much attention or refinement in the last 50 years), glass sniff bottles (Cheeseman and Townsend 1956, Doty et al.1986), plastic squeeze bottles (Amoore and Ollman 1983, Cain et al. 1988), air-dilution olfactometers (Punter 1983, Stone and Bosley 1965), and glass rods, wooden sticks, or strips of blotter paper dipped in odorant (Campbell and Gregson 1972, Semb 1968, Toyota et al. 1978) (Fig 1)[1]. Since sniff bottles are easy to use and reliable, they are the basis for many of the examples used in this Chapter. Although air-dilution olfactometers are the method of choice when exact stimulus specification is required, the odorant concentrations within the head space of sniff and squeeze bottles can also be quantified using analytical procedures, as can the outflows from the Zwaardemaker olfactometer. It should be noted that none of these procedures quantify the stimulus at the level of the receptors.

A. Classic Olfactory Detection Threshold Measurement

Since the pioneering work of Valentin (1848), Dibbits (1888), and Fischer and Penzoldt (1886), thousands of olfactory threshold values have appeared in the literature (Stahl 1973) and considerable effort has gone into the refinement of olfactory threshold testing. Unfortunately, threshold values are still treated by many as fixed biologic entities, and it is rarely appreciated that among "normal" individuals threshold measures fluctuate from moment to moment and that average threshold values differ by several orders of magnitude (Amoore 1971, Doty et al. 1986, Stevens et al. 1988). Although many investigators are now aware of the need to specify to a subject whether the focus of attention should be on the detection of a subtle difference in sensation or on the detection of a recognizable odor quality, others are not, leading to increased variability in the threshold values.

1. *Standard procedures for measuring olfactory thresholds*

The lowest concentration of an odor that can be perceived (or discerned from a blank stimulus or background noise) is commonly termed the odor detection threshold. As noted above, this concentration is not a fixed entity, but varies from one trial or moment to another and is somewhat elusive to measure. Therefore, mathematical estimates of an

[1] Another procedure for presenting odorants of which the reader should be aware is the blast injection technique of Elsberg and Levy (1935). In this procedure, a threshold is defined as the lowest volume of odorized air which, when injected into the nose, produces an odor sensation. Since this measure is unreliable, is based upon an unnatural stimulus presentation, and reflects not only stimulus concentration, but pressure and volume artifacts (Jerome 1942, Jones 1953, Wenzel 1948), it is not discussed in this Chapter.

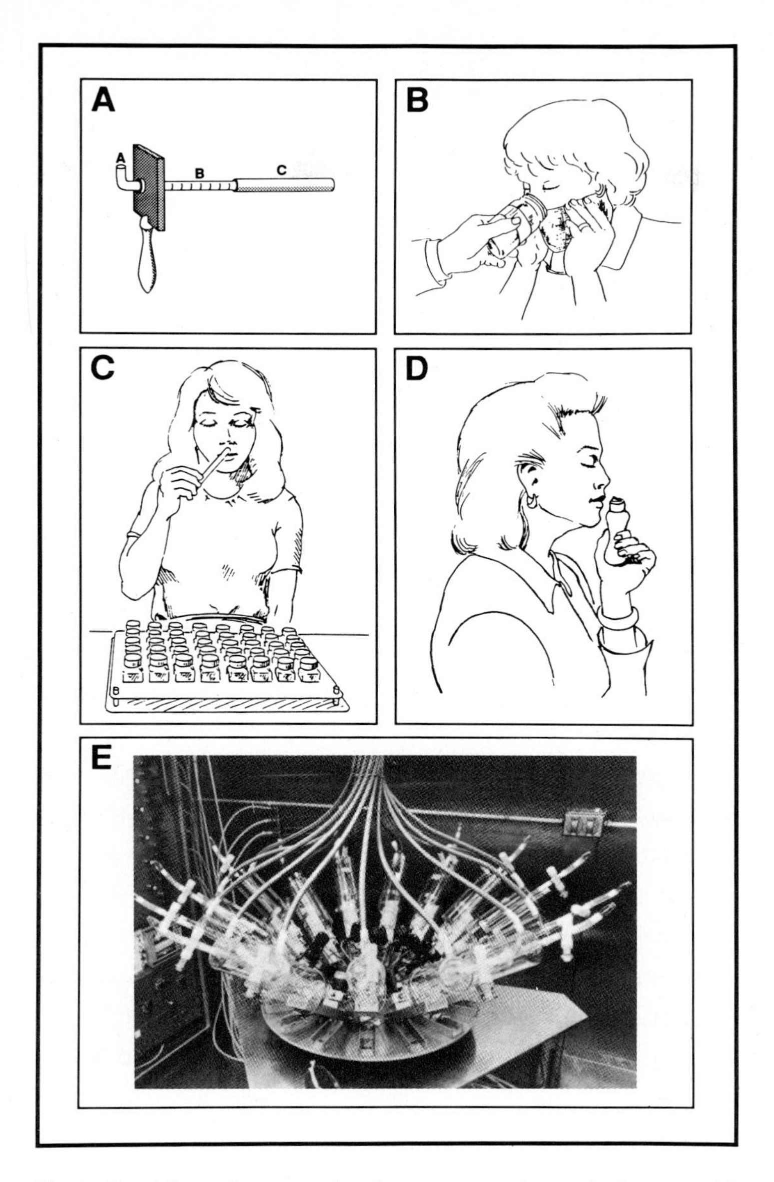

Fig 1. Procedures for presenting low concentrations of odors to subjects for assessment. A. An early draw-tube olfactometer of Zwaardemaker. Later models allowed mixtures to be presented from several parallel tubes. Adapted from Zwaardemaker (1925); B. Sniff Bottle. Reprinted with permission from Doty et al. 1978a; C. Perfumist's strip; D. Squeeze bottle; E. Sniff ports on a rotating table connected to the University of Pennsylvania's Dynamic Air-Dilution Olfactometer. Reprinted with permission from Doty et al. 1988.

average threshold are established. Sensations experienced by observers at such low odorant concentrations rarely, if ever, contain a recognizable qualitative odor component e.g., "rose like", reflecting only the experience of something more than nothing. The so-called recognition threshold, the lowest concentration of an odorant that has a recognizable quality, is typically higher than the detection threshold i.e., occurs at stronger odorant concentrations, and is more difficult to reliably measure. Although a distinction can be made between recognition thresholds and identification thresholds, which require the qualitative sensation to be identified in addition to being recognized as an odor quality (Dember 1965), it appears that this has never been done in the chemical senses.

The three psychophysical methods described in this section are the most popular for measuring basal detection olfactory thresholds. The first two were formally developed by Fechner, although not in relation to olfaction, and published in his 1860 treatise.

a. The method of constant stimuli

In this method, also termed the method of right and wrong cases and the constant method, the subject is presented with a series of concentrations of an odorant which encompass the imperceptible to clearly perceptible stimulus range. The order of presentations of concentrations is typically randomized, and hundreds of trials are usually made in an effort to obtain a reliable result. In the two-alternative forced-choice variant of this technique, a pair of stimuli (blank vs odorant) is presented at each concentration level in counterbalanced order, and the subject indicates which of the two stimuli elicits the stronger sensation. A response is required on each trial, even if no difference is discernible. The proportion of correct responses is plotted as a function of odorant concentration, and threshold is defined as that concentration corresponding to the 75% correct performance level i.e., that point located halfway between chance (50%) and perfect (100%) performance. In another variant of this technique, three rather than two choices are given at each concentration level (two blanks and an odorant). The threshold in this case is commonly defined as the odorant concentration corresponding to the point half way between chance (33.33%) and perfect (100%) performance i.e., at the 66.67% performance level.

Although traditionally such stimulus-response functions have been assumed to be monotonic and ogival in form, there is now evidence that, for many odorants, "notches", "dips" or "reversals" are present in these functions. When these reversals span the perithreshold region, more than one threshold value, as operationally defined above, can be present i.e., there can be more than one concentration for which 75% correct detection occurs in the two-choice task. This phenomenon is present in perithreshold data from humans (Fig 2), dogs, and rats (Marshall et al. 1981, Marshall and Moulton 1981, Moulton and Marshall 1976, Moulton et al. 1960). At least in some species, the location of the reversal on the stimulus-response curve varies as a function of a molecule's position within a homologous series. For example, in the case of n-aliphatic acids and acetates, reversals occur at successively higher odorant concentrations as carbon chain length increases (Moulton 1960, Moulton and Eayrs 1960).

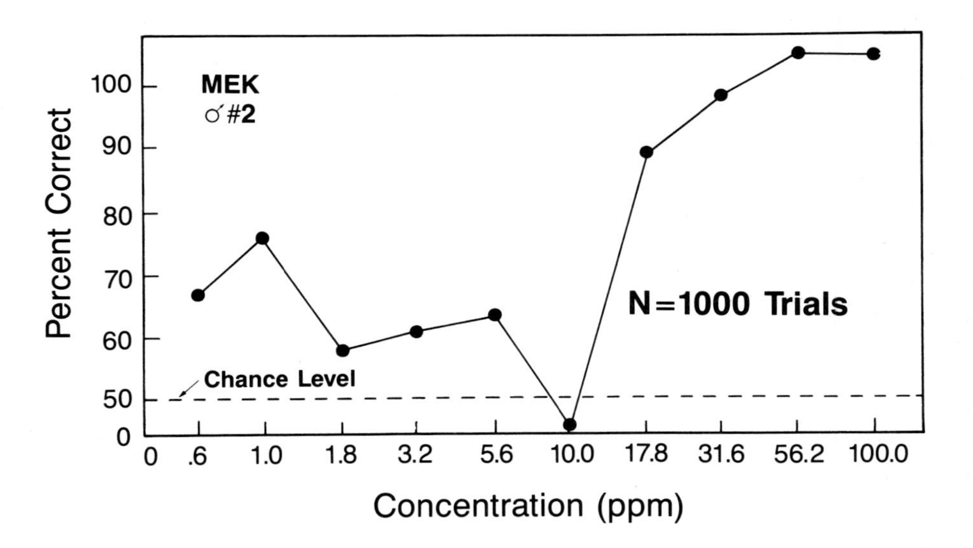

Fig 2. An example of a non-monotonic relationship between odor detection performance and odorant concentration. The curve represents data from a college student repeatedly tested in a two-choice task (odorant vs blank) using the olfactometer shown in Figure 3 (D). For this chemical (methyl ethyl ketone), such curves appear to be idiosyncratic. Doty, unpublished.

Not all odorants show reversals in the stimulus-response function and a number of others show only minor alterations (Doty, unpublished data). For these reasons, threshold estimates of practical value can be obtained for most odorants using the method of constant stimuli by ignoring the notches, when present, determining the best fitting curve to the data, and establishing the concentration corresponding to the 75% correct point on the function. However, the lack of monotonicity should not be forgotten, as it contributes to the variability in test measures and may provide clues to the nature of the underlying physiological mechanisms involved in odor quality coding.

In classical psychophysics, the method of constant stimuli is viewed as the quintessential measurement procedure. However, in studies of olfaction, it is less popular than the other procedures mentioned in this section, mainly because of the large number of trials and attendant amount of time required to obtain a meaningful threshold estimate. In addition, potential for subject fatigue and boredom, as well as for the introduction of adaptation, habituation, or facilitation, are other reasons why investigators have shied away from establishing olfactory thresholds by this method. Nevertheless, this procedure is preferred in cases where time is not at a premium, since it provides information about detection performance across the entire perithreshold concentration range, including differences among individuals and odorants in the shape of the function relating odorant concentration to performance.

b. The method of limits

The method of limits, also termed the method of minimal changes, or one of its variants (see next section), is the most widely used means of establishing an olfactory threshold. In the classical method of limits, alternating series of ascending and descending concentrations of a test odorant are presented. During an ascending trial series, an odorant is increased incrementally in concentration from an initially non-detectable level until it is reported as being detected by the subject. In a descending series, the stimulus is decreased incrementally from an initially detectable level until it is no longer discernible. The average concentration of the transition points where reports of non-detection and detection occur is used as the threshold estimate.

Largely as a result of concerns about adaptation from above-threshold stimuli (Pangborn et al. 1964), the classical procedure is rarely used in olfactory testing. More commonly, investigators present one or more ascending series only, which is termed the ascending method of limits. Although the presentation of a single ascending series is economical of time and has clinical utility e.g., Cain et al. (1988), this procedure provides a threshold estimate which is higher i.e., indicative of less sensitivity, than that obtained from averaging data from multiple ascending series or from other procedures which more thoroughly sample the perithreshold region.

In both the traditional and ascending series method of limits procedures, the starting point of the trial runs of a given type e.g., the ascending runs, is typically selected at a different concentration from run to run so that the subject does not consciously or unconsciously respond to the number of trials that occur before the category change is noticed. As described for the method of constant stimuli, a requirement that the subject selects the more intense stimulus from a set containing one or more blanks is preferred in order to minimize the bias associated with alterations in the response criterion.

c. The staircase or up-down method

This common variant of the method of limits technique (Cornsweet 1962, Dixon and Massey 1957, McCarthy 1949) results in a reliable measure of threshold that requires a minimum number of trials. An advantage of this procedure is that it concentrates the stimulus presentations near the threshold region and eliminates most of the ascending or descending trials of the method of limits. The staircase method is a more accurate and reliable test of threshold than the popular single series ascending method of limits, since it is less dependent upon early test trials in which practice or "warm-up" effects are present, and it uses a number of points within the perithreshold region to bracket the threshold value, thereby resulting in a threshold estimate based upon a sample of more than a single category change. Although there are numerous variations on the general procedure, the description below illustrates the key components of most staircase tasks.

In the two-alternative forced-choice single staircase detection threshold procedure used in our laboratory (Deems and Doty 1987), a trial consists of the presentation of two 120 ml volume glass sniff bottles to the subject in rapid succession. One bottle contains 20 ml of a given concentration of phenethyl alcohol (PEA), a rose-like smelling odorant with low intranasal trigeminal stimulative properties (Doty et al. 1978a, von Skramlik 1926), dissolved in USP-grade light mineral oil. The other bottle contains 20 ml of the mineral oil alone.[2] The sniff bottles are opened and immediately positioned over each subject's nose in a standardized manner (Fig 1B). The subject indicates which one of the two bottles smells stronger. Even if no difference is noticed or if no sensations are perceived, the subject is required to select one or the other bottle. The staircase is begun at the -6.50 log (liquid volume/volume) concentration step and moved upwards in full log steps. Successive blank-odorant pairs are presented at each step until an incorrect response is made or the responses to five sets of such pairs are correct. When a miss occurs, the next higher log concentration is presented. When five consecutive correct trials occur, the staircase is "reversed" and the subsequent pair of trials is presented at a concentration 0.5 log step lower. However, from this point on, only one or two trials are presented at each step i.e., if the first trial is missed, the second is not given and the staircase is moved to the next higher 0.5 log step concentration. The geometric mean of the last four staircase reversal points of a total of seven is used as the threshold estimate.

It should be noted that the threshold value produced by this method is not mathematically equivalent to that produced by such procedures as the method of constant stimuli or the traditional method of limits technique, since two correct trials are required for the staircase to descend and only one incorrect trial is required for the staircase to ascend. Thus, the threshold concentration determined by this method represents the value chosen correctly by the subject 71% of the time, not 50% of the time (Wetherill and Levitt 1965).

Examples of data obtained from six subjects using this procedure are shown in Figure 3 (Doty, unpublished data). All six cases illustrate that 1) the initial staircase reversal point, which is equivalent to the threshold value established in a single series method of limits procedure, is much higher than the subsequent reversal points, and 2) considerable between-subject variability occurs in the detection performance.

---

[2] Diluents other than mineral oil e.g., water, propylene glycol, diethyl phthalate, benzyl benzoate can also be used. Light mineral oil is preferred because it is odorless and is a good solvent for most organic compounds. Water tends to evaporate more quickly than oil (potentially altering the stimulus composition), is not a good solvent for many hydrocarbons, and readily harbors bacteria. Propylene glycol, benzyl benzoate, and diethyl phthalate often have weak odors which can vary from sample to sample, even after repeated distillations.

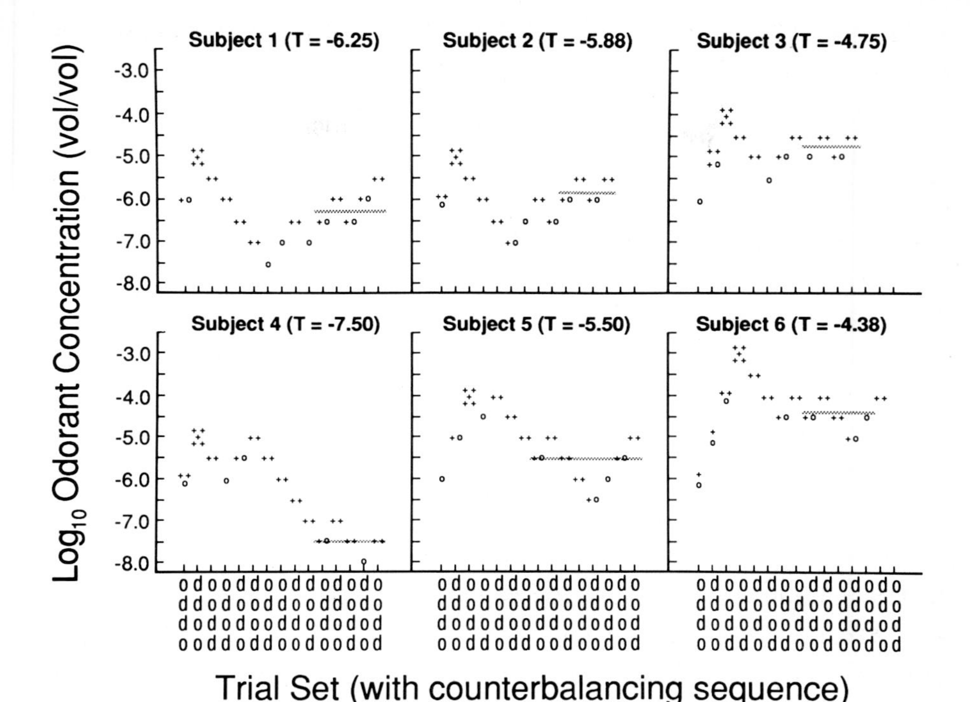

Fig 3. Data from single staircase detection threshold trials. Within the staircases, each + indicates a correct detection when an odorant vs a blank is presented, whereas each 0 indicates an incorrect report of odorant. T = threshold value (vol/vol in light mineral oil) calculated as the mean of the last four of seven staircase reversal points. O's and d's below the abscissa indicate the counterbalancing order of the presentation sequences for each set of trials and are read downward (O = odorant presented first, then diluent; d = diluent presented first, then odorant). For example, on the first trial, the odorant was always presented before the diluent. Subjects 1, 2, 4 and 6 responded correctly on this trial (as signified by the first +), whereas subjects 3 and 5 did not (as signified by the first 0). At the first reversal point, the fifth order sequence was determined by the first O or d of the next column of four order sequences.

B. Signal Detection Measurement of Olfactory Sensitivity

In the examples cited above, the use of a forced-choice procedure minimized the influence of subjects' response criteria on the detection measure. In essence, the response criterion is the internal rule a subject uses to decide whether or not to report the presence of a stimulus. For example, one subject may experience the same degree of sensation from a very weak stimulus as another subject and yet, because of uncertainty, report that no stimulus was perceived. In contrast, the other subject may be less cautious and report the presence of the stimulus. In both cases the stimulus was, in effect, perceived to the same degree, however, the two observers evidenced different response criteria for reporting its presence. Traditional non-forced choice threshold techniques would lead to the erroneous conclusion that the first of these subjects was less sensitive than the second. The theory of signal detection provides a means for distinguishing between the influences of the response criterion, sometimes termed the response bias, and sensory sensitivity in a detection situation, as is discussed in detail elsewhere (Doty 1976, Gescheider 1976, Green and Swets 1966).

In a simple olfactory signal detection experiment, the subject is presented with a relatively large number of trials of a single low concentration of odorant interspersed with blank trials (Doty et al. 1981, Semb 1968). The proportion of the total odor trials on which a subject reports detecting an odor (the "hit" rate) is calculated, as is the proportion of blank trials on which an odor is reported (the "false alarm" rate). The sensitivity measure, d', can then be computed by converting the proportions to z-scores via a normal probability table; d' equals the z-score for hits minus the z-score for false alarms. A more rapid procedure for determining d' for any combination of hit and false alarm proportions is to use the table developed by Elliot (1964). Measures of response criteria, such as the traditional measure of beta, can also be easily calculated from hit and false alarm rates by use of ordinate values from the normal curve, as discussed by Gescheider (1976).

There are procedures for testing the parametric assumptions of traditional signal detection analysis for a given set of data. These procedures are beyond the scope of this Chapter (for details see Gescheider 1976, Green and Swets 1966). Additionally, various non-parametric signal detection measures are available which are not dependent upon such assumptions (Brown 1974, Frey and Colliver 1973, Grier 1971, Hodos 1970, O'Mahony 1979, Pollack and Norman 1964).

IV. MEASUREMENT OF SUPRATHRESHOLD OLFACTORY FUNCTION

Numerous procedures have been developed to measure how persons respond to odor sensations that fall within the clearly perceptible range. Included in this class of measurement procedures are tests of 1) odor quality discrimination, 2) odor quality recognition, 3) odor quality identification, 4) odor intensity discrimination, 5) odor memory, and 6) sensory attribute scaling. The sensory processes measured by such tests are not independent. For example, if one cannot discriminate among odorants, then identification of odors is impossible. Similarly, if one has no short-term odor memory, a reliable determination of odor intensity can be difficult. It is important to be aware that no simple relationship between threshold and suprathreshold measures of olfactory function need be present, as evidenced by the fact that many odorants that are detectable at very low concentrations are perceived as being weak at high concentrations e.g., musks.

A. Odor Quality Discrimination

Odor quality discrimination is among the least demanding of olfactory tasks that a subject can be asked to perform, in that he or she merely needs to evidence the ability to distinguish among odors, not to recognize, identify, or remember them. In practice, however, the difficulty of this task is directly related to the similarity of the stimuli to be discriminated and to the number of trials presented. The three procedures described below provide a means for quantitatively assessing how well a subject discriminates among odors or, alternatively, how discriminable a set of odors is to a subject.

The most basic odor quality discrimination test requires a decision as to whether two odors are the same or different. A number of same- and different-odorant pairs are presented, and the proportion of the pairings that are correctly identified as same or different can be used as the measure of discrimination. Alternatively, data from this type of paradigm can be quantified using measures derived from signal detection theory. For example, a difference that is perceived correctly can be considered a hit, and a difference that is reported when the same odors are presented, a false alarm. Using the hit and false alarm rates, d' or an analogous non-parametric measure can serve as the test measure (O'Mahoney 1979, Potter and Butters 1980).

A more commonly used odor quality discrimination test requires the subject to choose an odd stimulus from a set of stimuli which, except for the odd stimulus, are identical to one another. Usually the stimuli are presented to the subject in counterbalanced order. When three choices are provided (two same and one different) and both of the stimuli occur on half the trials as the odd sample i.e., orders AAB, BAA, ABA, BBA, ABB, BAB, this procedure becomes the so-called triangle test, which is commonly used in applied settings. When only one of the stimuli serves as the odd sample e.g., AAB, ABA, BAA, the test is called a three alternative forced-choice test. Although the proportion of correct trials is a common measure of a subject's discrimination ability, measures derived from signal detection theory can also be used in data evaluation (see Frijters 1980, Frijters et al. 1980).

A third procedure for establishing an individual's ability to discriminate among odorants makes use of an odor confusion matrix (Koster 1972, 1975; see also Wright 1987, in Section C on odor identification). In Koster's variation of this procedure, each odorant of a set is presented in counterbalanced order to a subject. The subject's task is to determine which one of a set of codes previously assigned to each of the odorants best describes the odor sensation on a given trial. The odorants are randomly presented to a subject an equal number of times. If a wrong answer is given, the response is noted, the subject is told the correct response, and the next stimulus is presented. This task is repeatedly performed until the subject has made no mistakes or until some criterion of asymptotic performance is attained. The percentages of responses given to each alternative for each odorant are displayed in a rectangular matrix (stimuli making up rows and response alternatives making up equivalently-ordered columns) and subsequently converted to distance values and subjected to multidimensional scaling (MDS) analysis (Harshman et al. 1982). The values along the negative diagonal represent correct responses, whereas those that fall away from the diagonal are considered to represent confusions.

B. Odor Quality Recognition

Suprathreshold odor quality recognition tests can be divided into two general categories. In the first category, a small set of odorants (often two or three) are presented in succession, and the subject is asked to report in each case whether or not an odor is recognized. Identification of the stimuli is not required. This type of task is relatively unsophisticated, since it is very subjective, and only a "yes" i.e., the odor is recognized, or "no" i.e., the odor is not recognized, is obtained. Unfortunately, this procedure is used by many medical practitioners to ascertain whether a patient has normal smell function. In the second category, a subject is presented on a given trial with a "target" odorant and provided with a set of several odorants which includes the target stimulus. The subject's task is to report which odor of the set is the same as the previously presented target. The number correct in a series of such presentations provides the test score. This forced-choice test eliminates the subjectivity of a simple yes or no response. When multiple sets of odorants are presented in this fashion with different time intervals between the presentation of the target stimulus and the response set, this procedure becomes a test of odor recognition memory.

A variant of this general theme is termed odor matching. An example of an odor matching test is presented by Abraham and Matha (1983). In their procedure, four odors are contained in eight vials (two vials per odorant). The subject is simply required to pair up the equivalent two-vial containers, demonstrating the ability to match the odors. Scoring consists of giving a single point to each pair correctly matched (0 to 4). The test is given twice; if the first and second tests are discrepant, a third test is administered and the average of the three tests is taken. Abraham and Matha found this test to be sensitive to lesions of the right temporal lobe.

C. Odor Quality Identification

Odor quality identification testing requires a subject to either provide a name for each of a set of odors, signify whether or not a stimulus smells like a particular odorant e.g., does this smell like a rose?, or identify the odor from a list of odor names. The first of these tests is termed an odor naming test, the second a yes/no odor identification test, and the third a multiple-choice odor identification test.

Despite the fact that odor naming tests have found use in clinical settings (Gregson et al. 1981), many normal persons have difficulty in naming or identifying even familiar odors without cues. This has been called "the tip of the nose" phenomenon (Lawless and Engen 1977). Therefore, response alternatives are most commonly provided.

The yes/no odor identification test is relatively simple, in that the subject is required only to report whether or not each stimulus smells like that described by a target word supplied by the examiner. Even though this test requires that the subject is capable of keeping the odor in memory long enough to compare it with the target word, which, of course, must also be recalled from memory, it is conceivably less influenced by cognitive and memory demands than multiple-choice odor identification tests (next section). Since chance performance on this type of test is 50%, compared, for example, to 25% on a 4-alternative multiple-choice identification test, its range of discriminability is considerably lower and therefore more trials are needed to obtain the same power as that of the 4-alternative test.

Although several types of multiple-choice odor identification tests have been described (Cain et al. 1988, Doty 1979, Doty et al. 1984b, Wood and Harkins 1987, Wright 1987), they are conceptually similar, and in the few cases that have been examined, strongly correlated with one another (Cain and Rabin 1989, Smith 1988, Wright 1987). The most widely used of these tests is the University of Pennsylvania Smell Identification Test (UPSIT) which is commercially available as the Smell Identification Test™ (Sensonics, Haddonfield, NJ). This test focuses on the ability of subjects to identify each of 40 "scratch and sniff" odorants (Doty 1989, Doty et al. 1984a, Doty et al. 1984b). As shown in Figure 4, the UPSIT consists of four envelope-sized booklets, each containing ten odorants. The odorants are embedded in 10- to 50-μm urea-formaldehyde polymer microcapsules located on brown strips at the bottom of the pages of the booklets. The stimuli are released by scratching the strips with a pencil tip in a standardized manner. Above each odorant strip is a four-alternative multiple-choice question. For example, one of the items reads: "This odor smells most like 1) chocolate, 2) banana, 3) onion, or 4) fruit punch". The test is forced-choice, the subject having to mark one of the four alternatives, even if no smell is perceived. The criteria for the selection of the odorants and response alternatives are described elsewhere, as are the age- and gender-related norms which are based upon thousands of subjects (Doty et al. 1984a, Doty 1989). This test, which is scored as the number of items correct out of 40, has several unique features, including a means for detecting malingering and being amenable to self-administration.

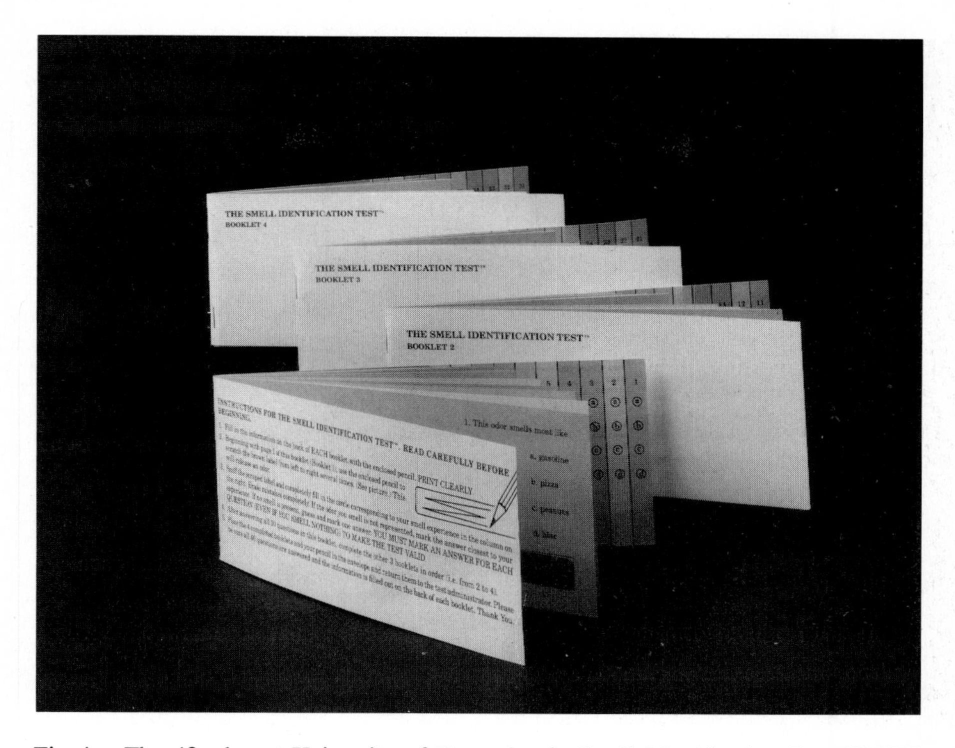

Fig 4. The 40-odorant University of Pennsylvania Smell Identification Test (UPSIT). Each page contains a microencapsulated odorant which is released by scraping the surface. Answers are marked on the columns of the last page of each booklet.

Wright (1987) has recently described a clinical confusion matrix odor identification test similar to the test described by Koster (1972, 1975) for assessing odor dissimilarities. Unlike Koster's test, however, the odors are familiar ones and therefore the response alternatives are the names of the 10 stimuli, rather than codes assigned to them in preliminary training. No feedback is given concerning the correctness or incorrectness of the subjects' responses. In this 100-trial test, each of the stimuli is presented 10 times in a counterbalanced order. An example of an odor confusion matrix obtained from the responses of five pseudohypoparathyroid patients deficient in Gs-protein, is presented in Figure 5. Wright argues that the off-diagonal responses i.e., the "confusions" of such matrices, provide meaningful clinical information in addition to that provided by the on-diagonal responses, which are a direct measure of identification ability.

Overall, forced-choice odor identification tests are highly reliable e.g., the test-retest reliability r of the UPSIT is 0.95 (Doty et al. 1989), and provide for convenient and accurate measurement of smell function. Strong correlations between odor identification and detection threshold tests suggest they are measuring, to a large degree, a common sensory domain e.g., Cain et al. (1988), Doty et al. (1984b), although notable exceptions occur, particularly in some brain-damaged patient groups e.g., Jones-Gotman and Zatorre (1988), Potter and Butters (1980).

RESPONSE

STIMULUS		AMMONIA 1	CINNAMON 2	LICORICE 3	MINT 4	MOTHBALLS 5	ORANGE 6	ROSE 7	RUB. ALCH. 8	VANILLA 9	VINEGAR 10	
AMMONIA	1	**.800**	—	—	—	—	—	—	.140	.020	.040	1
CINNAMON	2	.020	**.320**	.140	.120	—	.120	.120	.020	.100	.040	2
LICORICE	3	—	.220	**.380**	.140	.040	.020	—	.080	.100	.020	3
MINT	4	.020	.180	.220	**.160**	.040	.080	.060	.100	.040	.100	4
MOTHBALLS	5	.080	.020	.020	.040	**.420**	.040	.040	.160	.060	.120	5
ORANGE	6	.040	.140	.140	.120	.020	**.220**	.160	.040	.100	.020	6
ROSE	7	.040	.080	.040	.160	.120	—	**.200**	.080	.100	.180	7
RUB. ALCH.	8	.100	.060	.020	.060	.020	.020	.040	**.280**	.100	.300	8
VANILLA	9	—	.240	.040	.040	.020	.100	.120	.020	**.360**	.060	9
VINEGAR	10	.020	.160	.100	.080	.020	.080	.020	.060	.160	**.300**	10
		1	2	3	4	5	6	7	8	9	10	

Fig 5. Odorant confusion matrix from five female pseudohypoparathyroid patients with deficient Gs protein activity. The rows represent the alternative odorants and the columns represent the proportion of responses to each odorant. Note the incorrect responses which fall away from the negative diagonal. Reproduced with permission from Wright et al. (1987).

D. Odor Intensity Discrimination

Determination of a difference threshold is the classical procedure for assessing odor intensity discrimination. A difference threshold is that concentration of an odorant which is perceived to be just noticeably different or more intense than another suprathreshold concentration of the same odorant. It is typically calculated in the same manner as a detection threshold, except that the comparison stimulus is not a blank. Thus, the aim is to determine that concentration of an odorant which, on average, is reported as stronger than another on 75% of the trials. Since difference thresholds are dependent upon the magnitude of the standard stimulus, it is commonly assumed, at least within a middle range of odorant concentrations, that Weber's fraction (Weber 1834) holds reasonably well. That is, the size of the increment in odorant concentration (ΔI) required to produce a just noticeable difference (stronger intensity) on 75% of the trials relative to the concentration of the standard stimulus (I) is a constant (K). This fraction ($\Delta I/I = K$) has been used as an index of both the discriminative ability of a subject and the discriminative power of a given sensory system, or a factor representing a component of the system, such as represented by a specific odorant. The smaller the Weber fraction, the better the discrimination.

The size of the Weber fraction, K, is clearly dependent upon the test stimulus, the odorant presentation method, and other factors; consequently, considerable variation has been observed in the value of Weber fractions. Early workers who used the Zwaardemaker olfactometer, such as Gamble (1898) and Hermanides (von Skramlik 1926), found average Weber fractions around 0.33 across a number of odorants. Zigler and Holoway (1935) reported systematic average variation in the fraction for the odor of India rubber, with a value of 0.17 appearing at the upper end of the stimulus series and a value of 0.99 at the lower end of the series. These authors concluded that differential sensitivity is related to odor strength by a hyperbolic function.

Later workers, using stimulus presentation procedures other than the Zwaardemaker olfactometer, have found values usually falling at the lower end of the aforementioned range. Thus, Wenzel (1949) found an average Weber fraction of about 0.15 for phenethyl alcohol, Stone and Bosley (1965) reported an average fraction of 0.28 for acetic and propionic acid, which, unfortunately, also have considerable intranasal trigeminal stimulative properties (Doty et al. 1978a), and Slotnick and Ptak (1977) an average value of 0.32 for amyl acetate. Cain (1977a,b) reported, in two subjects, average Weber fractions for ethyl n-butyrate, n-amyl alcohol, n-butyl alcohol, and n-amyl butyrate of 0.30, 0.19, 0.07, and 0.09, respectively, arguing that larger fractions observed by most other investigators likely reflect fluctuations in stimulus concentrations. Despite such variability, however, most results are in accord with Wenzel's (1948, p 140) statement regarding human differential olfactory sensitivity: "... it appears that olfaction is not as sensitive in the detection of differences as are vision and audition, although it is not as far behind the latter as might be expected. Since the need for such acuity is rarely, if ever, forced upon us, we are apparently unaware of our capabilities".

V. SENSORY ATTRIBUTE SCALING

An odor can be described in terms of a number of psychological attributes. For example, one can denote an odor in terms of its strength, pleasantness, and quality e.g., mintiness, pungentness, floweriness, etc. In fact, no odor is perceived as having only one attribute and, therefore, the selection of attribute terms used to denote a sensation usually reflects an investigator's implicit or explicit theoretical perspective.

Procedurally, it is of value to divide sensory attribute scaling methods into two general classes: unidimensional and multidimensional. Unidimensional procedures assume that a given attribute is reliably discernible by a subject and can be ranked or scaled in some manner from less to more of the attribute. This does not preclude, however, the scaling of the same stimulus on each of several attribute dimensions e.g., intensity and pleasantness. Multidimensional procedures attempt to indirectly derive underlying attribute dimensions from the perceived similarities among all combinations of a set of stimuli. In effect, it is the number and identity of independent dimensions that is sought in most multidimensional scaling studies. Both of these approaches have theoretical and empirical strengths and weaknesses, depending upon the goal of the investigation, and need not produce equivalent outcomes. For example, it is often the case that some dimensions derived from multidimensional scaling procedures fail to correspond to identifiable unitary psychological attributes. It is also the case that some psychological attributes assumed to be unidimensional may, in fact, reflect the interplay of two or more attributes which are not immediately apparent to the subject.

A. Unidimensional Scaling Procedures

1. *Rating scales*

Rating scales, which were apparently introduced by Sir Francis Galton during the latter part of the 19th Century (Galton 1883), are among the easiest ways to quantify attributes of odors and have been widely used for this purpose. Their simplicity and flexibility, along with the fact that most persons in industrialized societies have used rating scales at one time or another in their lifetimes, make them straightforward. Although such scales have considerable merit in many basic and applied situations, they can have drawbacks when the goal is to obtain consistent mathematical relationships among scale values.

a. Numerical category scales

A numerical category scale is a rating scale with discrete response alternatives or categories, all of which are associated in one way or another with a number. The optimum number of categories in such a scale is a function of both the subject's ability to make fine distinctions along the stimulus continuum e.g., intensity continuum, and the amount of discriminability inherent in the stimulus continuum itself.

In practice, the use of more than 10 categories in the rating of the intensity of odors is rare, as exemplified by the widely used intensity category scale first applied by Katz and Talbert in 1930:

0 no odor
1 very faint odor
2 faint odor
3 easily noticeable odor
4 strong odor
5 very strong odor

As a general rule, it is better to have too many categories than too few, since the associated error will tend to be averaged out.

b. Line scales

A line scale, also termed a "graphic" or "visual analog" scale, consists of a line along which a subject places a mark to indicate the degree of the attribute that is experienced. The end points of the scale are "anchored" to descriptors e.g., "very weak odor", "very strong odor". In some cases, the descriptor on the end of the scale denoting the larger amount of the attribute (usually on the right) is located under the line a little towards the center near the end of the scale in an effort to help avoid constriction of scale values at high sensation levels (Marks et al. 1988). Relative to category scales, line scales are less susceptible to number preferences and to possible memory effects when stimuli are repeated (Anderson 1974). However, because of the limitation in spatial extent, they can fall prey to end effects, despite the fact that finer discriminations are made within the midscale sectors. One advantage of a line scale is that the responses can be assigned empirically to nearly any number of categories desired by the experimenter, although usually the distance measured from one end of the scale is used as the indicant.

An example of a simple bipolar line scale is as follows:

very unpleasant ____ very pleasant

2. *Within-modality intensity matching procedures*

Another procedure for quantifying the perceived strength of odors is to match each of their intensities to one of a set of concentrations from a standard reference odorant, a procedure analogous to that used in vision to obtain accurate scotopic visibility curves (Hecht and Williams 1922). This procedure, which has not been fully exploited, could lead to the development of intensity scales which would allow direct intensity comparisons across qualitatively different odorants. An example of this technique is provided by Kruger et al. (1955). These authors chose a binary dilution series of n-heptanol for the standard reference scale and determined the olfactory intensity of a number of aliphatic and aromatic compounds with widely different physico-chemical properties.

In general, scales of odor intensity obtained using a within-modality intensity matching procedure are very reliable (Beck et al. 1954). One limitation of this procedure, however, is that a relatively large number of odors are commonly smelled in succession to produce accurate matching and, therefore, adaptation can easily occur. For this reason, considerable time must be interspersed between trials to minimize distortion from adaptation.

3. *Cross-modal matching procedures: the method of magnitude estimation*

In cross-modal matching, an observer signifies the relative magnitudes of a sensory attribute, usually intensity, for each member of a set of stimuli using some other sensory or cognitive domain. Cross-modal continua that have been used in this paradigm include distance e.g., pulling a tape measure or drawing a line a distance proportional to the odor's intensity; brightness e.g., adjusting a rheostat-regulated light until its brightness is the same as the intensity of the odor, and strength of handgrip e.g., gripping a hand dynamometer in relation to the odor's intensity. In the most widely used of these procedures, termed magnitude estimation, the observer matches numbers to the relative amount of the attribute. Because of its popularity, only magnitude estimation, which was the first of the cross-modal matching procedures to be widely applied in studies of olfaction, is described in detail in this section. The reader is referred elsewhere for examples of studies using cross-modal matching procedures other than magnitude estimation (Burdach and Doty 1987, Stevens et al. 1960, Stevens and Marks 1965, Stevens 1966).

In magnitude estimation, an odor which smells three times as intense as one initially given an arbitrary value of 10 would be assigned the number 30, whereas an odor which smells half as strong would be assigned the value 5. No limits are placed on the size or the range of numbers observers can assign to the stimuli, so long as they reflect the relative magnitudes of the perceived intensities. In some cases, a standard to which the investigator has assigned a number, usually the middle stimulus of the series, is presented within the stimulus set in an attempt to keep the subject's responses reliable. This procedure is termed the "fixed modulus" method. In contrast, in the "free modulus method", no such standard is presented, and the subject is free to choose any number system, the only constraint being that the numbers be made proportional to the magnitude of the attribute. The key here is that the absolute value of the number is not important, only the ratios between the numbers are relevant.

Intensity data from magnitude estimation measures are most commonly analyzed by plotting the log magnitude estimates as a function of the log of the odorant concentrations. The best fit to the data is then determined by linear regression and is commonly expressed as

$$\log P = n \log \emptyset + \log k, \qquad (1)$$

where P = perceived intensity, k = the Y intercept, $\emptyset$ = stimulus concentration and n = the slope. The exponential form of this equation is a power function of the form,

$$P = k\emptyset^n, \qquad (2)$$

where the exponent n is equivalent to the slope above. In olfaction, n is nearly always less than one and reflects, on a linear-linear plot, an ascending concave downward function of physical concentration (Berglund et al. 1971). Numerous investigators have presented modifications of the power law to take into account such factors as adaptation and basal sensitivity. The reader is referred elsewhere for the details of such models e.g., Marks (1974), Overbosch (1986).

Magnitude estimation has the theoretical advantage of producing a ratio-like scale with minimal distortions due to boundary or category constraints. However, this procedure, like all psychophysical scaling procedures, has a number of problems. First, the task is rather complex, in that accurate responses to a stimulus requires a good memory for the prior stimulus. If too much time lapses between the stimuli, memory of the prior stimulus fades. On the other hand, if the trials are spaced too close in time, adaptation can distort the relationship. Second, there is a question as to whether subjects consistently provide ratio estimates of the stimuli. For example, some subjects revert to a covert category scale or to a combination category-ratio scale (Moskowitz 1977). A "round number bias" is common in which the overselection of 2, 5, 10, 20, 50, 100, etc. occurs (Baird et al. 1970). Third, the exponents are influenced by many non-sensory factors, including 1) the value of the numbers used in the instructions, 2) the number chosen for the modulus, 3) the range in quality and complexity of the stimuli, 4) the order of presentation of the stimuli, 5) the subjects' facility with numbers, and 6) the selection of the reference concentration in the fixed modulus paradigm (Cain 1969, Carlson 1977, Engen and Ross

1966, Poulton 1968, Pradhan and Hoffman 1963, Robinson 1976, Zwislocki and Goodman 1980).

Although it is often asserted that magnitude estimation is superior to other scaling procedures, this is likely true only for trained observers. Thus, Lawless and Malone (1986a) used nine-point rating scales, line scales, the method of magnitude estimation, and a hybrid of the category and line scales to assess the responses of housewives to visual stimuli (shininess of wood samples, degree of saturation of some green Munsell color swatches), tactile stimuli (roughness of sandpaper samples, thickness of stirred liquid silicone samples of different viscosities), and olfactory stimuli (intensity of citral and linalool samples). Comparisons of the degree to which the scales differentiated among the different stimulus values, as well as measures of variability, reliability, and ease of use, were determined. Using these criteria, category and line scales were found to be superior to magnitude estimation and the other procedures in these relatively untrained subjects. In a subsequent similar study in which the responses of college students were compared to those of housewives, Lawless and Malone (1986b) demonstrated that the previously observed inferiority of magnitude estimation was present only in the housewife group. All subjects performed the tasks with greater accuracy as they became familiar with the range of stimuli that were presented. The authors suggested that the differences observed between the college students and housewives may reflect differences in mathematical ability, comprehension of directions, memory for the reference standard, or other cognitive factors.

4. *The method of magnitude matching*

In the method of magnitude matching, subjects attempt to judge intensities of sensations from two or more modalities on a single common scale (Stevens and Marks 1980). Although this procedure has been used only for the attribute of intensity, it could be used to assess other attributes as well. If the assumption is made that the perception of the intensities of one of the modalities is invariant across subjects or groups of subjects, (an assumption that must be carefully considered in any specific case), then absolute comparisons of the scale values on the other modality can be made by normalizing the relative estimates to the invariant scale. In other words, if after such normalization one subject has a value of 10 for a given odorant concentration and another subject a value of

20, then the second subject is assumed to perceive the intensity of the odorant as twice as strong as does the first subject.

Recent studies demonstrate that this procedure, while useful in some circumstances, has a number of shortcomings. For example, the values resulting from magnitude matching are not independent of context effects when applied to chemical stimuli. Thus, in one of a series of exploratory studies, Marks et al. (1988) used sodium chloride taste solutions as the standard modality and butanol odor as the test modality. The concentration of the taste stimulus set ranged from 0.032 to 0.32 M in 0.25 log unit steps. The butanol stimuli, presented in squeeze bottles, were presented in two separate sessions using concentration sets that overlapped slightly (a low intensity set ranging from 0.049 to 0.78% in 0.3 liquid dilution unit steps and a high intensity set ranging from 0.39 to 6.25% in such steps). By using these two sets of concentrations, the investigators could determine the degree to which the judgment data, following normalization, were independent of the intensity context of the stimuli. These authors found that the judgments of the butanol concentrations common to the two stimulus sets were shifted by an average of 46% (in terms of log stimulus concentration) as a result of the stimulus context.

B. Multidimensional Scaling Procedures

All of the scaling procedures considered up to this point assume that the subject can discern a specific sensory attribute from a set of stimuli and can make reliable judgments of that attribute. For example, in the magnitude matching procedure, the subject is assumed to perceive the relative strengths of the stimuli from two different types of sensory continua and to be able to relate these strengths to one another on a common intensity scale. In multidimensional scaling (MDS), the goal is to ascertain underlying dimensions indirectly from ratings of the relative similarities or dissimilarities among stimuli of a given set. Obviously, the assumption is that such ratings can be reliably estimated. In the prototypical MDS paradigm, ratings are made for every possible pair of stimuli on a line scale anchored with descriptors like "exactly same" vs "completely different". The similarity or dissimilarity matrices are then converted to correlation matrices and analyzed by an MDS algorithm.

In essence, MDS mathematically represents the similarity of objects, in this case odors, in n-dimensional space. Often independent ratings of the stimuli are made on unidimensional rating scales in order to ascertain whether the MDS-derived dimensions correlate with them. The number of MDS dimensions ultimately chosen for representation is based upon changes in measures of goodness of fit to the model with changes in dimensionality. In most chemosensory studies, the data are statistically well represented in two or three dimensions, allowing for graphic representation and inspection of the relationships among the variables. There are numerous MDS programs which have been applied to olfactory data. Based upon a series of comparative studies, Schiffman et al.(1981) recommend, whenever possible, the use of programs that take into account individual differences among the responses of subjects e.g., NDSCAL, Carroll and Chang 1970; ALSCAL, Takane et al. 1977; MULTISCALE, Ramsay 1977).

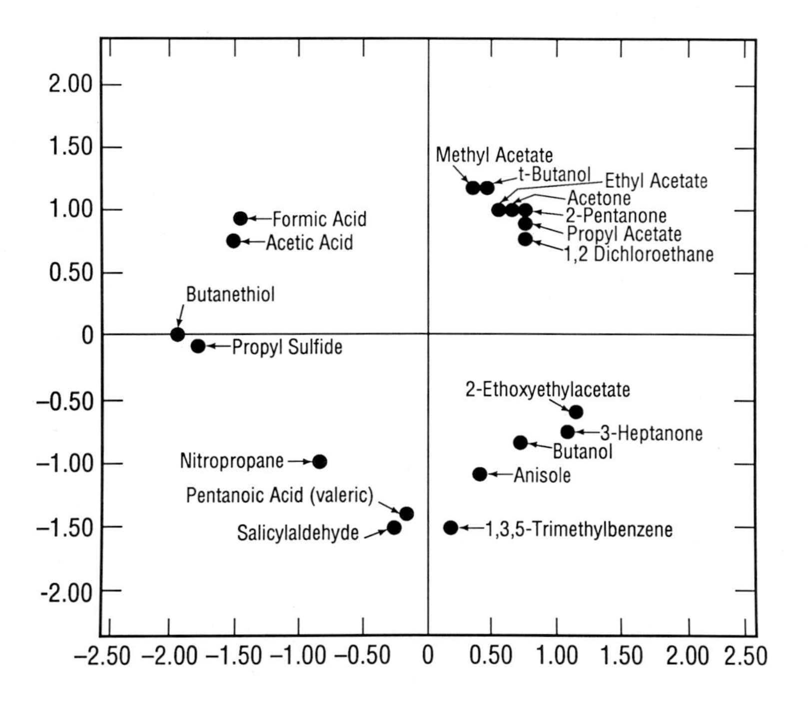

Fig 6. A two-dimensional solution from a multidimensional scaling analysis of 19 odorants using an individual differences ALSCAL model. Note that the more pleasant smelling chemicals are located on the right and the less pleasant ones on the left. From Schiffman et al. (1981) with permission.

An example of a two-dimensional space derived for 19 diverse stimuli using an individual differences model of ALSCAL is presented in Figure 6 (Schiffman et al. 1981). In this case, the more pleasant smelling chemicals fall on the right and the less pleasant smelling ones on the left. A trigeminal dimension appears to be present, in that the chemicals with sharp, pungent odors are in the top of the Figure and those with more aromatic smells at the bottom. In addition to its use in ascertaining underlying dimensions within a set of stimuli, MDS provides a convenient means to extract associations within the data which might otherwise not be observed. Thus, even in cases where it is not possible to identify psychological dimensions within the data, this procedure provides a pictorial representation of the data which is easier to grasp than a table of similarities. In this respect, it can be used in data reduction in the same manner as factor analysis. In addition, MDS is useful for comparing groups, as well as individuals, to determine the similarity of their odor perceptions. This is exemplified by MDS analyses which show that dimensional spaces for elderly persons are less distinct and reliable than those of younger persons, presumably reflecting an age-related decrement in ability to differentiate among odors (Schiffman and Leffingwell 1981, Schiffman and Pasternak 1979).

VI. SUBJECT FACTORS WHICH INFLUENCE OLFACTORY TEST MEASURES

Many factors influence the test measures described in this Chapter. Among the most salient subject factors are gender, smoking behavior, exposure to environmental chemicals, age, drug and alcohol usage, and various disease states. Since the latter three factors are reviewed in other Chapters of this book, comments here will be restricted to the influences of gender, smoking and environmental chemicals on such measures.

In general, women outperform men on tests of odor detection, discrimination and identification (Cain 1982, Deems and Doty 1987, Doty et al. 1984b, Griffiths and Patterson 1970, Le Magnen 1952, Schneider and Wolf 1955, Koelega and Koster 1974, Toulouse and Vaschide 1899). In addition, gender differences have been observed in the rating of the pleasantness of odors, with women rating the odors of such stimuli as human breath, axillary sweat, vaginal secretions, and 5-androst-16-ene-3-one, a component of sweat, as more intense and less pleasant than do men (Doty et al. 1975, 1978b, 1982,

Koelega and Koster 1974). At least in the case of odor identification, such sex differences do not appear to be culture specific (Doty et al. 1985) and are present by at least four years of age (Doty 1986), suggesting they may reflect an inborn sexually dimorphic trait.

The influence of smoking on olfactory function has been shown in most studies to be less important than the influence of gender or age (Doty et al. 1984a, Venstrom and Amoore 1968). Recently, Frye et al.(1989) demonstrated that the loss in olfactory ability associated with cigarette smoking is dose-related. Most noteworthy was the finding that past smokers also evidence decrements in function which reflect the amount and duration of past smoking, as well as the time since cessation of smoking (Fig 7). Thus, comparisons of smokers vs nonsmokers is problematic if neither smoking dose nor the influences of prior smoking are taken into account.

In addition to cigarette smoke, a wide variety of environmental agents, including industrial chemicals and dusts, has been associated with reversible and irreversible alterations in the general ability to smell (see Chapter 15 this book, and Amoore 1986). Chemicals for which decreased smell function has been reported following acute or chronic exposure include acetone, acetophenone, ammonia, benzene, cadmium, carbon disulfide, carbon monoxide, chlorine, cyclohexanone, chromium, formaldehyde, hydrazine, hydrogen cyanide, hydrogen selenide, hydrogen sulfide, lead, mercury, n-methylformimino-methyl ester, nickel, nitrogen dioxide, phosphorus oxychloride, silver, sulfur dioxide, sulfuric acid, sulfur dioxide, tetrahydrofuran, zinc chromate, and zinc sulfate. Unfortunately, only rarely have sound psychophysical procedures been used to document such exposure-induced decrements.

Recently, Schwartz et al. (1989) used the University of Pennsylvania Smell Identification Test and modern epidemiologic techniques to demonstrate an association between job exposure of workers to chemicals used in plastic manufacturing i.e., acrylates and methacrylates, and the ability to smell. In this study of over 700 chemical plant workers, individuals who had never smoked cigarettes but who had been exposed to such chemicals were six times more likely than their non-exposed counterparts to evidence olfactory dysfunction (see Chapter 15).

A final point that should be mentioned in this review is the influence of experience on measures of olfactory function. Surprisingly little is known about the degree to which the olfactory system can be trained to reach optimal performance, although at least three avenues of research have shed some light on this question. First, repeated testing within the perithreshold concentration range results in the lowering of thresholds and enhancement of signal detection sensitivity measures in both humans (Doty et al. 1981, Engen 1960, Rabin and Cain 1986, Wysocki et al. 1989), and rats (Doty and Ferguson-Segall 1989).

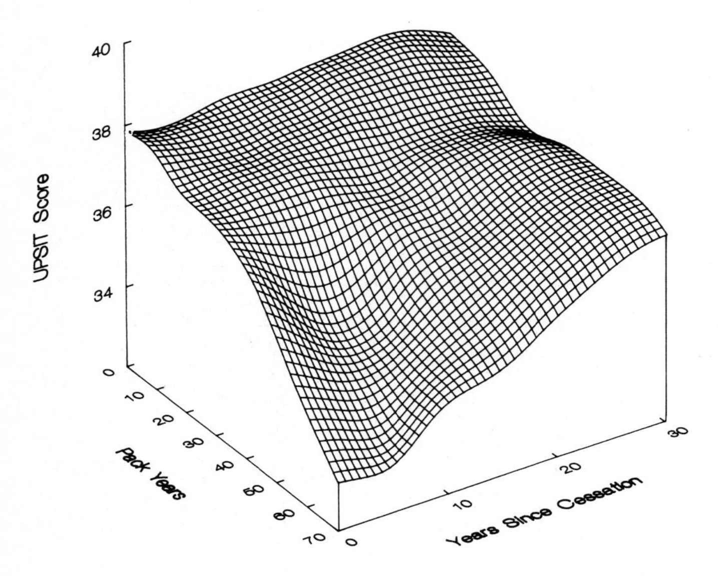

Fig 7. Relationship between past smoking dose in pack years, the time since cessation of smoking, and scores on the University of Pennsylvania Smell Identification Test in previous smokers of cigarettes. From Frye, et al. 1988 with permission.

Second, unpleasant odors become less unpleasant and pleasant odors become less pleasant following prior repeated exposure to them (Cain and Johnson 1978). Assuming that the latter phenomenon is not simply the result of adaptation, this would suggest that affective components of odors are able to habituate independently of odor intensity. Third, the ability to name odors is clearly influenced by practice in which feedback is provided (Desor and Beauchamp 1974, Engen and Ross 1973). Presumably such learning has evolved to a high degree in some perfumers, wine tasters and practitioners of the culinary arts who reportedly can identify many different odors by names or descriptors.

VII. OVERVIEW

In this Chapter psychophysical procedures that have been used to measure the sense of smell have been described and salient factors known to influence such measures have been briefly mentioned. As is apparent from the material presented in this Chapter, the assessment of olfactory function has been approached from many perspectives. Nevertheless, considerable gaps exist in our knowledge. A major challenge for olfactory psychophysicists within the next decade will be to answer several basic questions. High on the agenda are the following: To what degree do specific psychophysical tests measure independent physiological or psychological entities? What sets of odorants and techniques best describe human olfactory function? Can subtle differences in psychophysical responses to different types of odorants provide insight into the underlying quality coding mechanisms? Are some psychophysical measures more sensitive to peripheral than central olfactory system functioning and vice versa? Can psychophysical measures provide the type of information necessary to elucidate the biochemical and neurophysiological means by which the olfactory system recognizes and differentiates among seemingly thousands of disparate odor qualities? The answers to these questions will determine whether the infant field of olfactory psychophysics will prove as valuable to the field of olfaction in elucidating primary sensory mechanisms as have visual and auditory psychophysics to their respective parent disciplines.

REFERENCES

Abraham A, Matha KV (1983) The effect of right temporal lobe lesions on matching of smells. Neuropsychologia 21: 277-281

Amoore JE (1962a) The stereochemical theory of olfaction. I. Identification of the seven primary odours. Proc Sci Sect Toilet Goods Assoc No 37, Suppl, 1-13

Amoore JE (1962b) The stereochemical theory of olfaction. II. Elucidation of the stereochemical properties of the olfactory receptor sites. Proc Sci Sect Toilet Goods Assoc No 37, Suppl, 13-23

Amoore JE (1970) The molecular basis of odor. Charles C Thomas, Springfield

Amoore JE (1971) Olfactory genetics and anosmia. In: Beidler LM (ed) Handbook of sensory physiology, vol IV. Chemical senses. Springer-Verlag, New York, p 245

Amoore JE (1986) Effects of chemical exposure on olfaction in humans. In: Barrow CS (ed) Toxicology of the nasal passages. Hemisphere Publishing Corp, Washington DC, p 155

Amoore JE, Ollman BG (1983) Practical test kits for quantitatively evaluating the sense of smell. Rhinology 21: 49-54

Anderson NH (1974) Algebraic models in perception. In: Carterette EC, Friedman MP (eds) Handbook of perception, vol II. Psychophysical judgment and measurement. Academic Press, New York, p 215

Baird JC, Lewis C, Romer D (1970) Relative frequencies of numerical responses in ratio estimation. Percept Psychophys 8: 358-362

Beck LH, Kruger L, Calabresi P (1954) Observations on olfactory intensity: I. Training procedure, methods, and data for two aliphatic homologous series. Ann NY Acad Sci 58: 225-238

Beets MGJ (1971) Olfactory response and molecular structure. In: Beidler LM (ed) Handbook of sensory physiology, vol IV. Chemical senses. Springer-Verlag, New York, p 257

Berglund B, Berglund U, Ekman G, Engen T (1971) Individual psychophysical functions for 28 odorants. Percept Psychophys 9: 379-384

Boring EG (1933) The physical dimensions of consciousness. The Century Co Ltd, New York

Boring EG (1942) Sensation and perception in the history of experimental psychology. Appleton-Century-Crofts, New York

Brown J (1974) Recognition assessed by rating and ranking. Brit J Psychol 65: 13-22

Burdach K, Doty RL (1987) The effects of mouth movements, swallowing, and spitting on retronasal odor perception. Physiol Behav 41: 353-356

Cain WS (1969) Odor intensity: Differences in the exponent of the psychophysical function. Percept Psychophys 6: 349-354

Cain WS (1977a) Differential sensitivity for smell: "Noise" at the nose. Science 195: 796-798

Cain WS (1977b) Odor magnitude: Course versus fine grained. Percept Psychophys 22: 545-549

Cain WS (1982) Odor identification by males and females: Predictions vs. performance. Chem Sens 7: 129-142

Cain WS, Gent JP, Goodspeed RB, Leonard G (1988) Evaluation of olfactory dysfunction in the Connecticut Chemosensory Clinical Research Center. Laryngoscope 98: 83-88

Cain WS, Johnson F Jr (1978) Lability of odor pleasantness: Influence of mere exposure. Perception 7: 459-465

Cain WS, Rabin RD (1989) Comparability of two tests of olfactory functioning. Chem Sens 14: 479-485

Campbell IM, Gregson RAM (1972) Olfactory short-term memory in normal, schizophrenic and brain damaged cases. Aust J Psychol 24: 179-185

Carroll JD, Chang JJ (1970) Analysis of individual differences in multi-dimensional scaling via an n-way generalization of "Eckhart-Young" decomposition. Psychometrika 35: 283-319

Carlson VR (1977) Instructions and perceptual constancy judgments. In: Epstein W (ed) Stability and constancy in visual perception. Wiley, New York

Cheesman GH, Townsend MJ (1956) Further experiments on the olfactory thresholds of pure chemical substances using the "sniff-bottle method". Q J Exp Psychol 8: 8-14

Corbin A (1986) The foul and the fragrant: Odor and the French social imagination. Harvard University Press, Cambridge Massachussets

Cornsweet TN (1962) The staircase-method in psychophysics. Am J Psychol 75: 485-491

Coxon JM, Gregson RAM, Paddick RG (1978) Multidimensional scaling of perceived odour of bicyclo [2.2.1] heptane, 1,7,7-trimethylbicyclo [2.2.1] heptane and cyclohexane derivatives. Chem Sens Flav 3: 431-440

Crocker EC, Henderson LF (1927) Analysis and classification of odours. Am Perf Ess Oil Rev 22: 325-327

Deems DA, Doty RL (1987) Age-related changes in the phenyl ethyl alcohol odor detection threshold. Trans Penn Acad Ophthalmol Otolaryngol 39: 646-650

Dember WN (1965) The psychology of perception. Holt, Rinehart and Winston, New York

Desor JA, Beauchamp GK (1974) The human capacity to transmit olfactory information. Percept Psychophys 16: 551-556

Dibbits HC (1888) Jets ovez de geroeligheid van het reukvermogen ten oprichte van azynzuur. Festbundel Donder's Jubileum, p 497

Dixon WJ, Massey FJ (1957) Introduction to statistical analysis. McGraw Hill, New York

Doty RL (1976) Reproductive endocrine influences upon human nasal chemoreception: A review. In: Doty RL (ed) Mammalian olfaction, reproductive processes and behavior. Academic Press, New York, p 295

Doty RL (1979) A review of olfactory dysfunctions in man. Am J Otolaryngol 1: 57-79

Doty RL (1986) Gender and endocrine-related influences upon olfactory sensitivity. In: Meiselman HL, Rivlin RS (eds) Clinical measurement of taste and smell. MacMillan, New York, p 377

Doty RL (1989) The Smell Identification Test™ administration manual (2nd edn). Haddonfield, NJ: Sensonics, Inc.

Doty RL, Agrawal U, Frye R (1989) Evaluation of the internal consistency reliability of the fractionated and whole University of Pennsylvania Smell Identification Test. Percept Psychophys 45: 381-384

Doty RL, Applebaum SL, Zusho H, Settle RG (1985) A cross-cultural study of sex differences in odor identification ability. Neuropsychologia 23: 667-672

Doty RL, Brugger WE, Jurs PC, Orndorff MA, Snyder PJ, Lowry, LD (1978a) Intranasal trigeminal stimulation from odorous volatiles: Psychometric responses from anosmic and normal humans. Physiol Behav 20: 175-187

Doty RL, Deems DA, Frye R, Pelberg R, Shapiro A (1988) Olfactory sensitivity, nasal resistance, and autonomic function in the multiple chemical sensitivities (MCS) syndrome. Arch Otolaryngol Head Neck Surg 114: 1422-1427

Doty RL, Ferguson-Segall M (1989) Influence of castration on the odor detection performance of male rats. Behav Neurosci 103: 691-693.

Doty RL, Ford M, Preti G, Huggins G (1975) Human vaginal odors change in pleasantness and intensity during the menstrual cycle. Science 190: 316-1318

Doty RL, Gregor T, Settle RG (1986) Influences of intertrial interval and sniff bottle volume on the phenyl ethyl alcohol olfactory detection threshold. Chem Sens 11: 259-264

Doty RL, Kligman A, Leyden J, Orndorff MM (1978b) Communication of gender from human axillary odors: Relationship to perceived intensity and hedonicity. Behav Biol 23: 373-380

Doty RL, Kreiss DS, Frye, RE (1990) Human odor intensity perception: Correlation with frog epithelial adenylate cyclase activity and transepithelial voltage response. Brain Res 444: 95-103

Doty RL, Ram CA, Green P, Yankell S (1982) Communication of gender from breath odors: Relationship to perceived intensity and pleasantness. Horm Behav 16: 13-22

Doty RL, Shaman P, Applebaum SL, Giberson R, Sikorsky L, Rosenberg L (1984a) Smell identification ability: Changes with age. Science 226: 1441-1443

Doty RL, Shaman P, Dann M (1984b) Development of the University of Pennsylvania Smell Identification Test: A standardized microencapsulated test of olfactory function. Physiol Behav (Monogr) 32: 489-502

Doty RL, Snyder P, Huggins G, Lowry LD (1981) Endocrine, cardiovascular, and psychological correlates of olfactory sensitivity changes during the human menstrual cycle. J Comp Physiol Psychol 95: 45-60

Elsberg CA, Levy I (1935) The sense of smell: I. A new and simple method of quantitative olfactometry. Bull Neurol Inst NY 4: 5-19

Elliot PB (1964) Tables of d'. In: Swets JA (ed) Signal detection and recognition by human observers. Wiley, New York

Engen T (1960) Effect of practice and instruction on olfactory thresholds. Percept Motor Skills 10: 195-198

Engen T, Ross BM (1966) Effect of reference number on magnitude estimation. Percept Psychophys 1: 74-76

Engen T, Ross BM (1973) Long-term memory of odors with and without verbal descriptions. J Exp Psychol 100: 221-227

Fechner GT (1860) Elemente der psychophysik. Breitkopf and Harterl, Leipzig

Fischer E, Penzoldt F (1886) Ueber die Empfindlichkeit des Geruchssinnes. Sitzungsber Phys Med Soz Erlangen 18: 7-10

Frey PW, Colliver JA (1973) Sensitivity and responsivity measures for discrimination learning. Learn Motiv 4: 327-342

Frijters JER (1980) Three-stimulus procedures in olfactory psychophysics: An experimental comparision of Thurstone-Ura and three-alternative forced-choice models of signal detection theory. Percept Psychophys 28: 390-397

Frijters JER, Kooistra A, Vereijken PFG (1980) Tables of d' for the triangular method and the 3-AFC signal detection performance. Percept Psychophys 27: 176-178

Frye RE, Doty RL, Schwartz B (1989) Influence of cigarette smoking on olfaction: Evidence for a dose-response relationship. J Am Med Assoc 263: 1233-1236

Galton F (1883) Inquiries into human faculty and its development. Macmillan, London

Gamble EM (1898) The applicability of Weber's law to smell. Am J Psychol 10: 82-142

Gescheider GA (1976) Psychophysics: method and theory. Lawrence Erlbaum Associates, Hillsdale

Gescheider GA (1988) Psychophysical scaling. Ann Rev Psychol 39: 169-200

Green DM, Swets JA (1966) Signal detection theory and psychophysics. Wiley, New York

Gregson RAM, Free ML, Abbott MW (1981) Olfaction in Korsakoffs, alcoholics and normals. Brit J Clin Psychol 20: 3-10

Grier JB (1971) Nonparametric indexes for sensitivity and bias: Computing formulas. Psychol Bull 75: 424-429

Griffiths MN, Patterson RLS (1970) Human olfactory responses to 5α-androst-16-en-3-one, the principal component of boar taint. J Sci Food Agric 21: 4-6
Gross-Isseroff R, Lancet D (1988) Concentration-dependent changes of perceived odor quality. Chem Sens 13: 191-204
Guilford JP (1954) Psychometric methods. McGraw-Hill, New York
Haller A von (1763) Olfactus. elementa physiologiae corporis humani. Liber XIV. Tomus Quintus. Francisci Grasset, Lausanne, p 125
Harper R, Bate-Smith EC, Land DG (1968) Odour description and odour classification. American Elsevier, New York
Harshman R, Green P, Wind Y, Lundy ME (1982) A model for the analysis of asymmetric data in marketing research. Marketing Sci 1: 205-242
Hecht S, Williams RE (1922) The visibility of monochromatic radiation and the absorption spectrum of visual purple. J Gen Physiol 5: 1-7
Henning H (1916) Der geruch. Barth, Leipzig
Herbart JF (1824-1825) Psychologie als wissenschaft, neu gegrundet auf erfahrung, metaphysik und mathematik. Unzer, Konisburg
Hodos W (1970) Nonparametric index of response bias for use in detection and recognition experiments. Psychol Bull 74: 351-354
Ikeda K, Sakurada T, Sasaki Y, Takasaka T, Furukawa Y (1988) Clinical investigation of olfactory and auditory function in type I pseudo-hypoparathroidism: participation of adenylate cyclase system. J Larnygol Otol 102: 1111-1114
Jerome EA (1942) Olfactory thresholds measured in terms of stimulus pressure and volume. Arch Psychol 274: 1-44
Jones FN (1953) A test of the Elsberg technique of olfactometry. Am J Psychol 66: 81-85
Jones-Gotman M, Zatorre RJ (1988) Olfactory identification deficits in patients with focal cerebral excision. Neuropsychologia 26: 387-400
Katz SH, Talbert EJ (1930) Intensities of odors and irritating effects of warning agents for inflammable and poisonous gases. Tech paper 480, Bureau of Mines, US Dept of Commerce
Koelega HS, Koster EP (1974) Some experiments on sex differences in odor perception. Ann NY Acad Sci 237:234-246
Koster EP (1972) Odour similarities in nine odorous compounds. A methodological study. Report of the Psychological Laboratory, University of Utrecht
Koster EP (1975) Human psychophysics in olfaction. In: Moulton DG, Turk A, Johnston Jr JW (eds) Methods in olfactory research. Academic Press, New York, p 345
Kruger L, Feldzamen AN, Miles WR (1955) A scale for measuring supra-threshold olfactory intensity. Am J Psychol 68: 117-123
Laffort P, Patte F, Etcheto M (1974) Olfactory coding on the basis of physicochemical properties. Ann NY Acad Sci 237: 193-208
Lawless HT, Engen T (1977) Association to odors: Interference, memories, and verbal labeling. J Exp Psychol:Human Percept Perform 3: 52-59
Lawless HT, Malone GT (1986a) The discrimination efficiency of common scaling methods. J Sensory Stud 1: 85-98
Lawless HT, Malone GT (1986b) A comparision of rating scales: Sensitivity, replicates and relative measurement. J Sensory Stud 1: 155-174
Le Magnen J (1952) Les phenomenes olfacto-sexuels chez l'homme. Arch Sci Physiol 6: 125-160
Linnaeus C (1756) Odores medicamentorum. Amoenitates Academicae 3: 183-201

Lorry D (1784-1785) Observations sue les parties volatiles et odorants des medicamens tires des substances vegetales et animales. Hist Mem Soc Roy Med 7: 306-318

Mallevialle J, Suffet IH (1987) Identification and treatment of tastes and odors in drinking water. American Water Works Association, Denver

Marks LE (1974) Sensory processes: The new psychophysics. Academic Press, New York

Marks LE, Stevens JC, Bartoshuk LM, Gent JF, Rifkin B, Stone VK (1988) Magnitude-matching: the measurement of taste and smell. Chem Sens 13: 63-87

Marshall DA, Blumer L, Moulton DG (1981) Odor detection curves for n-pentanoic acid in dogs and humans. Chem Sens 6: 445-453

Marshall DA, Moulton DG (1981) Olfactory sensitivity to alpha-ionone in humans and dogs. Chem Sens 6: 53-61

McCarthy PJ (1949) A class of methods for estimating reaction to stimuli of varying severity. J Educ Psychol 40: 143-156

Meilgaard MC, Reid DS, Wyborski (1982) Reference standards for beer flavor terminology system. J Am Soc Brew Chem 40: 119-128

Moskowitz H (1977) Magnitude estimation: Notes on what, how, when, and why to use it. J Food Qual 3: 195-227

Moulton DG (1960) Studies in olfactory acuity. III. Relative detectability of n-aliphatic acetates by the rat. Q J Exp Psychol 12: 203-213

Moulton DG, Ashton EH, Eayrs JT (1960) Studies in olfactory acuity. 4. Relative detectability of n-aliphatic acids by the dog. Anim Behav 8: 117-128

Moulton DG, Eayrs JT (1960) Studies on olfactory acuity. II. Relative detectability of n-aliphatic alcohols by the rat. Q J Exp Psychol 12: 99-109

Moulton DG, Marshall DA (1976) The performance of dogs in detecting a-ionone in the vapor phase. J Comp Physiol 110: 287-306

Noble AC, Arnold RA, Buechsenstein J, Leach EJ, Schmidt JO, Stern PM (1987) Modification of a standardized system of wine aroma terminology. Am J Enol Vitic 38: 143-146

O'Mahony M. (1979) Short-cut signal detection measurements for sensory analysis. J Food Sci 44: 302-303

Overbosch P (1986) A theoretical model for perceived intensity in human taste and smell as a function of time. Chem Sens 11: 315-329

Pangborn RM, Berg HW, Roessler EB, Webb AD (1964) Influence of methodology on olfactory response. Percept Motor Skills 18: 91-103

Pike LM, Enns MP, Hornung DE (1988) Quality and intensity differences of carvone enantiomers when tested separately and in mixtures. Chem Sens 13: 307-309

Pollack R, Norman DA (1964) A non-parametric analysis of recognition experiments. Psychon Sci 1: 433-447

Potter H, Butters N (1980) An assessment of olfactory deficits in patients with damage to prefrontal cortex. Neuropsychologia 18: 621-628

Poulton EC (1968) The new psychophysics: Six models for magnitude estimation. Psychol Bull 69: 1-19

Pradhan PL, Hoffman PJ (1963) Effect of spacing and range of stimuli on magnitude estimation judgments. J Exp Psychol 66: 533-541

Punter PH (1983) Measurement of human olfactory thresholds for several groups of structurally related compounds. Chem Sens 7: 215-235

Rabin MD, Cain WS (1986) Determinants of measured olfactory sensitivity. Percept Psychophys 39: 281-286

Robinson GH (1976) Biasing power law exponents by magnitude estimation instructions. Percept Psychophys 19: 80-84
Schiffman SS (1974) Physiochemical correlates of olfactory quality. Science 185: 112-117
Schiffman SS, Leffingwell JC (1981) Perception of odors of simple pyrazines by young and elderly subjects: A multidimensional analysis. Pharmacol Biochem Behav 14: 787-798
Schiffman SS, Pasternak M (1979) Decreased discrimination of food odors in the elderly. J Gerontol 34: 73-79
Schiffman SS, Reynolds ML, Young FW (1981) Introduction to multidimensional scaling: Theory, methods, and applications. Academic Press, Orlando
Schneider RA, Wolfe S (1955) Olfactory perception thresholds for citral utilizing a new type olfactorium. J Appl Physiol 8: 337-342
Schwartz B, Doty RL, Monroe C, Frye RE, Barker S (1989) The evaluation of olfactory function in chemical workers exposed to acrylic acid and acrylate vapors. Am J Pub Health 1989 79: 613-618
Semb G (1968) The detectability of the odor of butanol. Percept Psychophys 4: 335-340
Skramlik E von (1926) Handbuch der physiologie der niederen sinne. I. Band: Die physiologie des geruchs- und geschmackssinnes. George Thieme, Leipzig
Slotnick BM, Ptak JE (1977) Olfactory intensity: Difference thresholds in rats and humans. Physiol Behav 19: 795-802
Smith DV (1988) Assessment of patients with taste and smell disorders. Acta Otolaryngol (Stockholm) Suppl 458: 129-133
Stahl WH (1973) (ed), Compilation of odor and taste theshold values data. American Society for Testing and Materials, Philadelphia
Stevens JC, Cain WS, Burke RJ (1988) Variability of olfactory thresholds. Chem Sens 13: 643-653
Stevens JC, Mack JD, Stevens SS (1960) Growth of sensation on seven continua as measured by force of handgrip. J Exp Psychol 59: 60-67
Stevens JC, Marks LE (1965) Cross-modality matching of brightness and loudness. Proc Nat Acad Sci USA 54: 407-411
Stevens JC, Marks LE (1980) Cross-modality matching functions generated by magnitude estimation. Percept Psychophys 27: 379-389
Stevens SS (1966) Matching functions between loudness and ten other continua. Percept Psychophys 1: 5-8
Stone H, Bosley JJ (1965) Olfactory discrimination and Weber's law. Percept Motor Skills 20: 657-665
Takane Y, Young FW, de Leeuw J (1977) Nonmetric individual differences multidimensional scaling: An alternating least squares with optimum scaling features. Psychometrika 42: 7-67
Tanner WP Jr, Swets JA (1954) A decision-making theory of visual detection. Psychol Rev 61: 401-409
Thurstone LL (1927) A law of comparative judgment. Psychol Rev 34: 273-286.
Toulouse E, Vaschide N (1899) Mesure de l'odorat chez l'homme et chez la femme. C R Soc Biol 51: 381-383
Torgerson WS (1958) Theory and methods of scaling. Wiley, New York
Toyota B, Kitamura T, Takagi SF (1978) Olfactory disorders, olfactometry and therapy. Igaku-Shoin, Ltd., Tokyo
Valentin G (1848) Lehrbuch der physiolgie des menschen. Braunschweig

Venstrom D, Amoore JE (1968) Olfactory threshold in relation to age, sex or smoking. J Food Sci 33: 264-265

Weber EH (1834) De pulsu, resorptione, auditu et tactu: Annotationes anatomicae et physiologicae. Koehler, Leipzig

Weinstock RS, Wright HN, Spiegel AM, Levine MA, Moses AM (1986) Olfactory dysfunction in humans with deficient guanine nucleotide-binding protein. Nature 322: 635-636

Wenzel B (1948) Techniques in olfactometry. Psychol Bull 45: 231-246

Wenzel B (1949) Differential sensitivity in olfaction. J Exp Psychol 39: 129-143

Wetherill GB, Levitt H (1965) Sequential estimation of points on a psychometric function. Brit J Math Stat Psychol 18: 1-9

Wood JB, Harkins SW (1987) Effects of age, stimulus selection, and retrieval environment on odor identification. J Gerontol 42: 584-588

Wright HN (1987) Characterization of olfactory dysfunction. Arch Otolaryngol Head Neck Surg 113: 163-168

Wysocki CJ, Dorries KM, Beauchamp GK (1989) Ability to perceive androstenone can be acquired by ostensibly anosmic people. Proc Nat Acad Sci USA 86: 7976-7978

Zigler MJ, Holoway AH (1935) Differential sensitivity as determined by the amount of olfactory substance. J Gen Psychol 12: 372-382

Zwaardemaker H (1895) Die physiologie des geruchs. W Engelmann, Leipzig

Zwaardemaker H (1925) L'Odorat. Doin, Paris

Zwaardemaker H (1927) The sense of smell. Acta Oto-Laryngol 11: 3-15

Zwislocki JJ, Goodman DA (1980) Absolute scaling of sensory magnitudes: A validation. Percept Psychophys 28: 28-38

6

HUMAN ELECTRO-OLFACTOGRAMS AND BRAIN RESPONSES TO OLFACTORY STIMULATION

GERD KOBAL

THOMAS HUMMEL

I. INTRODUCTION

Based on the experiments of Hosoya and Yoshida (1937), and Ottoson (1956), various attempts were made in the late sixties to record electrophysiological activities elicited by odorous stimuli in humans. The first to report a successful recording of an electro-olfactogram (EOG) from the human olfactory mucosa were Osterhammel et al. in 1969. Cerebral evoked potentials had already been obtained several years before from the electroencephalogram (EEG) by Finkenzeller (1966) and Allison and Goff (1967). Also, by the use of invasive techniques, spontaneous EEGs were recorded from the olfactory bulb as well as from the amygdala in patients (Hughes and Andy 1979).

Although some laboratories continued in their endeavours (Giesen and Mrowinski 1970, Herberhold 1973), the general tendency was to discontinue activities in this interesting field. Only since the introduction of more sophisticated stimulation techniques (Kobal and Plattig 1978, Kobal 1981) has it become possible to reliably record electrical correlates of odorous sensations in humans. Today, we are in the initial stage of utilizing improved electrophysiological techniques, in order to investigate the human sense of smell. In this Chapter a survey is presented of earlier attempts to record electro-olfactograms (EOG) from the peripheral receptor cells and olfactory evoked potentials (OEP) from the central nervous system.

Electro-olfactograms are generally considered to be summated generator potentials of olfactory receptor cells (Ottoson 1956). They are recorded with the aid of macroelectrodes from the surface of the olfactory mucosa. Therefore, it is of paramount importance that the process of applying the odorants does not produce artefacts at the mucosa. Consequently, an olfactometer which does not alter the mechanical or thermal conditions at the mucosa during stimulation, must be utilized. Similar requirements, albeit for different reasons, apply to the recording of event-related potentials. Since these potentials are highly integrated responses of the human brain, an additional activation of thermo- and mechanoreceptors would lead to synchronous cortical somatosensory responses, which would virtually be indistinguishable from the olfactory responses. However, even when causes responsible for artefacts are eliminated, distinguishing between responses elicited by activation of the chemoreceptors of the olfactory nerve and those of chemoreceptors of the trigeminal nerve (Silver and Maruniak 1981) remains a problem (Smith et al. 1971). The major technical problem in trying to record olfactory evoked potentials is that, in addition to the aforementioned requirements, steep stimulus onsets need to be guaranteed so that synchronous activation of a sufficient number of cortical neurons is accomplished to produce detection of the so-called event-related potentials (ERP).

Event-related potentials are changes in electrical fields generated by cortical neurons. They occur at a particular time before, during, or after an event has taken place in the physical or psychic world. Event-related potentials can be recorded non-invasively from the intact skull of a human subject, or directly from the surface of the brain. As early as 1890, the first ERPs evoked by chemical stimulation were obtained by Fleischl von Marxow, who presented ammonia to the nasal mucosa of a rabbit. This elicited a negative electrical deflection on the surface of the brain.

In order to measure an ERP it must be isolated from the overlapping potentials generated by cells that are not synchronized or time-locked to the event. This improvement of the signal-to-noise-ratio (ERP - to background activity) is generally achieved by averaging and/or filtering, using digital computers. For a detailed description of these techniques see Picton and Hillyard (1988). The recordings of the background activity are called electroencephalograms (EEG).

According to their time delay, or latency after onset of the event, ERPs are grouped into early, intermediate, late, and ultra-late potential components. Another way to classify ERPs is to differentiate between the distances of their generators from the recording electrode. In nearfield recordings, the active electrode is placed relatively close to the neural cortical generator. In farfield recordings, generators are located subcortically at much greater distances from the recording scalp electrode (Desmedt 1988). To date, in human olfaction, only nearfield late evoked potentials have been recorded.

II. STIMULATION

Different types of olfactometers were developed for the above applications in the past. The aim being to construct a device, which would deliver stimuli with a square wave shape, that is, with steep on- and offsets. Although earlier models were able to meet the requirement of steep stimulus onsets, they were not able to guarantee the equally important condition, namely, the absence of mechanical or thermal artefacts occurring synchronously with stimulation. Only recently have olfactometers been developed (Kobal and Plattig 1978, Kobal 1981, 1985, Tonoike and Kurioka 1979, 1985) which meet all of these demands. Since these methodological aspects are very important when recording electrophysiological correlates of olfactory sensations, the principal design of such a stimulator, which delivers the chemical stimulants without altering the mechanical or thermal conditions at the stimulated mucosa (Kobal 1981, 1985), will be briefly outlined.

Monomodal chemical stimulation is achieved by mixing pulses of the stimulants in a constantly flowing air stream with controlled temperature (36.5° C) and humidity (80% relative humidity). Figure 1 represents a diagram of the thermostabilized switching device. Essentially, two gaseous streams are being switched, comparable to the functioning of a railway switch. The switching is induced by applying two separate sources of vacuum so that either an odorant at a pre-established concentration (odor + dilution), or during interstimulus intervals, non-odorous air (control) reaches the nose. By carefully tuning the switching of the two sources of vacuum, the stimulus characteristics can be optimized in such a way that the rise time does not exceed 20 ms. Thus, it is guaranteed that the subjects have no additional cues, such as tactile, thermal or acoustic sensations, which would provide extraneous information about the timing of the stimulus presentation. The total flow rate and stimulus duration can be varied over wide ranges (Kobal 1981, 1985, 1987, Kobal and Hummel 1988).

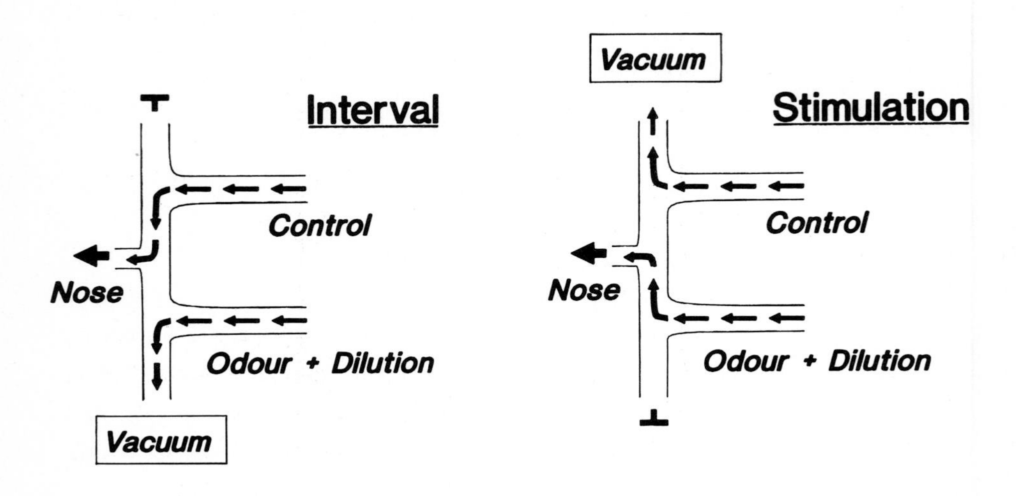

Fig 1. Schematic drawing of the flow switching device. No turbulence or changes in pressure or flow rate are detectable by the subject when the odorant is switched on or off. The carrier gas (air) is thermostabilized (36.5°C) and humidified (80% relative humidity).

This device made it possible to record EOGs from the nasal mucosa, as well as OEPs from the cerebral cortex, without eliciting artefacts. Another important aspect of applying olfactory stimuli is the necessity to control breathing. A prerequisite is that the respiratory air does not exert an undue influence on either the odorant's transportation to the olfactory mucosa, or on the responses to be measured. Moreover, the delivery of the stimuli should be independent from breathing, otherwise it would be impossible to investigate the temporal characteristics of the responses. Thus, to avoid flow of respiratory air in the nose during stimulation the technique of velopharyngeal closure (Fig 2) was introduced (Kobal 1981).

III. THE HUMAN ELECTRO-OLFACTOGRAM

The recording of electro-olfactograms (EOG) from the human nasal mucosa is still a difficult problem for, and a challenge to, those concerned with electrophysiology of the human sense of smell. The placing of the electrodes alone is no easy task, since the intrusion of foreign matter into the nose nearly always leads to sneezing and to an excessive discharge of mucus. Extensive local anaesthesia has also to be avoided since it might affect olfactory fibres and render the subject temporarily anosmic. This is the reason, why only a few publications dealing with this subject exist. In one study Osterhammel et al. (1969) discovered that the negative electrical potential recorded from the olfactory mucosa of two patients increased in relation to an incremental flow rate of coffee-saturated air. In more recent experiments,where the odorants amyl acetate, hydrogen sulphide (H_2S), and eugenol, were presented to four subjects, Kobal (1981) utilized the stimulation method described in this paper so that mechanical or thermal alterations of the stimulated mucosa due to different flow rates of the stimulant were eliminated. The responses, all of which characteristically showed negative electrical potentials at the surface of the nasal mucosa, were dependent on the concentration of the stimulus. Where stimuli (H_2S) of longer durations were applied, temporal integration (Cometto-Muniz and Cain 1984) over a period of 10 s was observed.

The same phenomenon also applied to stimuli presented in pairs, in which one stimulus, the duration of which was 1 s, succeeded the other by a short interval. The response to the second stimulus had a smaller amplitude and was superimposed upon the remnant of the preceding one (Fig 3). However, overall there was very little adaptation of the EOG. As a consequence the subjectively perceived decrease in intensity of the sensation, after repeated or continuous application of an odorant, must have been due, at least in part, to habituation.

While recording the EOGs from the nasal mucosa, a second electrode, placed on the regio respiratoria, registered additional negative potentials after application of odorants with strong trigeminal components. These negative deflections were not correlated with olfactory sensations, but clearly corresponded to painful sensations. There is strong

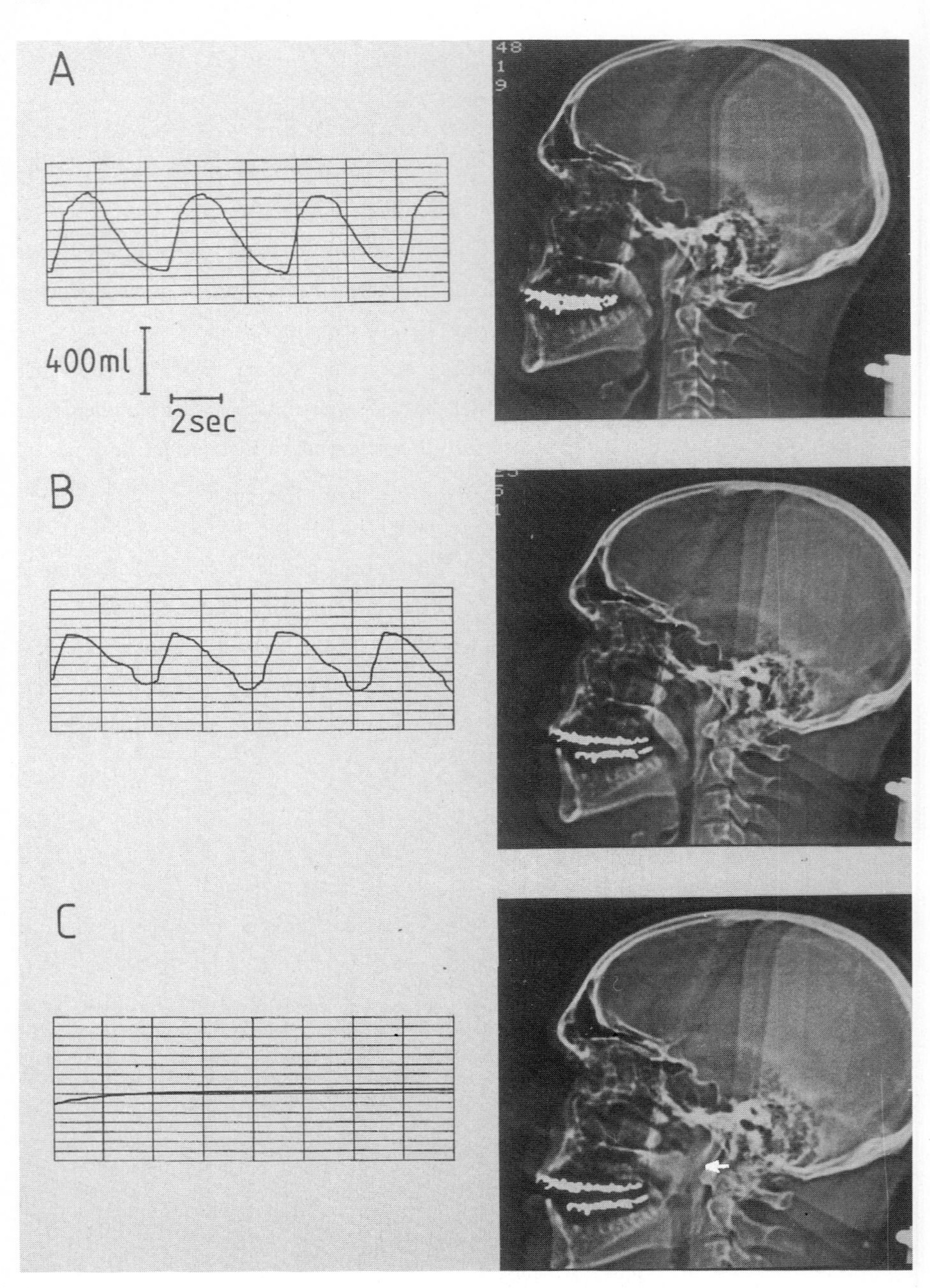
A
400ml
2sec
B
C

evidence that they represent peripheral events, which are related to the activation of nociceptors of the trigeminal nerve (Kobal 1985).

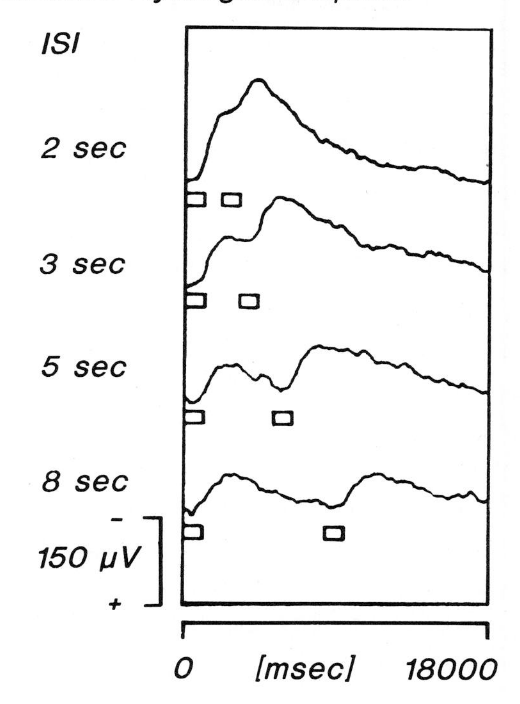

Fig 3. Electro-olfactograms recorded from human olfactory mucosa. The electrode consisted of Teflon tubing (1 mm outer diameter) filled with Ringer-agar (1%), inside of which was a silver chloride-silver wire (impedance <3 kOhm). A silver chloride-silver electrode at the contralateral medial eye angle served as a reference (filter: 0-40 Hz). Open boxes indicate the stimuli, which were presented in pairs with different inter-stimulus intervals ISI (hydrogen sulphide, 14000 $\mu g/m^3$, 1 s duration).

Fig 2. Positions of soft palate during and with no velopharyngeal closure. Right: Radiographs were obtained using a Siemens Somatotom. Left: Recordings of the flow rates of respiratory air in front of one nostril. A: Closed mouth; all respiratory air escaped through the nasal cavity. B: Open mouth; respiratory air escaped through the nasal and oral cavities. C: Open mouth with simultaneous velopharyngeal closure; all respiratory air escaped through the oral cavity since the connection between the nasal cavity and the trachea is blocked by the soft palate (arrow). Prof G Rettinger of the ENT-Clinic, University of Erlangen, kindly provided the radiographs.

Meanwhile, it has been shown in animal experiments (Kobal, Friedl, Thürauf, unpublished results) that these responses can be eliminated by premedication with capsaicin (Silver et al. 1986), which implies a relationship to the excitation of nociceptors. Thus, there are two possible explanations for their existence: either they represent a direct activation of the nociceptors themselves, that is, they are generator potentials, or they indicate a peripheral reaction e.g., based on an axon reflex, to the activation of nociceptors.

IV. THE HUMAN OLFACTORY EVOKED POTENTIAL

The earliest reports of human olfactory evoked potentials were those of Allison and Goff (1967) and Finkenzeller (1966). However, they were adversely commented on by Smith et al. (1971), who contested the olfactory nature of these responses on the grounds that they were not obtainable in patients who had lost trigeminal sensitivity. After introduction of a new stimulation technique, Kobal (1982) tested anosmic patients who yielded no evoked potentials after stimulation with 2-phenethyl alcohol or vanillin. These findings were concordant with the results of Doty et al. (1978), who reported that these substances were not perceived by anosmics. Thus, it can be concluded that olfactory evoked potentials indeed exist and can be recorded if precise and well controlled stimulation techniques are used (Kobal and Hummel 1988). So far, non-odorous carbon dioxide has been perceived by all anosmics investigated, and has elicited chemosomatosensory evoked potentials (CSSEP) which were mediated by the trigeminal nerve. An attempt to determine the differences in the shape of the potentials elicited by substances which, to different degrees, excite chemoreceptors of either the olfactory or the trigeminal nerve, was only partly successful (Kobal and Hummel 1988). However, the degree of excitation of the trigeminal nerve strongly influences the time of appearance and the amplitudes of chemosensory evoked potentials. The more the trigeminal nerve is involved in the processing of a stimulant, the shorter become the latencies, the amplitudes are larger and the intensity ratings of the overall (olfactory plus somatosensory) sensation are higher.

Based upon earlier findings concerning the topographical distribution of evoked potentials (Plattig and Kobal 1979), recent investigations indicate that different stimulants effect different topographical patterns of chemosensory evoked potentials (Kobal et al. 1987).

Maximal amplitudes of potentials evoked by substances (high concentrations of CO_2, menthol, acetaldehyde) which partly or exclusively excite the trigeminal nerve were found at the vertex, and are defined as chemosomatosensory evoked potentials (CSSEP) (Kobal and Hummel 1988).

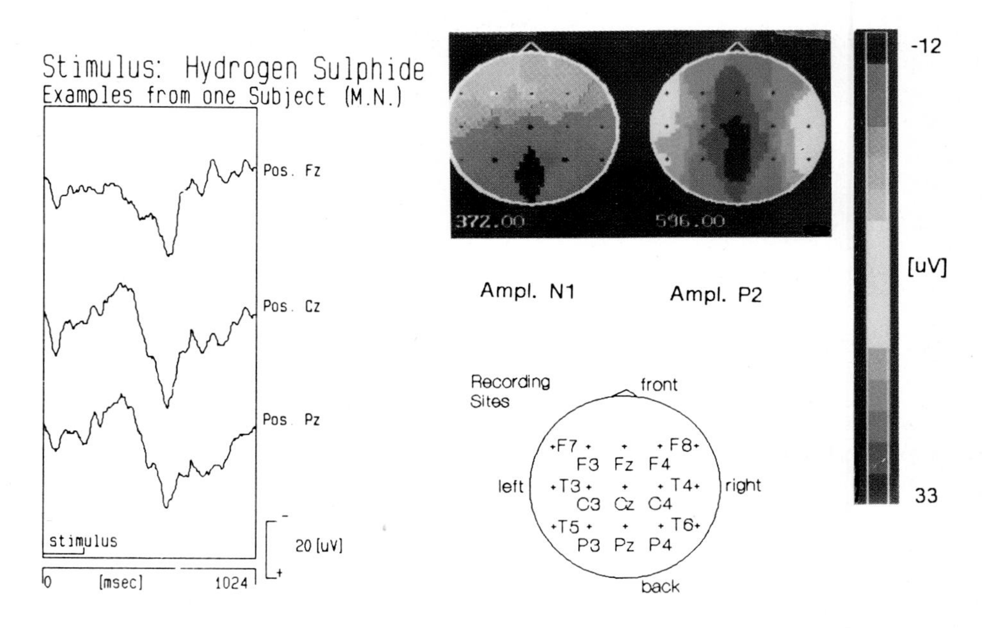

Fig 4. Topographical distribution of the olfactory evoked potentials induced by H_2S (200 ms duration, left nostril). Base-to-peak amplitudes N1 (322 ms), the main negative deflection (upwards), and P2 (596 ms), the main positive deflection (downwards), are mapped (top right-inset) showing a trend to maximal amplitudes at central (C) and parietal (P) leads. The EEGs (bandpass 0.2-70 Hz, impedance 1-2 kOhm) were recorded from 14 positions of the international 10/20 system referenced to A1-A2. EEG records of 1024 ms duration (left inset) were digitized (sampling frequency 250 Hz) and averaged. EEG records contaminated by eye blinks (Fp2) or motor artefacts were discarded.

Substances which exclusively, or to a great extent, excite the olfactory nerve (H_2S, vanillin) effected maximal responses in Pz (a parietal lead in the mid-line), and were defined as olfactory evoked potentials (OEP). In Figures 4 and 5 maps of these two patterns are presented. Evoked magnetic field recordings by Huttunen et al. (1986) demonstrated that responses to CO_2 in all probability are generated in the secondary

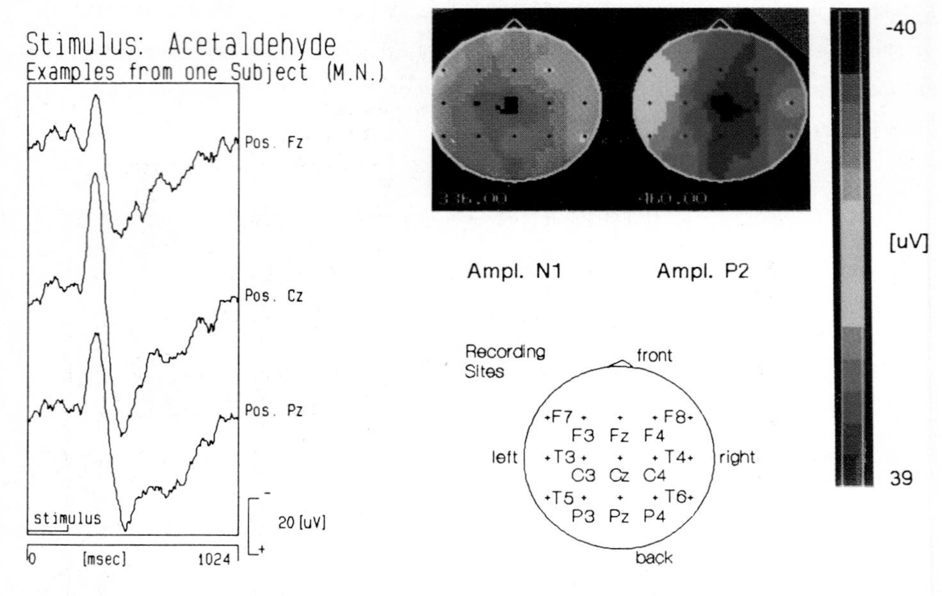

Fig 5. Topographical distribution of the chemosensory evoked potentials (mix of olfactory evoked potential and chemosomatosensory evoked potential) induced by acetaldehyde (200 ms duration, left nostril) in a concentration that additionally activated the trigeminal nerve. Base-to-peak amplitudes N1 (336 ms) and P2 (460 ms) are mapped showing a trend to maximal deflections at central leads.

somatosensory cortex (SII), whereas the generators of the responses to olfactory stimuli are not located at this site. These findings reveal that different generators must be responsible for the potentials evoked by "pure" odorants, and for those evoked by substances that excite chemoreceptors of the trigeminal nerve. However, since the distribution of these potentials is not totally different from the distribution of potentials evoked by CO_2, it may be assumed that they are also generated in the frontal cortex (Kobal and Hummel 1988).

Recently, an additional clue to distinguishing the responses to olfactory and somatosensory stimuli was found (Kobal et al. 1989). After randomized dichorhinic stimulation of both nostrils, it was possible to distinguish between two kinds of chemical stimulants; 1) those which subjects were able to localize to the stimulated side, and 2) those which the subjects were unable to localize to the stimulated nostril.

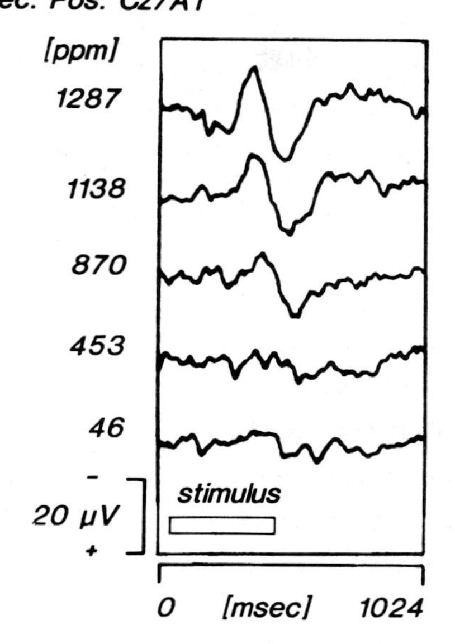

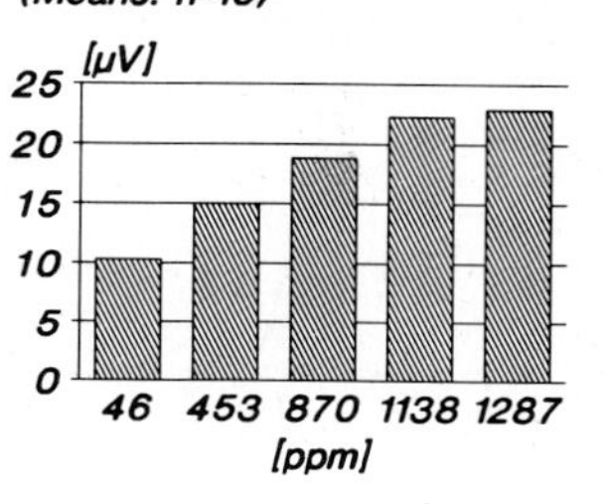

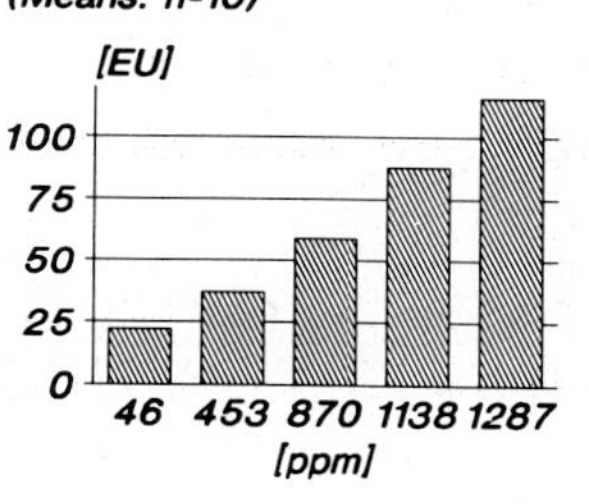

Fig 6. Influence of different concentrations of linalool on the OEP. Five subjects. Two separate sessions. Stimulus duration = 400 ms. Different concentrations were randomly presented 10 times to the left nostril in each one of 2 sessions. Since amplitudes of the evoked potentials were averaged (10 presentations), estimates were also averaged. EU = estimation units according to the length of a visual analog scale. 100 EU = standard stimulus (1138 ppm) that was presented as the first stimulus in the experiments. Means (right bar graphs) were calculated for the averaged measurements of 2 sessions with 5 subjects. The EEG recording technique was as is shown in Figure 4. Recording position was vertex (Cz) versus left ear lobe (A1).

Carbon dioxide, menthol (Kobal et al. 1987), ammonia, and sulphur dioxide belong to the first group (Kobal and Hummel, unpublished), while vanillin, acetaldehyde (in low concentrations), and hydrogen sulphide, none of which can be perceived by anosmics, belong to the second group. The differences in the topography of the chemosensory evoked potentials were concordant with the grouping of the stimulants in relation to their property of being locatable. Non-locatable stimuli yielded a typical olfactory distribution, whereas locatable stimuli yielded a typical somatosensory distribution (Kobal et al. 1987). Before employing a novel biological potential, it is expedient to study some of its more important characteristic features e.g., its relation to stimulus intensity. In olfaction the

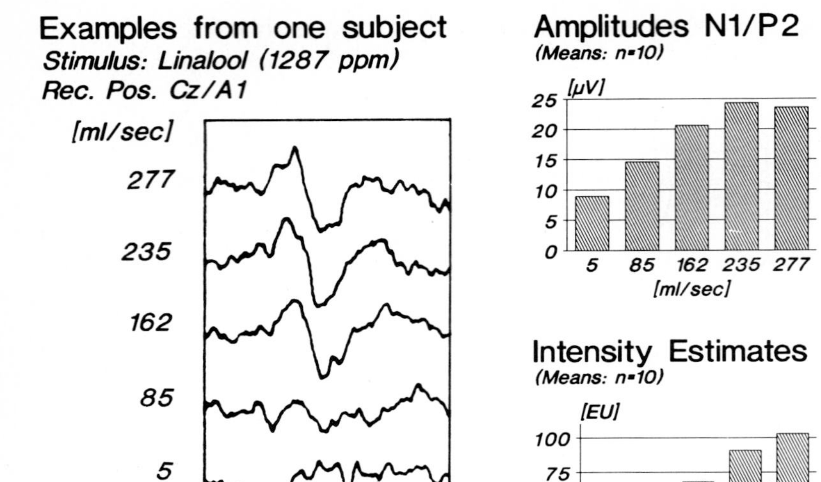

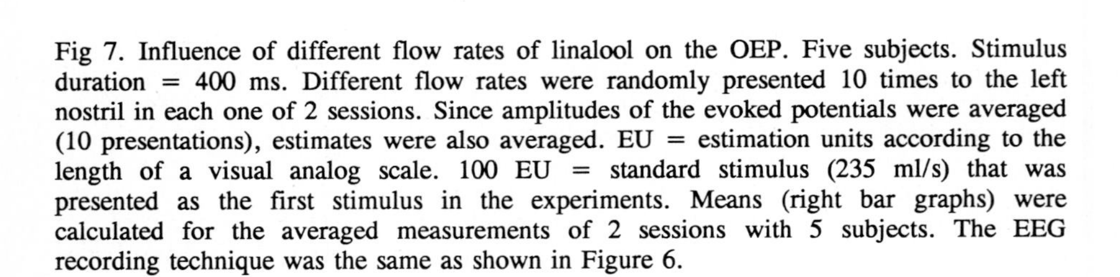

Fig 7. Influence of different flow rates of linalool on the OEP. Five subjects. Stimulus duration = 400 ms. Different flow rates were randomly presented 10 times to the left nostril in each one of 2 sessions. Since amplitudes of the evoked potentials were averaged (10 presentations), estimates were also averaged. EU = estimation units according to the length of a visual analog scale. 100 EU = standard stimulus (235 ml/s) that was presented as the first stimulus in the experiments. Means (right bar graphs) were calculated for the averaged measurements of 2 sessions with 5 subjects. The EEG recording technique was the same as shown in Figure 6.

stimulus intensity can be described as the number of molecules presented during a defined time segment. In practice this can be achieved by varying the concentration or the flow rate of an odorant. Figures 6 and 7 show the results of experiments in which the concentration and flow rate of linalool were varied (Kobal 1981). Suprathreshold stimuli were used in order to distinguish OEPs from background activity, indicating that an objective measurement of thresholds cannot be achieved by simply recording OEPs. On the other hand, it was equally apparent that the amplitudes of OEPs increase and their latencies decrease with an increment of the stimulus intensity (concentrations: amplitude N1/P2: $r=0.703$ ($p<0.05$); intensity estimates: $r=0.94$ ($p<0.01$), $n=5$; flow rates: amplitude

N1/P2: $r=0.780$ ($p<0.01$); intensity estimates: $r=0.972$ ($p<0.01$), $n=5$, Fischer's Z-transformed data). Where flow rates did not exceed 85 ml/s, no responses could be obtained. The conditions that limited recording of OEPs in these experiments were also valid for other odorants.

In another experiment it was demonstrated (Kobal 1981) that OEPs to repeated stimulation (200 ms duration, eucalyptol) showed considerable habituation. In 16 subjects interstimulus intervals (ISI) of 12, 22, 32, 42 and 52 s were used. While a shortening of the ISI from 52 to 32 s caused a reduction of the amplitudes by approximately 15%, ISIs of durations shorter than 32 s occasioned a considerable reduction in amplitudes. With an ISI of 12 s only late positive potentials were present in not more than 25% of the subjects. Thus, the ISI should be longer than 30 s, when repeated stimulation with short odorant pulses is necessary.

Currently, a small number of publications exist which are concerned with the employment of OEPs in different scientific fields. Using OEPs, Westhofen and Herberhold (1987), for example, investigated patients who suffered from diverse lesions of the CNS. They found changes in cortical responses to odorous stimuli, and assumed that the earlier peaks were related to trigeminal and the later ones to olfactory activity.

A further, very important application of OEPs, lies in clinical olfactometry for diagnostic purposes. Figure 8 depicts the results from a patient with transient anosmia which is typical for viral infections of the upper respiratory tract. From these and similar measurements it is clear that the olfactory and trigeminal systems are affected in different ways. Hence under certain circumstances these two functional systems can be studied separately.

Another type of induced brain wave activity is the endogenous component, called the contingent negative variation (CNV) (Walter et al. 1964). These components are termed endogenous, because their presence depends upon subjective response strategies rather than on stimulus characteristics. Usually the test procedure starts with an initial warning stimulus S1, which is followed by a second stimulus S2, indicating to the subjects that

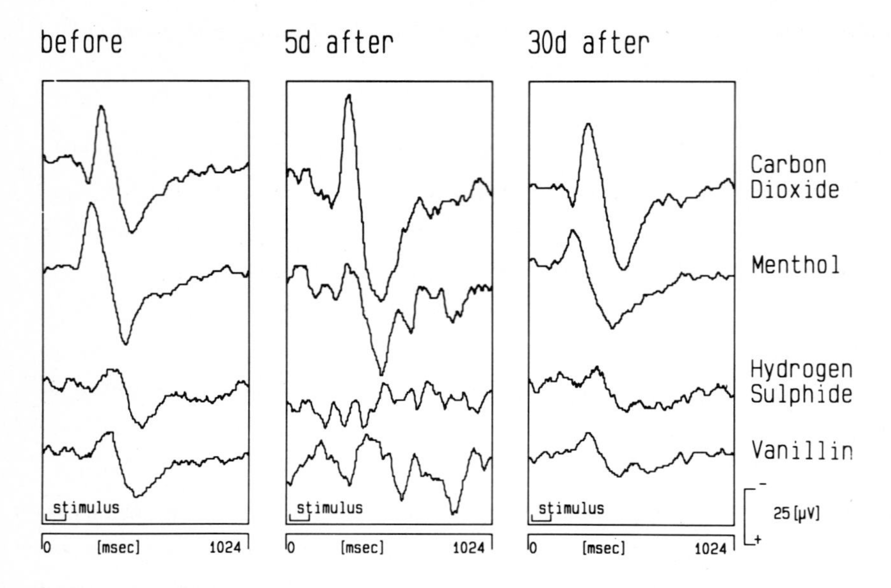

Fig 8. Transient anosmia due to viral infection of the upper respiratory tract in a patient (C.H.). Five days after the infection the chemosomatosensory evoked potential elicited by CO_2 was the only response that was not diminished by the viral infection (the nociceptors may have been even more sensitive because of the augmentation in the amplitude). Olfactory evoked potentials to H_2S and vanillin were no longer discernible. Because of the missing sensory input there was an increased slow wave activity indicating a low level of arousal. Thirty days later, the olfactory sensitivity had almost recovered according to subjective and objective measures. The responses to the trigeminal stimulus CO_2 were not diminished by the viral infection. Stimulus duration 200 ms. Vertex (Cz) versus left ear lobe (A1).

they are to fulfill a task e.g., press a button when S2 is detected. Between the occurrence of S1 and S2 a negative potential slowly builds up and breaks down after S2. When S1 has been an odorant, these components have been used for diagnostic purposes (Gerull et al. 1981), and to study the effects of pleasant and unpleasant odors (Torii et al. 1988). In order to study arousal or sedative effects, Kanamura et al. (1988) presented odorants as a continuous background stimulus during the CNV recording.

Background EEG activity has also been investigated. For this purpose frequency analysis was employed (Brandl et al. 1980), the aim being to find parameters which could be indicators of pleasantness (Lorig and Schwartz 1987, 1988, Yoshida et al. 1989), or of unconscious odor information processing (Lorig and Schwartz 1988). Although the recording of the background EEG activity is much simpler than recording olfactory event-related potentials, these measurements are not specifically related to activities in the olfactory system. Nevertheless, they show in what way odors can modulate general cerebral functions such as arousal, mood, etc., and evidently are relevant to the understanding of the functioning of the human brain.

V. SUMMARY

Scientists engaged in electrophysiological investigations of the human sense of smell find themselves faced with considerable technical difficulties. On the one hand an appropriate stimulation technique is required for generating reproducible temporally precise stimuli with a square wave characteristic. On the other hand it should be possible to non-invasively record the neuronal activity of the olfactory system. Notwithstanding these difficulties, two kinds of olfactory bio-potentials have been recorded to date; 1) the peripherally recordable electro-olfactogram (EOG) which represents a summated potential generated by olfactory sensory cells, a fact learned from experimental animals, and 2) the olfactory evoked potential (OEP) which is recorded from the surface of the skull. This potential is assumed to be related to the slow dendritic potentials of cortical neurons. From several studies it has become evident that the OEP is quantitatively correlated with the olfactory stimulus. Apart from these specific olfactory responses, several non-specific signals have been recorded e.g., contingent negative variations, and the background activity in the electroencephalogram. This paper has reviewed results obtained in several studies. In particular it has discussed: the basic principle of an olfactometric method suitable for employment in human electrophysiology; a specific breathing technique, which permits the presentation of stimuli independent of the respiratory cycle; the fact that EOGs show inconsiderable adaptation (although habituation of the intensity estimates of the subjects can be observed), and that EOGs sum when the olfactory sensory cells are repeatedly stimulated; specific differences in the topographical distribution of evoked potentials elicited by either the sensory endings of the olfactory or the trigeminal nerve;

the relation between the administered dose of an odorant and the amplitudes and latencies of the OEPs; and last but not least the employment of this technique in diagnosing olfactory deficits.

REFERENCES

Allison T, Goff WR (1967) Human cerebral evoked potentials to odorous stimuli. Electroenceph clin Neurophysiol 14: 331-343

Brandl U, Kobal G, Plattig K-H (1980) EEG-Correlates of olfactory annoyance in man. In: Van der Starre H (ed) Olfaction and taste, vol VII. IRL Press, London, Washington, p 401

Cometto-Muniz JE, Cain WS (1984) Temporal integration of pungency. Chem Sens 8: 325-329

Desmedt JE (1988) Somatosensory evoked potential. In: Picton TW (ed) Human event-related potentials. Elsevier, Amsterdam, p 245

Doty RL, Brugger WPE, Jurs PC, Orndorff MA, Snyder PJ, Lowry LD (1978) Intranasal trigeminal stimulation from odorous volatiles: psychometric responses from anosmic and normal humans. Physiol Behav 20: 175-185

Finkenzeller P (1966) Gemittelte EEG-Potentiale bei olfaktorischer Reizung. Pfluegers Arch 292: 76-80

Fleischl von Marxow E (1890) Mittheilung, betreffend die Physiologie der Hirnrinde. Zbl Physiol 4: 537-540

Gerull G, Mielke G, Mrowinski D (1981) Contingent negative variation bei olfaktorischer Reizung. Z EEG-EMG 12: 125-127

Giesen M, Mrowinski D (1970) Klinische Untersuchungen mit einem Impuls-Olfactometer. Arch Hals-Nas-Kehlk Heilk 196: 377-380

Herberhold C (1973) Nachweis und Reizbedingungen olfaktorisch und rhinosensibel evozierter Hirnrindensummenpotentiale sowie Konzept einer klinischen Computer-Olfactometrie. Westdeutscher Verlag, Opladen

Hosoya Y, Yoshida H (1937) Uber die bioelektrischen Erscheinungen an der Riechschleimhaut. Jap J med Sci III Biophysics 5: 2

Hughes JR, Andy OJ (1979) The human amygdala. I. Electrophysiological responses to odorants. Electroenceph clin Neurophysiol 46: 428-433

Huttunen J, Kobal G, Kaukoranta E, Hari R (1986) Neuromagnetic responses to painful CO2 stimulation of nasal mucosa. Electroenceph clin Neurophysiol 64: 347-349

Kanamura S, Kawasaki M, Indo M, Fukuda H, Torii S (1988) Effects of odors on the contingent negative variation and the skin potential level. Chem Sens 13: 327

Kobal G (1981) Elektrophysiologische Untersuchungen des menschlichen Geruchssinns. Thieme, Stuttgart

Kobal G (1982) A new method for determination of the olfactory and the trigeminal nerve's dysfunction: olfactory (OEP) and chemical somatosensory (CSEP) evoked potentials. In: Rothenberger A (ed) Event-related potentials in children. Elsevier, Amsterdam, p 455

Kobal G (1985) Pain-related electrical potentials of the human respiratory nasal mucosa elicited by chemical stimulation. Pain 22: 151-163

Kobal G (1987) Process for measuring sensory qualities and apparatus therefor. United States Patent Number 4681121.

Kobal G, Hummel C (1988) Cerebral chemosensory evoked potentials elicited by chemical stimulation of the human olfactory and respiratory nasal mucosa. Electroenceph clin Neurophysiol 71: 241-250

Kobal G, Hummel Th, Van Toller C (1987) Olfactory and chemosomatosensory evoked potentials from stimuli presented to the left and right nostrils. Chem Sens 12: 183

Kobal G, Van Toller S, Hummel Th (1989) Is there directional smelling? Experientia 45: 130-132

Kobal G, Plattig K-H (1978) Methodische Anmerkungen zur Gewinnung olfaktorischer EEG-Antworten des wachen Menschen (objektive Olfaktometrie). Z EEG-EMG 9: 135-145

Lorig TS, Schwartz GE (1987) EEG activity during fine fragrance administration. Psychophysiol 24: 599

Lorig TS, Schwartz GE (1988) Brain and odor: I. Alteration of human EEG by odor administration. Psychobiol 16: 281-284

Osterhammel P, Terkildsen K, Zilstorff K (1969) Electro-olfactograms in man. J Laryngol 83: 731-733

Ottoson D (1956) Analysis of the electrical activity of the olfactory epithelium. Acta Physiol Scand 35 Suppl 122: 1-83

Plattig K-H, Kobal G (1979) Spatial and temporal distribution of olfactory evoked potentials and techniques involved in their measurement. In: Lehmann D, Callaway E (eds) Human evoked potentials. Plenum Press, New York, p 285

Picton TW, Hillyard SA (1988) Endogenous event-related potentials. In: Picton TW (ed) Human event-related potentials. Elsevier, Amsterdam, p 361

Silver WL, Maruniak JA (1981) Trigeminal chemoreception in the nasal and oral cavities. Chem Sens 6: 295-305

Silver WL, Mason JR, Marshall DA, Maruniak JA (1986) Rat trigeminal, olfactory, and taste responses after capsaicin desensitization. Brain Res 333: 45-54

Smith DB, Allison T, Goff WR, Princitato JJ (1971) Human odorant evoked responses: Effects of trigeminal or olfactory deficit. Electroenceph clin Neurophysiol 30: 313-317

Tonoike M, Kurioka Y (1979) Correlation analysis of the waveforms of olfactory evoked brain potentials recorded from the human scalp. Bull Electrotech Lab 43: 14-22

Tonoike M, Kurioka Y (1985) Odor measured by catching changes in brain wave. Technocrat 18: 65

Torii S, Fukuda H, Kanemoto H, Miyanchi R, Hamazu Y, Kawasaki M (1988) Contingent negative variation (CNV) and the psychological effects of odour. In: Van Toller S, Dodd GH (eds) Perfumery. Chapman and Hall, London, New York, p 107

Walter WG, Cooper R, Aldrige VJ, McCallum WC, Winter AL (1964) Contingent negative variation: an electric sign of sensorimotor association and expectancy in the human brain. Nature 203: 380-384

Westhofen M, Herberhold C (1987) Trigeminal and olfactory synergism in the perception of smell. Acta Oto-Rhino-Laryngol Belg 41: 66-71

Yoshida T, Saito S, Iida T, Yamamura M, Kanamura S (1989) Effect of odors on frequency fluctuation of brain-waves. Chem Sens 14: 311

PART 3

DEVELOPMENT AND SENESCENCE

7

OLFACTORY FUNCTION IN NEONATES

RICHARD L DOTY

I. INTRODUCTION

In most mammals, a close social relationship exists between a mother and her young which begins at birth and extends to the time of weaning. During this early period the infant comes to recognize the odor of the mother, the species, the nesting area, and the other siblings. In turn, the mother learns to recognize the odor of her infant or infants and, in the extreme case, will not accept offspring for nursing unless the appropriate smell is present. Although recent research suggests that humans also evidence an olfactory infant-mother association, its relative importance for physiological and psychological development is not clear. What is clear, however, is that the human neonate has a functional olfactory system at birth that is responsive not only to maternal odors, but to a wide range of airborne stimulants.

The purpose of this Chapter is to present an overview of the developmental anatomy of the human olfactory system and to discuss what is known about the behavioral responses of neonates to olfactory stimuli, including those present during nursing. The reader is referred elsewhere for more comprehensive and general recent treatises which also present data on the olfactory function of neonates (Doty 1986) and of older children (Mennella and Beauchamp, this book; Moncrieff 1966, Porter 1991, Schall 1988).

II. OLFACTION IN THE NEONATE

A. Anatomical Considerations

A summary of the prenatal developmental stages of the human olfactory system is presented in Table 1. The human fetus has a well developed ciliated olfactory epithelium by 9 weeks of age. Completely differentiated olfactory cells are observed by 11 weeks (Pyatkina 1982), while adult-like lamination of the olfactory bulb and a clearly-defined glomerular layer within the extent of the olfactory formation are present by 18.5 weeks (Humphrey 1940). Both periglomerular and interglomerular cells are also observed at this time. By 32 weeks, a modest amount of olfactory marker protein (a probable cytoplasmic enzyme which is unique to olfactory receptor cells) is detectable in the peripheral olfactory nerve layer (Chuah and Zheng 1987).

TABLE 1 : Pre-natal ontogeny of nasal chemoreceptors in humans. Modified from Schall (1988).

Gestational Age (post-ovulatory week)	Event
3.5 - 5	Formation of the olfactory placode
4.5 - 6	Formation of the olfactory grooves
4.5 - 7	Differentiation of the olfactory nerves
5 - 8	Formation of the vomeronasal grooves
5.5 - 13	Presence of olfactory-like cells in the vomeronasal regions
5.8 - 8	Differentiation of terminal-vomeronasal pathways
6 - 6.5	Formation of the main olfactory bulbs
6.5	Formation of the accessory olfactory bulbs
7 - 8	Characteristic structure of the main olfactory bulbs delineated
7.5 - 9.5	Differentiation of ophthalmic and maxillary divisions of the trigeminal nerve
9.5	Increase of mitral cell size in the main olfactory bulbs
11	Presence of ciliated olfactory neuroreceptors suggesting they are ready for reception
11 - 18.5	Mitral cell layer clearly delineated in the main olfactory bulbs
16 - 24	Presence of epithelial nasal plugs in the external nares
32 - 35	Olfactory marker protein present in the olfactory epithelium, olfactory nerve, and bulbar glomerular layer

The trigeminal nerve (cranial nerve V), which mediates intraoral and intranasal somatosensory responses to chemicals (including coolness, warmth, sharpness, and irritation), is well formed in utero (Gasser and Hendrickx 1969, Hogg 1941). Indeed, the first region of the embryo to be sensitive to cutaneous stimulation (circa 7.5 weeks) is the perioral area supplied by the mandibular and maxillary divisions of this nerve. The ophthalamic division, which innervates the nasal mucosa, is present in month-old embryos and is functional by 10.5 weeks (Brown 1974, Humphrey 1966, Streeter 1908).

Humans possess a well developed vomeronasal organ in utero and, by five weeks of age, olfactory-like sensory cells have been observed in this organ (Bossy 1980, Humphrey 1940, Nakashima et al. 1984, Pyatkina 1982). The lumen of this organ is present in a relatively large number of adults (Johnson et al. 1985) and developed vomeronasal structures have been observed in older fetuses and in the newborn (MacCotter 1915, Read 1908, Kreutzner and Jafek 1980). However, it is unlikely that the vomeronasal system functions postnatally, as the accessory formation, the first relay station of the vomeronasal pathway, undergoes degenerative changes during the second trimester of gestation (Humphrey 1940). Whether the vomeronasal system serves a role in prenatal development or other intrauterine processes is unknown.

B. Neonatal Olfactory Perception

1. *Responses to general odors*

A number of researchers in the late 19th Century and early 20th Century reported that odors elicited movements and facial expressions in human newborns, including premature infants as young as seven months (Genzmer 1873, Kroner 1881, Kussmaul 1859, Peterson and Rainey 1910-1911). Although the results indicated that neonates are responsive to odorants, they are inconclusive, since proper controls, such as the use of blank stimuli and blind scoring procedures, were not employed (Disher 1934).

Perhaps the most noteworthy of these studies was that performed by Peterson and Rainey (1910-1911). These investigators tested the behavioral responses of 207 normal term babies, as well as several premature ones, to the odors of asafoetida, oil of rose geranium, compound spirits of orange, tincture of gentian and mother's milk. Sucking was commonly observed following the presentation of the more pleasant (to adults) odorants (orange extract, oil of geranium), whereas grimaces and head movements were commonly observed following the presentation of the unpleasant odorant asafoetida. In accordance with more recent work (Sarnat 1978), these authors noted remarkable nasal chemosensitivity in premature infants.

In an effort to overcome the apparent subjectivity of such studies, Engen et al. (1963) used electronic transducers to measure the leg withdrawal, general body activity, respiration and heart rate of 20 neonates (32 to 68 h old) before, during and following the presentation of odorants on cotton swabs. Ten odor trials were alternated with 10 control trials, where non-odorized swabs were presented. In the first of two experiments, significant increases were observed in one or more of the dependent measures after the presentation of either acetic acid or phenethyl alcohol. More responses were observed for the acid than for the alcohol, suggesting that stronger or perhaps more irritating stimulants may be more effective in producing such changes. In the second experiment, anise and asafoetida were presented. Again, both stimuli elicited increases in the responses, with a greater increase occurring for asafoetida, the less pleasant of the two stimuli.

In another series of studies, Lipsitt et al. (1963) sought to explore potential changes in olfactory sensitivity over the first four days of life. They determined an average threshold value each day by presenting successively higher concentrations of asafoetida until a response was elicited. The thresholds decreased as the infants got older, although it is not clear whether this reflected changes in learning, sensitization, olfactory responsiveness, trigeminal responsiveness or motor responsiveness.

Using stimuli they believed to be free from trigeminal activity (anise, asafoetida, valerian and water), Self et al. (1972) evaluated changes in respiration and body movements in 32 newborns before and following odor presentation. Eight subjects were tested daily over three consecutive days, whereas the remainder were tested only once. Although odor-related responses were noted in most of the infants on at least one of the test days, large individual differences were present and a relatively high rate of responding occurred for the water condition. In addition, the same infant rarely gave uniform responses across test days, and the two dependent measures were often in disagreement. Nonetheless, odor-related responses were observed in both irregular and deep sleep, and subjects tended to consistently fall into low, moderate, or high levels of responding.

Following in the tradition of a number of early studies which noted that odors elicited facial responses in newborns (Ciurlo 1934, Peterson and Rainey 1910-1911, Stirnimann 1936), Steiner (1974, 1977, 1979) found that odorants such as banana extract, vanilla extract and butter produced smile-like expressions accompanied by sucking movements in neonates less than 12 h old, whereas shrimp and rotten egg odors elicited rejection-like responses, such as a depression of the mouth angles and lip pursing. Steiner recorded such responses cinemagraphically and observed that they also occurred in anencephalic infants.

Overall, these studies imply that neonates respond to odorants. However, since many of the stimuli used in a number of these studies likely had considerable trigeminal stimulative properties (Doty et al. 1978), some of these findings may reflect stimulation of the latter nerve rather than stimulation of the olfactory system. Evidence in support of the hypothesis that many of the behavioral responses are due to olfactory, rather than trigeminal stimulants , should include the demonstration of 1) clear-cut responsiveness to relatively non-irritating stimuli, and 2) differential responsiveness to qualitative, rather than intensive, aspects of the stimuli. Assuming that infants have a trigeminal system of similar sensitivity to that of the adult, the reliable elicitation of sucking and other behaviors in the neonate by odorants such as butter, vanilla, phenethyl alcohol and anise would appear to be in accord with point (1). The differential responding noted to several other types of stimulants, along with more recent studies (described in detail below), which demonstrate that newborn infants can distinguish their own mothers' breast and axillary odors from those of other mothers, provide examples of point (2).

C. Responses to Odors in the Maternal Environment

In two simple but elegant pioneering studies, Macfarlane (1975) demonstrated that neonates could distinguish the odor of their own mother's breasts from 1) a blank control and 2) that of other lactating females. In the first study, Macfarlane hung, side by side, a clean breast pad and a breast pad worn previously by its lactating mother in a test crib where a neonate was lying supine. The pads just touched the baby's cheeks. Videotape records revealed that 17 of the 20 subjects, who were between 2 and 7 days of age, spent more time orienting towards the mother's breast pad than towards the clean breast pad.

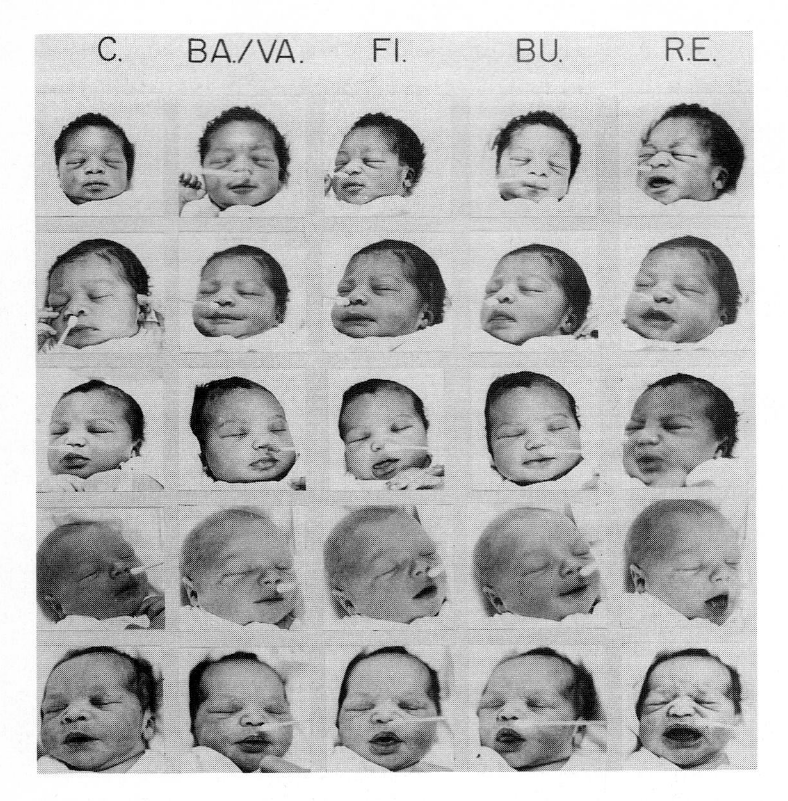

Fig 1. "Perinatal infants' facial reactions to different food-related odors presented to them, prior to their first food intake experience" (between birth and the first breast or bottle-feeding 4-10 h of age). Each horizontal row is the same infant. C = Control (applicators/Q-tips without odorants were presented). BA/VA = response to artificial banana or vanilla flavor. FI = response to odor of artificial flavor of shrimps, concentrated. BU = response to artificial butter flavor. R.E. = response to an artificial, non-poisonous "rotten egg" odor. Photograph kindly donated by Prof J E Steiner, Hebrew University, Jerusalem, Israel.

The majority of the time spent orienting towards either pad was directed towards the mother's pad.

In the second study 32 neonates were similarly tested on days 2 and 6 post-partum for their reactions to breast pads previously worn by their mothers vs pads previously worn by a strange lactating mother. A separate group of 32 infants within the age range of 8 to 10 days was also tested. The percentage of time spent turned towards one or the other breast pad was 72.3 on day 2, 82.6 on day 6, and 89.4 on days 8-10. The percentage of time spent oriented towards the breast pad of their own mother, as a percentage of the total time spent with both breast pads, of the three age groups was 51.8, 60.3 and 68.2% respectively. Fifty-three percent of the babies turned more towards their own mother's pad at 2 days of age (not significantly different from chance), whereas 69% and 78% did so at 6 and 8-10 days of age, respectively (respective ps < 0.01 and 0.001).

In a less structured test situation, Russell (1976) presented breast-fed neonates with 1) a clean moist breast pad, 2) a pad previously worn by the infant's mother, and 3) a pad previously worn by another lactating mother. Although reliable differences in responding were not observed at 2 days of age, by 2 weeks of age 8 of the 10 infants responded to the odor of their own mother and 7 of the 10 to that of the strange mother. Three responded solely to the odor of their own mother. By 6 weeks of age, 6 babies responded only to their own mother's odor and one responded to both the odor of its own mother and to that of the unfamiliar mother.

Using a test paradigm similar to that devised by Macfarlane (1975), Cernoch and Porter (1985) demonstrated that preferences for maternal odors are not confined solely to stimuli from breast pads of the mother. Thus, two-week old breast-fed neonates preferentially oriented towards axillary odors from their mother relative to axillary odors from 1) another lactating mother, or 2) a non-lactating unfamiliar mother. Since bottle-fed infants failed to show this phenomenon, it would appear that either breast and axillary odors have elements in common, or the act of breast feeding brought the infants into closer proximity to the axillary stimuli. No preference was observed when the axillary odors of the father were compared to those of an unfamiliar male.

Support for the role of learning in producing such preferences comes from a study by Schleidt and Genzel (1990). Lactating women applied perfume to their breasts, but not to their nipples, before their babies nursed. When tested at 1 and 2 weeks of age, the infants oriented more towards the familiar perfume than towards a novel one. At 4 weeks of age (2 weeks after the mothers stopped perfuming their breasts), most of the infants failed to demonstrate a preference for the originally-presented odorant, suggesting that the preference may extinguish without repeated presentation or conditioning.

Balogh and Porter (1986) found that exposure to odors without any known conditioned reinforcers is sufficient to induce at least some types of odor preferences in neonates. These investigators taped pads odorized with either ginger or cherry inside the bassinets of babies less than 22 h after birth. On the following day, female neonates oriented more towards a pad impregnated with the exposure odor than with a control odor. The males evidenced no such preference and exhibited a bias in turning their heads to the right regardless of the position of the stimulants. The authors suggest that this difference in the behavior of the sexes may be a reflection of the sex difference in the performance of a number of olfactory tasks that is well-documented in older infants and in adults (Doty et al. 1984, 1985, Deems and Doty 1987, Koelega and Koster 1974, Wallace 1977).

The results of the above studies, along with observations that maternal odors can quiet restless babies (Schall et al. 1980), provide strong evidence that newborn infants have a keen sense of smell. Whether there is any meaningful post-natal development in the perception of the quality of odors is unknown, although there is evidence that sensitivity to odors may be enhanced in adulthood by repeated psychophysical testing or odor exposure (see Chapter 5, this book).

D. Responses to Irritating Vapors

Currently there appears to be only one study that has systematically examined the responsiveness of neonates to irritating odors (Rieser et al. 1976), concluding that newborns are capable of responding directionally to the presence of irritating vapors. These investigators videotaped movements of infants ranging in age from 16 to 131 h following the brief presentation of two open-ended glass sleeves in front of their nares, one containing a low concentration of ammonium hydroxide and the other no odor. On 64% of 304 trials the newborns turned away from the side with ammonia, and on 30% they turned towards that side. As in studies of responsiveness to touch and sound (Turkewitz et al. 1966), the infants exhibited a bias in turning to the right, conceivably reflecting neonatal cerebral asymmetries (Wada et al. 1975).

III. SUMMARY AND CONCLUSIONS

Six general conclusions can be made from the studies reviewed in this Chapter: 1) The human olfactory system is functional at birth; 2) human neonates exhibit facial responses resembling those of pleasure following the presentation of odorants which are judged by most adults as pleasant and of displeasure following stimulation with odorants perceived by adults as unpleasant; 3) human neonates can detect, by inhalation, air-borne irritants such as ammonia, and can localize the direction from which they are presented; 4) newborn babies learn to distinguish between, and come to prefer, odors of their own mothers to those of other lactating mothers; 5) maternal odors can decrease the movements of a baby, and 6) exposure to odorants without traditionally paired reinforcers can influence odor preferences early in life. The degree to which long-lasting odor preferences can be induced, however, is unknown and may well depend upon the age of the infant and the nature of the stimulus presented.

REFERENCES

Balogh RD, Porter RH (1986) Olfactory preferences resulting from mere exposure in human neonates. Infant Behav Devel 9: 395-401

Bossy J (1980) Development of olfactory and related structures in staged human embryos. Anat Embryol (Berlin) 161: 225-236

Brown, JW (1974) Prenatal development of the human chief sensory trigeminal nucleus. J Comp Neurol 156: 307-335

Cernoch JM, Porter RH (1985) Recognition of maternal axillary odors by infants. Child Devel 56: 1593-1598

Ciurlo, L (1934) Sulla funzione olfattoria nel neonato. Valsalva 10: 22-34

Chuah MI, Zheng DR (1987) Olfactory marker protein is present in olfactory receptor cells of human fetuses. Neuroscience 23: 363-370

Deems DA, Doty RL (1987) The nature of age-related olfactory detection threshold changes in man. Trans Penn Acad Ophalmol Otolaryngol 39: 646-650

Disher DR (1934) The reactions of newborn infants to chemical stimuli administered nasally. In: Dockeray FC (ed) Studies of infant behavior. Ohio State University Press, Columbus, p 1

Doty RL (1986) Ontogeny of human olfactory function. In: Breipohl WB (ed) Ontogeny of olfaction. Springer-Verlag, Berlin, p 3

Doty RL, Applebaum SL, Zusho H, Settle RG (1985) A cross-cultural study of sex differences in odor identification ability. Neuropsychologia 23: 667-672

Doty RL, Brugger WE, Jurs PC, Orndorff MA, Snyder PJ, Lowry LD (1978) Intranasal trigeminal stimulation from odorous volatiles: Psychometric responses from anosmic and normal humans. Physiol Behav 20: 175-185

Doty RL, Shaman P, Applebaum SL, Giberson R, Sikorski L, Rosenberg L 1984) Smell identification ability: Changes with age. Science 226: 1441-1443

Engen T, Lipsitt LP, Kay H (1963) Olfactory responses and adaptation in the human neonate. J Comp Physiol Psychol 56: 73-77

Gasser RF, Hendrickx AG (1969) The development of the trigeminal nerve in baboon embryos (Papio sp.). J Comp Neurol 136: 159-182

Genzmer A (1873) Untersuchungen ueber die Sinneswahrnehmungen des neugeborenen Menschen. Inaugural Dissertation, Halle

Hogg ID (1941) Sensory nerves and associated structures in the skin of human fetuses of 8 to 14 weeks of menstrual age correlated with functional capability. J Comp Neurol 75: 371-410

Humphrey T (1940) The development of the olfactory and the accessory olfactory formations in human embryos and fetuses. J Comp Neurol 73: 431-468

Humphrey T (1966) The development of trigeminal nerve fibers to the oral mucosa, compared with their development to cutaneous surfaces. J Comp Neurol 126: 91-108

Johnson A, Josephson R, Hawke M (1985) Clinical and histological evidence for the presence of the vomeronasal (Jacobson's) organ in adult humans. J Otolaryngol 14: 71-79

Koelega HS, Koster EP (1974) Some experiments on sex differences in odor perception. Ann NY Acad Sci 237: 234-246

Kreutzner EW, Jafek BW (1980) The vomeronasal organ of Jacobson in the human embryo and fetus. Otolaryngol Head Neck Surg 88: 119-123

Kroner T (1881) Ueber die Sinnesempfindungen der Neugeborenen. Breslauer Aerzliche Zeitschrift. Cited in Peterson and Rainey, 1910-1911

Kussmaul A (1859) Untersuchungen ueber das Seelenleben des neugeborenen Menschen. C.F. Winter'sche Verlagsbuchhandlung, Leipzig und Heidelberg

Lipsitt LP, Engen T, Kaye H (1963) Developmental changes in the olfactory threshold of the neonate. Child Devel 34: 371-376

MacCotter RE (1915) A note on the course and distribution of the nervus terminalis in man. Anat Rec 9: 243-246

Macfarlane A (1975) Olfaction in the development of social preferences in the human neonate. In: Macfarlane A (ed) Ciba Found Symp 33: 103-117

Moncrieff RW (1966) Odour preferences. Wiley, New York

Nakashima T, Kimmelman CP, Snow JB Jr (1984) Structure of human fetal and adult olfactory neuroepithelium. Arch Otolaryngol 110: 641-646

Peterson F, Rainey LH (1910-1911) The beginnings of mind in the newborn. Bull Lying-in Hosp 7: 99-122

Porter RH (1991) Human reproduction and the mother-infant relationship: The role of odors. In: Getchell TV, Doty RL, Bartoshuk LM, Snow JB (eds) Smell and taste in health and disease. Raven Press, New York

Pyatkina GA (1982) Development of the olfactory epithelium in man. Z Mikrosk Anat Forsch 96: 361-372

Read EA (1908) A contribution to the knowledge of the olfactory apparatus in dog, cat and man. Am J Anat 8: 17-47

Rieser J, Yonas A, Wikner K (1976) Radial localization of odors by newborns. Child Devel 47: 856-859

Russell MJ (1976) Human olfactory communication. Nature 260: 520-522

Sarnat HB (1978) Olfactory reflexes in the newborn infant. J Pediatr 92: 624-626

Schaal B (1988) Olfaction in infants and children: developmental and functional perspectives. Chem Sens 13: 145-190

Schaal B, Montagner H, Hertling E, Bolzoni D, Moyse A, Quichon R (1980) Les stimulations olfactives dans les relation entre l'enfant et la mere. Reprod Nutr Devel 20: 843-858

Schleidt M, Genzel C (1990) The significance of mother's perfume for infants in the first weeks of their life. Ethol Sociobiol 11: 145-154

Self PA, Horowitz FD, Paden LY (1972) Olfaction in newborn infants. Devel Psychol 7: 349-363

Steiner JE (1974) Innate, discriminative human facial expressions to taste and smell stimulation. Ann NY Acad Sci 237: 229-233

Steiner JE (1977) Facial expressions of the neonate indicating the hedonics of food-related chemical stimuli. In: Weiffenbach JM (ed) Taste and development. DHEW Publication No. (NIH) 77-1068, p 173

Steiner JE (1979) Human facial expressions in response to taste and smell stimulation. Adv Child Devel Behav 13: 257-295

Stirnimann F (1936) Versuche uber Geschmack und Geruch am ersten Lebenstag. Jhb Kinderhlk 146: 211-227

Streeter GL (1908) The peripheral nervous system in the human embryo at the end of the first month (10 mm). Am J Anat 8: 285-302

Turkewitz G, Birch HG, Moreau T, Levy L, Cornwall AC (1966) Effect of intensity of auditory stimulation on directional eye movements in the human neonate. Anim Behav 14: 93-101

Wada JA, Clarke R, Hamm A (1975) A cerebral hemispheric asymmetry in humans: cortical speech zones in 100 adults and 100 infant brains. Arch Neurol 32: 239-246

Wallace P (1977) Individual discrimination of humans by odor. Physiol Behav 19: 577-579

8

OLFACTORY PREFERENCES IN CHILDREN AND ADULTS

JULIE A MENNELLA

GARY K BEAUCHAMP

I. INTRODUCTION

The hedonic valence of an odor comprises a fundamental component of olfactory perception (Schiffman 1974). The present Chapter will discuss selected aspects of the development of olfactory preferences from infancy to adulthood. Such selected topics include the origin of olfactory preferences and developmental changes in affective reactions to selected odors. No attempt has been made to be encyclopedic.

II. THE ORIGINS OF OLFACTORY PREFERENCES

A. Infants

A key question in the area of human olfaction is whether hedonic responses to any odors are present at birth or whether all are acquired postnatally, presumably as a function of experience (Engen 1982). Two fundamental problems exist, however, in addressing this question. First, how does one measure affective reactions to odors in the pre-verbal infant? And second, what odors should be used?

Some investigators have measured movements towards or away from an odor and have concluded that such responses indicate olfactory preference and aversion, respectively. If this interpretation is warranted, then responses to odors are evidenced shortly after birth (Engen et al. 1963, Peterson and Rainey 1910-1911, Reiser et al. 1976, Self et al. 1972, Sarnat 1978). However, it is not certain whether the infant's approach or avoidance response was the result of olfactory or trigeminal stimulation because few of the odors used in these studies were pure olfactory stimulants (Doty 1986, Schaal 1988). Moreover, the behaviors observed may not be mediated by affective reactions to the odors. That is, a tendency to avoid novel or strong odors or to approach familiar or weak odors does not prove the odors elicited hedonic experiences.

Facial expressions also have been used as a measure of early hedonic response because such stereotypic, reflex-like behaviors are thought to be indicators of various internal states (Steiner 1973, 1974, 1976, but see Gilbert et al. 1987). A number of taste studies in newborns have demonstrated distinctive facial responses to strong bitter, sour and sweet tastes (Steiner 1977, 1979, Rosenstein and Oster, in press). Although some preliminary data indicate positive and negative facial responses to odors (Datta et al. 1982, Peterson and Rainey 1910-1911, Steiner 1973, 1974, 1977, 1979, Steiner and Finnegan 1975), these studies suffer from interpretational problems. Again, the odors used in some of the studies included substances with potentially significant trigeminal effects, thus confounding the issue as to whether the responses were elicited by the common chemical sense. Moreover, blind scoring procedures and blank controls were not always implemented (Doty 1986).

A logical first approach when studying the infant's response to odors, is to use odors that, a priori at least, have potential biological significance to infants. Examples of such odors are those originating from the mother. Indeed, studies have shown that 2- to 3-day old infants spent more time orienting towards a breast pad previously worn by their lactating mother than one worn by an unfamiliar lactating female (Macfarlane 1975, Russell 1976, Schaal 1986). Moreover, the incidence of head and arm movements decreased (Schaal 1986) and the incidence of sucking increased (Russell 1976) during exposure to their mother's breast odors but not during exposure to the breast odors of an unfamiliar lactating female. This ability of breast-fed infants to discriminate odors from their mothers is not limited to odors emanating from the breast region because they also orient towards odors originating from their mother's underarm (Cernoch and Porter 1985) and neck (Schaal 1986).

While breast-fed infants discriminated their mother's breast and axillary odors, no discrimination was observed with bottle-fed infants when these odors were paired with same-source odor stimuli from an unfamiliar bottle-feeding mother (Cernoch and Porter 1985, Schaal 1986). It has been suggested that breast-fed infants are able to discriminate these odors because they, unlike bottle-fed infants, have prolonged periods of skin contact with their mother, and their nostrils are in close proximity to their mother's breasts and underarm during feeding (Cernoch and Porter 1985).

But are these data indicative of an affective reaction to maternal odors or are newborns merely preferring the familiar and/or avoiding the unfamiliar odors? Recent studies suggest that bottle-fed infants spend more time orienting towards the breast odors of an unfamiliar, lactating female than that of an unfamiliar, non-parturient female (Makin 1987). Thus breast odors, or perhaps the volatile components of breast milk, may be particularly attractive to all newborn infants prior to post-natal exposure to these odors. But even if this conclusion is warranted, it does not prove that the infant's response is unlearned; attraction may be due to a familiarity with the odor given the similarities in volatiles found in breast milk and amniotic fluid (Stafford et al. 1976). Moreover, the animal literature indicates pre-natal exposure to certain odors influences post-natal responses to these odors (Hepper 1987). Whether one can ever determine if a pre-verbal child "likes" one odor more than the other is uncertain.

B. Young Children

The few published studies that have focused on the issue of olfactory preferences and aversions in the verbal child indicate that they indeed have preferences and aversions to a range of odors but that their hedonic experience of odors may be different from that of adults. Because children less than 5 years of age appear to respond positively to some odors, for example, the odor of synthetic sweat and feces that adults dislike, some investigators have concluded that they do not have aversions to odors that adults and older children find offensive and are generally more tolerant of odors than are adults (Engen 1974, Lipsitt et al. described in Engen 1982, Peto 1936, Stein et al. 1958).

Some of the discrepancies as to when olfactory preferences and aversions arise, can be traced to methodological and technical difficulties in evaluating odor hedonics in young children. Children younger than six years of age tend to answer a positively-phrased question in the affirmative (Engen 1974) and they have a shorter attention span than older children (Kniep et al.1931). Their responses, therefore, may be biased and not reflect their actual reaction to the odor. By using methodologies that are sensitive to these behavioral limitations of the child, two recent studies have demonstrated that olfactory preferences and aversions are evidenced in children as young as three years of age. In both studies, the olfactory task was embedded in the context of a game.

In one of the studies the responses of 3-, 4-, and 5-year old children to the odors of benzaldehyde (almond) and dimethyl disulfide (sweat) (Strickland et al. 1988) were investigated. Children were asked to smell a jar which contained an odor and then to indicate how the odor made them feel by placing the jar either on the smile face ("makes you feel good") or frown face ("does not make you feel good"). The data demonstrated that the two odors elicited different hedonic responses at all ages tested.

The other study investigated the response of 3-year old children and adults to a wider range of odor qualities and hedonic values (Schmidt and Beauchamp 1988). Children were told they were playing a "smell game" and that "good" smells (things the subject liked) should be given to Big Bird and "bad/yucky" smells (things the subject disliked) to Oscar the Grouch. Both characters are from "Sesame Street", a popular children's television program with which all of the subjects were familiar. Prior to the start of the game, a teaching session was implemented to be certain the child understood the task involved.

Overall, children and adults exhibited a similar pattern of preferences and aversions (Fig 1). The majority of adults and children rated C-16 aldehyde (strawberry), phenylethyl--methylethyl carbinol (PEMEC:floral), L-carvone (spearmint) and methyl salicylate (wintergreen) as 'good', and butyric acid (strong cheese/vomit) and pyridine (spoiled milk) as 'bad'. In addition, although children were more likely to rate amyl acetate (banana) as 'bad', there was no significant difference between the age groups. There was a significant difference, however, in the proportion of children and adults liking spearmint and disliking pyridine. Differences were also evidenced in affective judgments of eugenol (cloves) and androstenone (variable quality: urinous, sweaty, sandalwood, musky, or no odor). Despite the differences, analyses revealed that the odor was a substantially better predictor of its affective quality than was the age of the subject. Of special interest was the difference between children and adults in their affective judgments

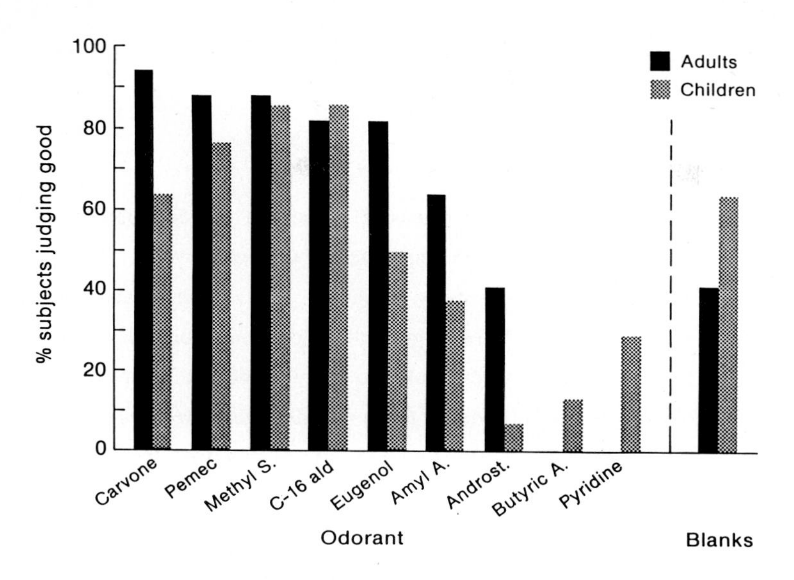

Fig 1. The percentage of adults and 3-year old children who rated each odor as "good". (From Schmidt and Beauchamp 1988).

of androstenone (92% of the children and 59% of the adults rated this odor as 'bad'). Because approximately half of the adult population is anosmic to the odor of androstenone (Wysocki and Beauchamp 1984), the high percentage (92%) of children rating the odor as bad, suggests that most or perhaps all 3-year old children can detect the odor of androstenone (Schmidt and Beauchamp 1988). The difference in sensitivity between children and adults is not due to a general response bias as indicated by the hedonic ratings of pyridine, since all of the adults but only 71% of the children rated pyridine as bad.

In summary, it can be concluded from the aforementioned studies that children as young as 3 years of age have distinct odor preferences and aversions, some of which are similar to that reported for adults. The ontogeny of these preferences and aversions remains a subject for future research.

III. DEVELOPMENTAL CHANGES IN OLFACTORY PREFERENCES: OLDER CHILDREN AND ADULTS

A. The Studies of Moncrieff

Moncrieff (1966) conducted one of the most extensive studies on the development of olfactory preferences to date. In his work, two approaches were employed. The first approach entailed testing a few subjects with many odorants. Twelve individuals, males and females ranging in age from 10 to 77 years, ranked 132 odorants from "best liked" (1) to "least liked" (132). This task took several hours to a day or more to complete, and four of the individuals ranked the odors twice. From the data generated, one of the conclusions was that children, aged 10, 12 and 14 years (n=3) preferred fruity smells more and floral smells less than did adults.

The second approach entailed testing many individuals (559), ranging in age from 4 to 82 years, with only ten odors (almond, chlorophyll, lavender, musk lactone, naphthalene, neroli, rape oil, spearmint, strawberry, vanillin). Again, subjects were asked to rank the odors on the basis of pleasantness. Moncrieff reported that substantial changes in the ranking occurred as a function of the subject's age with the greatest changes occurring during the first twenty years of life (Fig 2). In particular, strawberry was more likely to be judged the most pleasant odor by subjects under 20 years of age than by those older. In contrast, lavender odor and neroli odor (orange blossom) increased in ranking among subjects over 20 years of age. These changes were consistent with that observed by Moncrieff in the smaller subject group.

Other developmental generalizations were that odors commonly disliked by adults (rape oil, chlorophyll) were best tolerated by children younger than eight years and that there were substantial age-related changes during adolescence and adulthood to the odor of musk-lactone, a presumed 'sexual smell'.

The ranking method used by Moncrieff has both strengths and weaknesses. Because this method is essentially a forced-choice procedure, it avoids the tendency of children to answer in the affirmative (Engen 1982). However, there is no absolute scale of pleasantness.

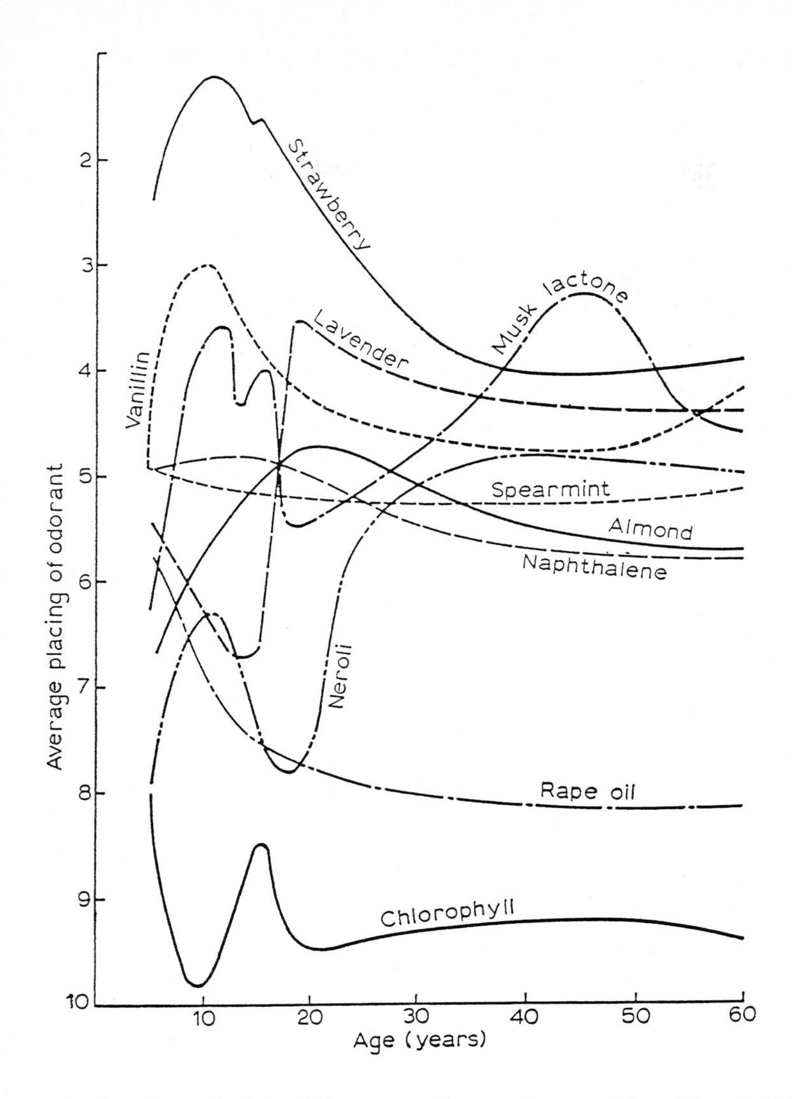

Fig 2. The age/relative liking curves for ten odorants. (From Moncrieff 1966).

Hence, if the ranking of one odor went up and another down, there is no way of determining whether the former became more pleasant and/or the latter became less pleasant. In spite of this problem, Moncrieff's work remained for over twenty years the most extensive attempt to examine developmental changes in the reported pleasantness of odors. Recently, an even broader examination of this topic was conducted in the context of a world-wide study of the sense of smell (Gilbert and Wysocki 1987).

B. The National Geographic Smell Survey

In 1986, the National Geographic Society and the Monell Chemical Senses Center in Philadelphia, conducted the largest smell survey in history (Gilbert and Wysocki 1987). The survey, completed by approximately 1.4 million people worldwide, consisted of a 'scratch and sniff' test with questions regarding the detection, identification, intensity rating and perceived pleasantness of six odorants, namely androstenone (musky, urinous, like sandalwood or no odor), eugenol (clove), iso-amyl acetate (odor of banana), galaxolide (musky or no odor), a mixture of mercaptans (sulfur compounds added to natural gas as a warning agent) and rose (floral). Here we discuss only the age-related changes in odor intensity and hedonic ratings (see Gilbert and Wysocki 1987, Wysocki and Gilbert 1989, for further details).

The authors reported on data generated by 1.2 million questionaires completed by individuals from the United States, who ranged in age from 10 to 99 years (Wysocki and Gilbert 1989). They concluded that although detection and intensity ratings of odorants declined with age, this decline was not uniform across odors. Pleasantness ratings also showed some variation as a function of age and odorant, but there was remarkable consistency in the overall ordering of odorants across age groups. That is, the same ranking of the most pleasant to the least pleasant odor (rose, eugenol, iso-amyl acetate, galaxolide, androstenone and mercaptans) was evidenced in virtually all of the age groups (Fig 3). There were in fact few crossovers of the sort obtained by Moncrieff (compare Figure 2 with Figure 3). This difference between the two studies in the age-related changes in odor hedonics may be due to the difference in tasks, the use of ranking versus absolute ratings, as well as the difference in the amount and kind of odors used. Nevertheless, these data suggest that while there is heterogeneity in the perception of odor pleasantness across the lifespan, there is considerable consistency as well.

The ratings for mercaptans presented an interesting developmental profile. Perceived intensity ratings for this odor steadily declined and pleasantness ratings steadily increased across the lifespan, possibly reflecting the fact that some odors are perceived as more pleasant at low concentrations and less pleasant at higher concentrations (Doty 1975). This finding is alarming given the fact that this odorant is often used as a warning agent in natural gas. Similarly, there was a small but steady decline in the intensity ratings for eugenol as a function of age and a steady but slight increase in preference ratings from individuals in their teens to those in their 50's. This negative association between intensity and pleasantness ratings was not found with three of the odors. Although intensity ratings for the odor of banana gradually declined with age, pleasantness ratings remained relatively stable over the decades. Furthermore, while no age-related changes were evidenced in intensity ratings for the odor of galaxolide, an increase in pleasantness rating for this odor was seen in individuals from the second to third decade of life. In addition, intensity and hedonic ratings for the odor of rose remained stable for the first six decades, but both steadily decreased during the last three decades.

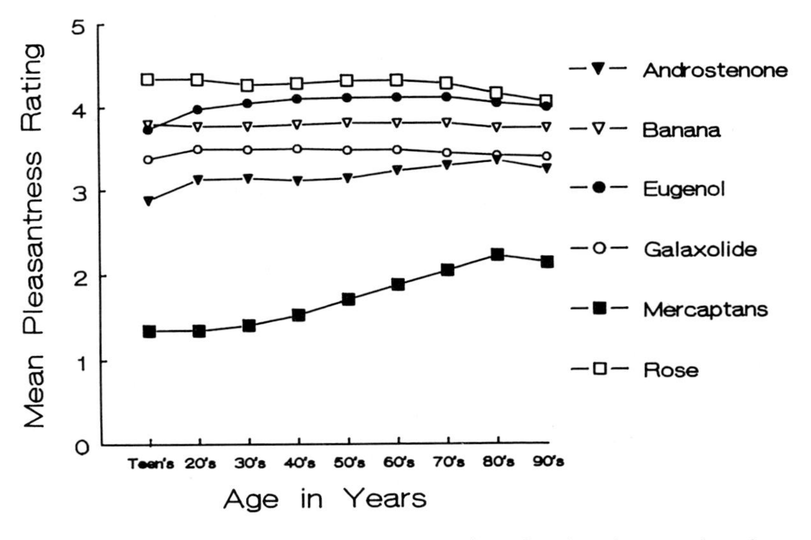

Fig 3. The mean perceived pleasantness ratings for six odorants plotted as a function of age. Subjects were asked to rate odor pleasantness on a 5-point scale (1, unpleasant; 5, pleasant). (Data provided by C. J. Wysocki from Wysocki and Gilbert 1989)

With regard to the odor of androstenone, intensity ratings and the percentage of individuals detecting androstenone declined throughout adolescence with the most precipitous drop occurring in males between the ages of 10 to 20 years. Conversely, pleasantness ratings increased during this time span (Fig 3). As discussed previously, striking differences between children and adults were found in affective responses to the odor of androstenone (Schmidt and Beauchamp 1988). These observations stimulated a more in-depth examination of age-related changes in the perception of this odor.

C. The Odor of Androstenone

Androstenone, a steroid found abundantly in the urine of male pigs, stimulates behaviors associated with sexual receptivity in female conspecifics (Melrose et al. 1974). It is also found in humans, with concentrations being higher in adult males than females (Brooksbank et al. 1974, Gower 1972). The ability to smell the odor of androstenone may in part be genetically determined, as suggested by the finding that sensitivity to this odor is more similar among identical than fraternal twins (Wysocki and Beauchamp 1984). For the 50% of the adult population that can perceive androstenone, a great amount of individual variation exists as to the hedonic quality experienced; its odor can be described as smelling urinous (bad), sweaty (bad), musky (good or bad) or like sandalwood (good).

Apparently, almost all the 3-year old children tested in a previous study detected androstenone and perceived it as an unpleasant odor (see above discussion and Schmidt and Beauchamp 1988). Sensitivity and hedonic response to this odor, however, changed with age, and a pronounced sex difference in the timing of this change existed (Dorries et al. 1989). With regard to changes in sensitivity, the proportion of males insensitive to androstenone tripled between the ages of 9-14 and 15-20 years, whereas the proportion of females insensitive to this odor remained the same during this time period and only doubled in subjects aged between 15-20 years and over 21 years of age. Moreover, adult females had greater olfactory sensitivity to the odor of androstenone than did adult males (Koelega and Koster 1974). With regard to changes in hedonic responses, individuals with high sensitivity to the odor were more likely to rate it as unpleasant, whereas those with low sensitivity rated it as neutral or pleasant. Accordingly, more adult females than males tended to rate androstenone as unpleasant. These data, combined with the National Geographic Smell Survey (Gilbert and Wysocki 1987, Wysocki and Gilbert 1989), support the hypothesis that the ability of humans to detect and dislike androstenone declines with age; this decline is most pronounced for males during early adolescence.

Of further interest is the recent finding that it is possible to induce an ability to perceive androstenone in approximately half of the adults who are initially anosmic to it by exposing them to the odor for an extended period of time (Wysocki et al. 1989). No work has been reported, however, on the hedonic qualities of androstenone in those with induced perception.

IV. DISCUSSION

Why some odors are judged pleasant and others unpleasant is not readily apparent. The simplest hypothesis, that objects producing bad odors are likely to be dangerous and those producing good odors beneficial, is difficult to support or refute. Although some unpleasant odors often arise from substances considered to be harmful e.g., ammonia and odors associated with fecal decomposition, such odors often have a significant trigeminal component. Therefore, it is difficult to determine if the negative hedonic tone is attributable to the trigeminal and/or the olfactory sense.

It could also be hypothesized that odors that play a role in sexual interaction should be inherently pleasant. However, the odors often considered in this category (musk odors) are not species specific. Furthermore, androstenone and androstenol, the two odors most often considered to be mammalian sexual signals, are perceived as unpleasant by approximately 30% of the adult population (Dorries et al. 1989), and the evidence that these odors play a role in human psychosexual interactions is weak indeed (Black and Biron 1982, Cowley et al. 1977, Filsinger et al. 1984, Kirk-Smith et al.1978). This does not negate the possibility that these odors play a role in sexual interaction, but if they do have such a function, it appears that in some cases it is in spite of, rather than because of, the hedonic tone of the odor. Engen (1982) has argued that one need look no further for the explanation of odor hedonics than the contingencies of reinforcement. This strong behaviorist hypothesis suggests that any odor, given the appropriate reinforcement or punishment, can be rendered good or bad, respectively. Two arguments were used to support this viewpoint. First, there are individual as well as cultural differences as to what odors are judged pleasant or unpleasant. And second, odor preferences are not evidenced until childhood, suggesting that the liking or disliking for an odor is acquired.

The first argument appears to be a matter of emphasis. Indeed, substantial individual and cultural differences in olfactory preferences do exist. But there is also general agreement across individuals, in individuals over time and across cultural groups, as to which odors are pleasing and which are not (Gilbert and Wysocki 1987, Moncrieff 1966, and Pangborn et al. 1988). These agreements suggest that some odors are likely to be perceived as good or bad, or that the contingencies of reinforcement are similar across cultures.

As regards the second argument, recent data suggest that odor hedonics can be evaluated in children younger than those reported by Engen. As discussed earlier, children as young as three years of age have distinct olfactory preferences and aversions, some of which are similar to those reported in adults (Schmidt and Beauchamp 1988, Strickland et al. 1988). Moreover, day-old infants exhibit preferences for the breast odors of an unfamiliar lactating females to those of an unfamiliar non-lactating female (Makin 1987). Whether this preference is due to familiarization with the odor and/or to some inherent connection between the odor and positive hedonic tone remains uncertain, and due to the problems involved in controlling in utero exposure to odors, is exceedingly difficult to ascertain.

V. CONCLUSIONS

Future studies in this area should include longitudinal studies on odor hedonics combined with experimental approaches for assessing the ontogeny of odor pleasantness and unpleasantness. Until then, the issue remains as open as it was when Moncrieff (p 77, 1966) wrote: "How wonders the author, who smelled honeysuckle this same afternoon whilst cutting the hedge, can children be so indifferent to this heavenly perfume? The question cannot be answered yet".

REFERENCES

Black SL, Biron C (1982) Androstenol as a human pheromone: No effect on perceived physical attractiveness. Behav Neural Biol 34: 326-330

Brooksbank BWL, Brown R, Gustafsson JA (1974) The detection of 5α-Androst-16-en-3-one in human male axillary sweat. Experientia 30: 864-865

Cernoch JM, Porter RH (1985) Recognition of maternal axillary odors by infants. Child Devel 56: 1593-1598

Cowley JJ, Johnson AL, Brooksbank BWL (1977) The effect of two odorous compounds on performance in an assessment- of- people test. Psychoneuroendocrinol 2: 159-172

Datta T, Prasad SS, George S (1982) Response to olfactory stimulation in normal newborn infants. Indian Pediatr 19: 591-595

Dorries KM, Schmidt HJ, Beauchamp GK, Wysocki CJ (1989) Changes in sensitivity to the odor of androstenone during adolescence. Devel Psychobiol 22: 423-435

Doty RL (1975) An examination of relationships between the pleasantness, intensity, and concentration of 10 odorous stimuli. Percept Psychophys 17: 492-496

Doty RL (1986) Ontogeny of human olfactory function. In: Breipohl W (ed) Ontogeny of olfaction in vertebrates. Springer-Verlag, Berlin, p 3

Engen T (1974) Method and theory in the study of odor preferences. In: Turk A, Johnston JW, Moulton DG (eds) Human responses to environmental odors. Academic Press, New York, p 121

Engen T (1982) The perception of odors. Academic Press, New York

Engen T, Lipsitt LP, Kaye H (1963) Olfactory responses and adaptation in the human neonate. J Comp Physiol Psychol 56: 73-77

Filsinger EE, Braun JJ, Monte WC, Linder DE (1984) Human (Homo sapiens) responses to the pig (Sus scrofa) sex pheromone 5α-androst-16-en-3-one. J Comp Psychol 96: 220-223

Gilbert AN, Fridlund AJ, Sabini J (1987) Hedonic and social determinants of facial displays to odors. Chem Sens 12: 355-363

Gilbert AN, Wysocki CJ (1987) National Geographic smell survey: The results. Nat Geogr Mag 122: 514-525

Gower DB (1972) 16-unsaturated c19 steroids: A review of their chemistry, biochemistry and possible physiological role. J Steroid Biochem 3: 45-103

Hepper PG (1987) The amniotic fluid: An important priming role in kin recognition. Anim Behav 35: 1343-1346

Kirk-Smith M, Booth DA, Carroll D, Davies P (1978) Human social attitudes affected by androstenol. Res Commun Psychol Psychiat Behav 3: 379-384

Kniep EH, Morgan WL, Young PT (1931) Studies in affective psychology: XI. Individual differences in affective reactions to odors. Am J Psychol 43: 406-421

Koelega HS, Koster ED (1974) Some experiments on sex differences in odor perception. Ann NY Acad Sci 237: 234-246

Macfarlane AJ (1975) Olfaction in the development of social preferences in the human neonate. Ciba Found Symp 33: 103-117

Makin JIW (1987) Bottle-feeding infants' responsiveness to lactating maternal breast and axillary odors. Unpublished doctoral dissertation, Vanderbilt University

Melrose DR, Reed HCB, Patterson RLS (1974) Androgen steroids as an aid to the detection of oestrus in pig artificial insemination. Brit Vet J 130: 61-67

Moncrieff RW (1966) Odour preferences. John Wiley, New York

Pangborn RM, Guinard JX, Davis RG (1988) Regional aroma preferences. Food Qual Pref 1: 11-19

Peto E (1936) Contribution to the development of smell feeling. Brit J Med Psychol 15: 314-320

Peterson R, Rainey LH (1910-1911) The beginnings of mind in the newborn. Bull Lying-in Hosp 7: 99-122

Reiser J, Yonas A, Wikner K (1976) Radial localization of odors by human newborns. Child Devel 47: 856-859

Rosenstein D, Oster H (in press) Differential facial responses to four basic tastes in newborns. Child Devel

Russell MJ (1976) Human olfactory communication. Nature 260: 520-522

Sarnat HB (1978) Olfactory reflexes in the newborn infant. J Pediatr 92: 624-626

Schall B (1986) Presumed olfactory exchanges between mother and neonate in humans. In: Le Camus J, Conier J (eds) Ethology and psychology. Privat-IEC, Toulouse, p 101

Schaal B (1988) Olfaction in infants and children: Developmental and functional perspectives. Chem Sens 13: 145-190

Schiffman S (1974) Physiochemical correlates of olfactory quality. Science 185: 112-117

Schmidt HJ, Beauchamp GK (1988) Adult-like preferences and aversions in three-year old children. Child Devel 59: 1138-1143

Self PAF, Horowits FD, Paden LY (1972) Olfaction in newborn infants. Devel Psychol 7: 349-363

Stafford M, Horning MC, Zlatkis A (1976) Profiles of volatile metabolites in body fluids. J Chromatogr 126: 495-502

Stein M, Ottenberg P, Roulet N (1958) A study on the development of olfactory preferences. AMA Arch Neurol Psychiatr 80: 264-266

Steiner JE (1973) The gustofacial response: Observation on normal and anencephalic newborn infants. In: Bosma JF (ed) Oral sensation and perception. NIH-DHEW, Bethesda MD, p 254

Steiner JE (1974) Discussion paper: Innate discriminative facial expressions to taste and smell stimulation. Ann NY Acad Sci 237: 229-233

Steiner JE (1976) Further observations on sensory motor coordinations induced by gustatory and olfactory stimuli. Israel J Med Sci 12: 1231

Steiner JE (1977) Facial expressions of the neonate indicating the hedonics of food-related chemical stimuli. In: Weiffenbach JM (ed) Taste and development. NIH-DHEW, Bethesda MD, p 173

Steiner JE (1979) Human facial expressions in response to taste and smell stimulation. Adv Child Devel Behav 13: 257-295

Steiner JE, Finnegan L (1975) Innate discriminative facial expression to food-related odorants in the neonate. Israel J Med Sci 11: 858-859

Strickland M, Jessee PO, Filsinger EE (1988) A procedure for obtaining young children's reports of olfactory stimuli. Percept Psychophys 44: 379-382

Wysocki CJ, Beauchamp GK (1984) Ability to smell androstenone is genetically determined. Proc Natl Acad Sci 81: 4899-4902

Wysocki CJ, Dorries KM, Beauchamp GK (1989) Ability to perceive androstenone can be acquired by ostensibly anosmic people. Proc Natl Acad Sci 86: 7976-7978

Wysocki CJ, Gilbert AN (1989) The National Geographic Smell Survey: Effects of age are heterogenous. Ann NY Acad Sci 561: 12-28

9

INFLUENCES OF AGING ON HUMAN OLFACTORY FUNCTION

RICHARD L DOTY

I. INTRODUCTION

The ability to smell, like other primary sensory abilities, diminishes as we grow older. Olfactory decrements are especially prominent in the elderly. Indeed, over three-fourths of individuals 80 years of age and older are anosmic or nearly anosmic, and over one half of those between the ages of 65 and 80 years are similarly deficient (Doty et al. 1984b). Such a decline may be universal, as it appears to occur in persons from all walks of life and in all cultures (Gilbert and Wysocki 1987).

Although overlooked by society at large, age-related olfactory deficits have caused millions of dollars of damage to persons and property. If left unappreciated, such deficits will continue to jeopardize the safety and well-being of people. For example, hundreds of butane and propane gas explosions occur each year which can be linked directly to the inability to smell or recognize the odors added to them as warning agents (Cain and Turk 1985), and a disproportionate number of the elderly die in accidental gas poisonings (Chalke and Dewhurst 1957). Such problems will likely gain even greater significance in the future. Thus, in the United States alone, the number of persons over the age of 65 years has increased from 3.1 million in 1900 to 24.1 million in 1978 and is expected to exceed 32 million by the year 2000 (Brody and Brock 1985).

In addition to protecting individuals from fire and leaking gas, the sense of smell plays a primary role in the determination of the flavor and palatability of foods and beverages. For this reason, some of the nutritional problems of the elderly may stem from a loss of motivation to eat as a result of a decreased ability to appreciate flavor, an ability largely dependent upon the stimulation of the olfactory receptors via the retronasal route during deglutition (Burdach and Doty 1987, Mozell et al. 1969). Clearly, understanding age-related changes in the olfactory system and developing strategies to minimize them should be a societal priority.

The purpose of the present Chapter is to provide an overview of the scientific literature related to the influences of aging on the human olfactory system. The alterations in smell function associated with age-related diseases such as Alzheimer's disease are discussed elsewhere in this book (Chapter 14) and will only be mentioned briefly here. As in all studies of aging, it must be kept in mind that it is often difficult, if not impossible, to differentiate between the biological influences of aging, per se, and the influence of other factors associated with simply having led a long life such as, long-term cumulative exposure to airborne toxins.

II. AGE-RELATED CHANGES IN OLFACTORY PERCEPTION

Age-related changes occur in the ability to perceive odors at both threshold and suprathreshold levels. Such changes are robust and include deficits in absolute olfactory sensitivity (for reviews see Doty 1990, Doty and Snow 1988, Murphy 1986, Schiffman 1979, Weiffenbach 1984), odor identification or recognition (Cain and Gent 1986, Doty et al. 1984a, Doty et al. 1984b, Murphy 1985, Schemper et al. 1981, Schiffman 1977), suprathreshold odor discrimination (Schiffman and Pasternak 1979, Stevens and Cain 1985, Stevens et al. 1982), the perception of odor pleasantness (Murphy 1983, Springer and Dietzmann, cited in Engen 1977), and odor memory (Stevens et al. 1990). Furthermore, aging is associated with a greater susceptibility to olfactory adaptation and a slower rate of recovery from such adaptation (Stevens et al. 1989).

Examples of the changes which are observed across the age span on tests of odor detection and odor identification are shown in Figures 1 and 2, respectively. In the case of the widely-used University of Pennsylvania Smell Identification Test, 1) peak performance occurs in the third through the fifth decade of life and markedly declines after the seventh decade, 2) nonsmokers outperform smokers at all ages, and 3) women outperform men, particularly in the later years of life (Fig 1). Interestingly, the sex difference is present to the same relative degree in several cultural groups, including American Blacks, American Caucasians, American Koreans and Native Japanese (Doty et al. 1985).

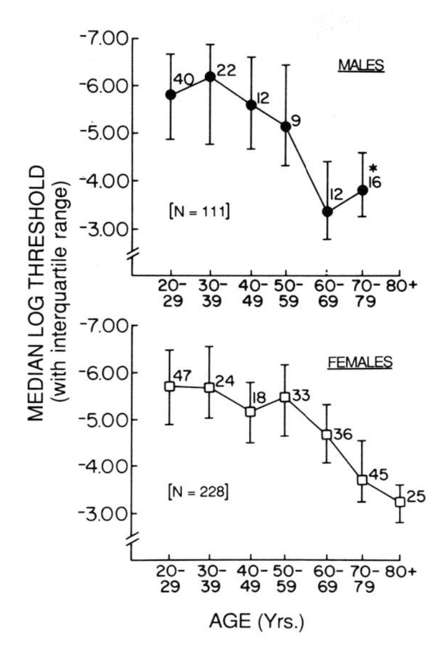

Fig 1. Phenethyl alcohol detection thresholds as a function of age and gender in non-smoking subjects. Numbers by data points indicate sample sizes. Concentration values are plotted inversely on the ordinate. *Because of small sample sizes, the 7 men over 79 years of age have been included in the 70 to 79 year old age point. Reproduced with permission from Deems and Doty (1987).

Age-related decrements in smell perception have also been reported for odorants presented retronasally i.e., from inside the oral cavity, as occurs during chewing and swallowing. Thus, Stevens and Cain (1986) noted that young subjects (18-24 yrs of age) perceived the overall intensity of orally-sampled solutions of ethyl butyrate to be much weaker when the nose was occluded than when it was open. This disparity was not evident in a group of elderly subjects (67-83 yrs), presumably because of age-related alterations in the olfactory system proper.

There is some controversy as to whether the decline in olfactory function observed in the elderly is the result of a gradual loss of smell ability throughout life or whether it reflects a precipitous loss which occurs after the age of 60 (Cain and Stevens 1989, Wysocki and Gilbert 1989). Although most studies report an accelerating decline in the ability to smell after the age of 60, many of the tests that have been administered have ceiling effects which may underestimate the ability of young persons to smell and thereby mask the detection of changes in smell ability across the earlier years of life. Although different

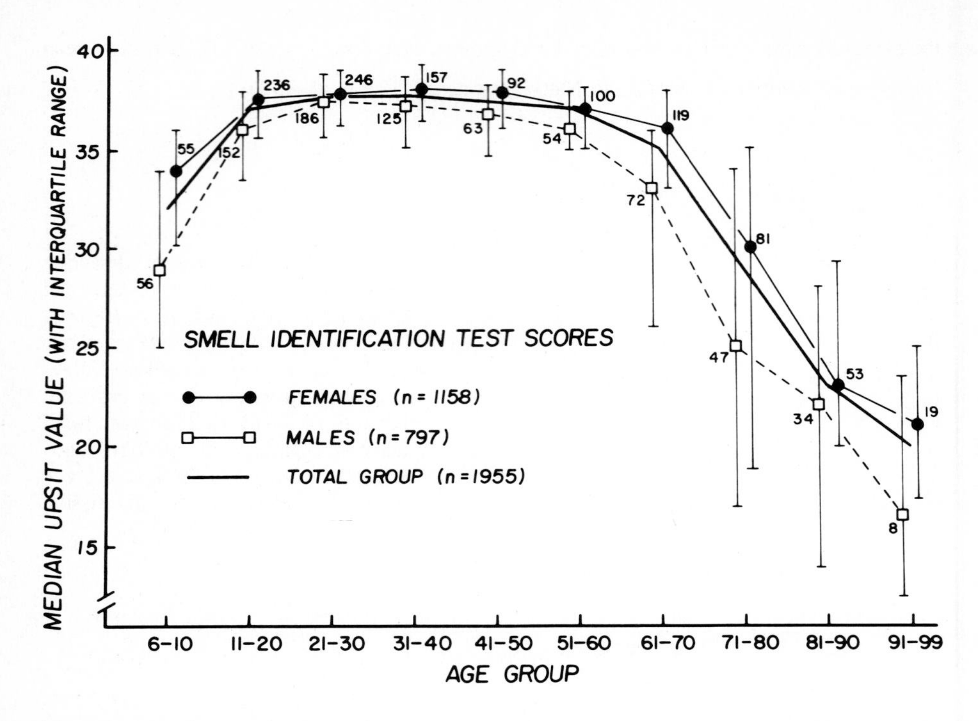

Fig 2. Scores on the University of Pennsylvania Smell Identification Test (UPSIT) (see Chapter 5) as a function of age in a large heterogeneous group of subjects. Numbers by data points indicate sample sizes. Reproduced with permission from Doty et al. (1984b).

odors exhibit different age-related performance functions, in part, because they exhibit different concentration-response functions across their dynamic range, it is of interest to note that persons who evidence low sensitivity to one odorant typically evidence low sensitivity to others. Conversely, those who evidence high sensitivity to a given odorant typically evidence high sensitivity to others. This suggested to Yoshida (1984) that a "general olfactory acuity" factor exists which is independent of odor type, analogous to the general intelligence factor derived from items of intelligence tests (Koelega and Koster 1974, Punter 1983, Stevens and Cain 1985). These observations imply that loss of perception to one odor can often be used as an index of loss of sensitivity to others, an assumption that is implicit in most clinical tests of olfactory function.

In general, most functions relating age to various olfactory test scores are well described by curves with both linear and quadratic components (Deems and Doty 1987). However, more research is needed to determine to what degree such functions are dependent upon task demands, stimulus range, odorant type, and statistical artefacts.

III. PHYSIOLOGICAL BASIS FOR AGE-RELATED CHANGES IN ODOR PERCEPTION

A number of factors are likely to be responsible for age-related changes in human odor perception. Although viral damage to the receptor epithelium is perhaps the most common basis for such loss (Deems et al. in press), other factors must be considered, including age-related atrophy of the olfactory receptors e.g., as a result of vascular or metabolic insufficiency, and loss of specific neurotrophic factors (Appel 1981). Complex interactions among environmental and metabolic factors may also be present. For example, hypothyroidism is implicated in olfactory dysfunction as well as in alterations within the olfactory membrane (for a review see Deems et al. in press), and nutritional deficiencies, including ones associated with some types of dementia, may influence the integrity of epithelial tissue. However, at the present time the relationship between olfactory function and nutrition is obscure (Ferris and Duffy 1989).

In the following section, changes that have been described in the anatomy and physiology of the olfactory pathways, the nasal cavity, and the cribriform plate of older individuals are discussed in detail.

A. Age-related Changes in the Nasal Cavity

Before a smell can be experienced, the odorant molecules responsible for the smell must reach the olfactory receptor cells. Such cells are located in the uppermost recesses of the nasal cavity and occupy regions of the cribriform plate, superior turbinate, and nasal septum. Alterations in the highly vascularized nasal respiratory epithelium, particularly the epithelium located on the nasal turbinates, can alter airflow patterns within the nose and presumably the amount of odorized air reaching the receptors through the small (< 1 mm wide) superior meatus. Thus, age-related alterations in factors which influence airway patency, such as decreased mucosal thickness, nasal polyposis, turbinal engorgement, and inflammation, could influence odor perception in elderly individuals (Doty and Frye 1989).

Although much is understood about changes that occur in aging human skin and in the respiratory system (Fowler 1985, Kligman et al. 1985), surprisingly little is known about age-related alterations in the nasal epithelium and their potential influence upon the olfactory mucosa or upon airflow patterns to the mucosa. However, the few data available suggest that age-related alterations may be profound. For example, Bende (1983) observed a -0.51 correlation between age and blood flow within the nasal mucosa using the ^{133}Xe wash-out method, and Hasegawa and Kern (1977) noted increased nasal patency and a less prominent nasal cycle in persons over the age of 40 (see also Nishihata 1984). Sakakura et al. (1983) noted a decline in the nasal mucocilliary clearance rate in some elderly individuals, and others have reported age-related atrophy and decreases in the vascular elasticity of the nasal epithelium e.g., Somlyo and Somlyo (1968).

According to Koopman (1989), elderly persons evidence, on average, a 7% decrease in body water which results in an increase in nasal mucus viscosity, a problem which is often perceived as "postnasal drip" (Janzen 1986), as well as atrophy of mucus secreting glands and lymphatics. As a result, mucosal dryness, less effective ciliary clearance and the sensation of nasal obstruction can occur. How such changes impact on the olfactory mucosa and on odor perception is unknown.

B. Calcification of the Cribriform Plate

The axons of the olfactory nerve i.e., the collection of bipolar receptor neurons within the olfactory epithelium, form into fascicles and enter the anterior cranial fossa via foramina in the cribriform plate. Therefore, processes which block the transit of these axons through the cribriform plate would be expected to alter olfactory function. In a pioneering study, Krmpotic-Nemanic (1969) demonstrated calcification of the foramina of the cribriform plate in some elderly individuals, implying that the connections between the olfactory epithelium and olfactory bulb were severed, or at least impaired. This important observation, however, does not appear to have received further attention. Clearly, studies examining the proportion of skulls from older persons which evidence such occlusion would be of considerable value in understanding the importance of this phenomena for producing age-related olfactory deficits. Importantly, a number of the alterations in the olfactory epithelium discussed below would be expected to occur as a result of such calcification.

C. Alterations in the Olfactory Epithelium

Since Nagahara's report in 1940 of mitotic activity within the basal regions of the olfactory epithelia of adult mice, it has been widely assumed that olfactory receptor cells are replaced at regular intervals throughout the lifespan of the organism (Moulton 1974, Graziadei and Monti-Graziadei 1978). Such plasticity has been suggested to be an adaptation to the fact that the apical processes of these neurons are exposed almost directly to the outside environment, making them very susceptible to insult from bacteria, viruses and airborne toxins. However, several lines of evidence suggest that this plasticity is altered by age-related processes and that the concept of continuous turnover and replacement of the olfactory receptor cells by new cells is probably incorrect (Breipohl et al. 1986). If, for example, continuous replacement of the olfactory receptor cells occurred throughout life, a constant ratio of dead or dying cells to the number of receptor cells would be observed in animals of all ages. However, as noted by Breipohl, in mice and salamanders there appears to be an age-related increase in this ratio, suggesting that olfactory receptor cell populations from older individuals may have less need for mitotic activity than those from younger ones. Furthermore, following chemical destruction of the olfactory receptor cells with zinc sulfate or methyl-formimino-methylester, morphological repair is slower or non-existent in old mice, unlike the case with younger ones (Rehn et al. 1986, Matulionis 1982). Thus, the neurogenic process appears to be altered with age and may explain, in part, why so many older individuals evidence permanent olfactory loss.

In humans, histological studies of the olfactory epithelium suggest that the plasticity in receptor cell turnover fails to protect all sectors of the olfactory epithelium from destruction, even in young individuals. For example, Nakashima et al. (1984) examined, by light microscopy, the olfactory epithelia of five aborted human fetuses and 21 adults ranging in age from 20 to 91 years at autopsy, and noted zonal degeneration of receptor cells in specimens of all ages, although the effect was more marked in the older speci-mens. The distribution of the basal, supporting, and sensory receptor cells was often disturbed. Metaplasia of the respiratory epithelium was evident within the olfactory epithelium, suggesting that regions of the damaged olfactory epithelium were replaced with respiratory epithelium. In an earlier study, Naessen (1971) noted a regression of the vessels and the loss of cellularity of the lamina propria adjoining the basement membrane in neuroepithelia from older persons. Furthermore, pigment granules were found to have accumulated in the supporting cells of adults, but not in the supporting cells of fetuses, infants, and young children. This suggested to Naessen that these granules may represent the inability of the cells to deal effectively with their own metabolic wastes and the products of neuronophagic activity.

Analogous age-related alterations have been observed in the olfactory epithelia of species other than humans. For example, Dodson and Bannister (1980) noted that, across the 6-to 30-month age range of albino housemice, the supporting cells of the older individuals also accumulated considerable cellular debris in their basal processes. A gradual reduction in cell body size, a reduction in the per cell amounts of granular and agranular endoplasmic reticulum, and an increase in the numbers of secondary lysosomes was observed. While the number of receptors per unit of epithelial volume did not fall appreciably, the number of cells observed in mitosis were less frequent in the older animals, in accordance with the observations of Breipohl et al. (1986). An appreciable decrease in the rate of cell migration away from the basal layer was in evidence. Overall, age-related reduced cellular metabolism and protein synthesis within olfactory epithelia was in evidence.

D. Alterations in the Olfactory Bulb

Smith (1942) examined the number and form of glomeruli present in 205 olfactory bulbs of 121 individuals at autopsy and found a considerable age-related decrease and alteration in these structures. Under the assumption that these alterations were secondary to age-related losses within the olfactory epithelium, he inferred that olfactory nerve loss begins soon after birth and continues throughout life at approximately 1% per year. However, a re-evaluation of Smith's data (Figure 1 in Smith's paper), using medians rather than means, suggests that the major loss may occur after the fifth decade of life, implying that this widely cited statistic may be misleading. Although considerable variability among bulbs was seen at all ages examined, sex differences were not apparent.

In a descriptive study, Liss (1956) examined olfactory bulbs (stained with silver carbonate) from an unreported number of humans ranging from 18 months to 80 years of age and concluded that no age-related effects were present. However, an increase in astroglia and corpora amylacea in those specimens from persons whose ages fell between 70 and 80 years was observed. In a similar non-quantitative study published two years later, Liss and Gomez (1958) reported that the olfactory bulbs and tracts from persons over the age of 70 years exhibited moderate loss of neurons and nerve fibers, as well as increased numbers of glial elements. These authors argued, as had Smith (1942), that the bulbar degeneration is secondary to the destruction of the olfactory receptor cells.

In a recent quantitative study, Bhatnagar et al. (1987) counted the number of mitral cells in the mitral cell and external plexiform layers of both the left and right olfactory bulbs from eight women between the ages of 25 and 102 years. Although no differences were found between the left and right bulbs, the mean number of mitral cells per olfactory bulb decreased from 50,935 at 25 years of age to 32,718 at 60 years and 14,501 at 95 years of age. The corresponding mean bulb volumes were 50.02, 43.35, and 36.68 mm^2, respectively. Thus, a 36% decrease in the number of mitral cells was observed between the ages of 25 and 60 years, whereas a 56% decrease was observed from 60 to 95 years. The corresponding decreases in bulb volume values were 13% and 15%, respectively. Although these data suggest that the decline across age in the number of mitral cells is non-linear, the small sample size precludes a definitive determination of this point.

The influence of age on olfactory bulb structures has also been the subject in animal studies. For example, Hinds and McNelly (1977), using rats of the Sprague Dawley strain, measured the volume of components of the main olfactory bulb, including the glomerular, external plexiform, internal granular and olfactory nerve layers, at 3, 12, 24, 27 and 30 months of age (see also Hinds and McNelly 1979). In addition, the number and size of mitral cells were measured in both the main and accessory areas of the bulbs. Although developmental increases in the layer volumes were noted during the first 24 months, decreases occurred after that time. Of note was a sharp decrease in mitral cell numbers, along with an increase in the volume of dendritic trees of individual mitral cells and in the size of the cell body and nucleus. In a subsequent study using Charles River rats, Hinds and McNelly (1981) observed most of the latter changes, although no loss in mitral cell number was found in older animals. Importantly, Hinds and McNelly examined the alterations in the olfactory neuroepithelium at the same time that they made the other measurements. A comparison of regression lines for changes in the number of olfactory receptor cells with that of the size of mitral cell bodies suggested that the decline in receptor cell number began several months before the increase in mitral cell size. Thus, the latter alteration may have been due to the decline in neuroepithelial receptor cells. An increase in the number of synapses per receptor cell was present in the oldest group evaluated (33 months), possibly reflecting a compensatory increase in the relative numbers of synapses per cell in the surviving receptor cells.

It should be noted that age-related structural changes in the main olfactory bulb may differ considerably among rat strains. Contrary to the above findings with Sprague-Dawley and Charles River rats, Forbes (1984) found no evidence in Fischer 344 rats of changes in the size of mitral cell perikarya and nuclei over the age range of 18 to 36 months. Whether such strain differences reflect differential susceptibility to epithelial damage from viruses or toxins is not known.

E. Alterations in Higher Olfactory Centers

Few studies have examined the olfactory cortices of older humans, although there is now a relatively large literature on changes within the higher sectors of the olfactory system of persons with Alzheimer's disease (Ferreyra-Moyano and Barragan 1989). As noted in Chapter 14, Alzheimer's disease is associated, even at early stages, with major changes in olfactory function which could reflect alterations in cortical olfactory regions. It is now well established, for example, that Alzheimer's disease is accompanied by severe neuro-chemical and anatomical changes in higher olfactory centers, such as structures within the ventromedial temporal lobe (Pearson et al. 1985, Reyes et al. 1987). Whether aging, per se, results in significant changes in higher olfactory regions is not yet established, although it should be noted that 1) some plaques and tangles are present in a number of the regions of the olfactory system of many older individuals who evidence no apparent symptoms of dementia, and 2) neuronal loss occurs in selected layers and regions of the aging human cerebral cortex (Duara et al. 1985, Tomlinson et al. 1968).

In contrast to the hypothesis that aging is associated with changes in central olfactory structures, studies of the piriform cortices of older rats suggest that comparatively few alterations are present, unlike the situation within the olfactory epithelium and bulb. For example, Curcio et al. (1985), in a study of the cells and synapses of the piriform cortices of rats aged 3, 12, 18, 24, 30 and 33 months, found no significant changes in the volumes of cortical laminae Ia and Ib, or in the numerical and surface densities of the synaptic apposition zones in layer Ia, which are formed mainly by mitral cell axons. However, a modest (18%) decline in the proportion of layer Ia occupied by dendrites and spines was observed. This decrease was accompanied by an increase in the proportion of glial processes but not by any alteration in the proportion of axons and terminals. Age-related changes in soma volume, nuclear volume, or numerical density of layer II neurons, were not observed.

IV. SUMMARY AND QUESTIONS FOR THE FUTURE

It is clear from the material reviewed in this Chapter that olfactory function decreases significantly in most older persons. Such decline can be detected by a variety of psychophysical tests, including ones which assess odor detection, identification, discrimination, memory and adaptation. Furthermore, evidence has been presented which suggests that such perceptual alterations are accompanied by changes in the anatomy and physiology of the olfactory epithelium and bulb. However, it is not yet clear whether or to what degree age-related changes occur in higher olfactory centers, although, if Alzheimer's disease serves as a model for such changes, they are most likely to be profound. Questions begging for empirical resolution include: Is there continuous and cumulative destruction of the olfactory epithelium throughout life which, when a critical point is reached, leads to noticeable perceptual deficits in later years of life? Are non-demented older individuals who evidence olfactory dysfunction more likely to develop symptoms of Alzheimer's disease than non-demented older individuals who do not evidence such dysfunction? Is the destruction of the olfactory epithelium in older persons caused solely by environmental insults which ultimately cause changes in metabolic processes within the epithelium, or do age-related metabolic changes set the stage for destruction by such environmental agents? To what degree do metabolic, genetic and environmental factors interact to produce the age-related changes observed in odor perception and in the olfactory pathways? Hopefully, answers to these questions will be forthcoming in the not-too-distant future.

REFERENCES

Appel SH (1981) A unifying hypothesis for the cause of amyotrophic lateral sclerosis, parkinsonism, and Alzheimer disease. Ann Neurol 10: 499-505

Bende M (1983) Bloodflow with ^{133}Xe in human nasal mucosa in relation to age, sex, and body position. Acta Otolaryngol 96: 175-179

Bhatnagar KP, Kennedy RC, Baron G, Greenberg RA (1987) Number of mitral cells and the bulb volume in aging human olfactory bulb: A quantitative morphological study. Anat Rec 218: 73-87

Breipohl W, Mackay-Sim A, Grandt D, Rehn B, Darrelmann C (1986) Neurogenesis in the vertebrate main olfactory epithelium. In: Breipohl W (ed) Ontogeny of olfaction. Springer-Verlag, Berlin, p 21

Brody JA, Brock DB (1985) Epidemiologic and statistical characteristics of the United States elderly population. In: Finch CE, Schneider EL (eds) Handbook of the biology of aging. Van Nostrand Reinhold, New York, p 3

Burdach KJ, Doty, RL (1987) Retronasal flavor perception: Influences of mouth movements, swallowing and spitting. Physiol Behav 41: 353-356

Cain WS, Gent JF (1986) Use of odor identification in clinical testing of olfaction. In: Meiselman HL, Rivlin RS (eds) Clinical measurement of taste and smell. Macmillan, New York, p 170

Cain WS, Stevens JC (1989) Uniformity of olfactory loss in aging. Ann NY Acad Sci 561: 29-38

Cain WS, Turk A (1985) Smell of danger: An analysis of LP-gas odorization. Am Ind Hyg Assoc J 46: 115-126

Chalke HD, Dewhurst JR (1957) Accidental coal-gas poisoning. Brit Med J 2: 915-917

Curcio CA, McNelly NA, Hinds JW (1985) Aging in the rat olfactory system: Relative stability of piriform cortex contrasts with changes in olfactory bulb and olfactory epithelium. J Comp Neurol 235: 519-528

Deems DA, Doty RL (1987) The nature of age-related olfactory detection threshold changes in man. Trans Penn Acad Ophthalmol Otolaryngol 39: 87-93

Deems DA, Doty RL, Settle RG, Moore-Gillon V, Shaman P, Mester AF, Kimmelman CP, Brightman VJ, Snow JB Jr (in press) Smell and taste disorders: A study of 750 patients from the University of Pennsylvania Smell and Taste Research Center. Arch Otolaryngol Head Neck Surg

Dodson HC, Bannister LH (1980) Structural aspects of ageing in the olfactory and vomeronasal epithelia in mice. In: van der Starre H (ed) Olfaction and taste, vol VII. London, IRL Press, p 151

Doty RL (1990) Aging and age-related neurological disease: Olfaction. In: Goller F, Grafman J (eds) Handbook of neuropsychology. Elsevier, Amsterdam, p 459

Doty RL, Applebaum SL, Zusho H, Settle RG, (1985) A cross-cultural study of sex differences in odor identification ability. Neuropsychologia 23: 667-672

Doty RL, Frye RE (1989) Nasal obstruction and chemosensation. Otolaryngol Clin North Am 22: 381-384

Doty RL, Shaman P, Dann M (1984a) Development of the University of Pennsylvania Smell Identification Test: A standardized microencapsulated test of olfactory function. Physiol Behav (Monogr) 32: 489-502

Doty RL, Shaman P, Applebaum SL, Giberson R, Sikorski L, Rosenberg L (1984b) Smell identification ability: Changes with age. Science 226: 1441-1443

Doty RL, Snow JB Jr (1988) Age-related changes in olfactory function. In: Margolis RL, Getchell TV (eds) Molecular neurobiology of the olfactory system. Plenum Press, New York, p 355

Duara R, London ED, Rapoport S (1985) Changes in structure and energy metabolism of the aging brain. In: Finch CE, Schneider EL (eds) Handbook of the biology of aging. Van Nostrand Reinhold, New York, p 595

Engen T (1977) Taste and smell. In: Birren JE, Schaie KW (eds) Handbook of the psychology of aging. Van Nostrand Reinhold, New York, p 554

Ferreyra-Moyano H, Barragan E (1989) The olfactory system and Alzheimer's disease. Int J Neurosci 49: 157-197

Ferris AM, Duffy VB (1989) Effect of olfactory deficits on nutritional status. Ann NY Acad Sci 561: 113-123

Forbes WB (1984) Aging-related morphological changes in the main olfactory bulb of the Fischer 344 rat. Neurobiol Aging 5: 93-99

Fowler RW (1985) Ageing and lung function. Age and Ageing 14: 209-215

Gilbert AN, Wysocki CJ (1987) The smell survey. Nat Geogr Mag 172: 514-525

Graziadei PPC, Monti Graziadei GA (1978) Continous nerve cell renewal in the olfactory system. In: Jacobson M (ed) Handbook of sensory physiology, vol IX. Springer, New York, p 55

Hasegawa M, Kern EB (1977) The human nasal cycle. Mayo Clin Proc 52: 28-34

Hinds JW, McNelly NA (1977) Aging of the rat olfactory bulb: growth and atrophy of constituent layers and changes in size and number of mitral cells. J Comp Neurol 171: 345-368

Hinds JW, McNelly NA (1979) Aging in the rat olfactory bulb: Quantitative changes in mitral cell organelles and somato-dendritic synapses. J Comp Neurol 184: 811-820

Hinds JW, McNelly NA (1981) Aging in the rat olfactory system: Correlation of changes in the olfactory epithelium and olfactory bulb. J Comp Neurol 203: 441-454

Janzen VD (1986) Rhinological disorders in the elderly. J Otolaryngol 15: 228-230

Kligman AM, Grove GL, Balin AK (1985) Aging human skin. In: Finch CE, Schneider EL (eds) Handbook of the biology of aging. Van Nostrand Reinhold, New York, p 820

Koelega HS, Koster EP (1974) Some experiments on sex differences in odor perception. Ann NY Acad Sci 237: 234-246

Koopmann CF Jr (1989) Effects of aging on nasal structure and function. Am J Rhinol 3: 59-62

Krmpotic-Nemanic J (1969) Presbycusis, presbystasis, and presbyosmia as consequences of the analagous biological process. Acta Otolaryngol 67: 217-223

Liss L (1956) The histology of the human olfactory bulb and the extra-cerebral part of the tract. Ann Otol Rhinol Laryngol 65: 680-691

Liss L, Gomez F (1958) The nature of senile changes of the human olfactory bulb and tract. Arch Otolaryngol 67: 167-171

Matulionis DH (1982) Effects of the aging process on olfactory neuron plasticity. In: Breipohl W (ed) Olfaction and endocrine regulation. IRL Press, London, p 299
Moulton DG (1974). Dynamics of cell populations in the olfactory epithelium. Ann NY Acad Sci 237: 52-61
Mozell MM, Smith BP, Smith PE, Sullivan RL, Swender P (1969) Nasal chemoreception in flavor identification. Arch Otolaryngol 90: 131-137
Murphy C (1983) Age-related effects on the threshold, psychophysical function and pleasantness of menthol. J Gerontol 38: 217-222
Murphy C (1985) Cognitive and chemosensory influences on age-related changes in the ability to identify blended foods. J Gerontol 40: 47-52
Murphy C (1986) Taste and smell in the elderly. In: Meiselman HL, Rivlin RS (eds) Clinical measurement of taste and smell. Macmillan, New York, p 343
Naessen R (1971) An enquiry on the morphological characteristics and possible changes with age in the olfactory region of man. Acta Otolaryngol 71: 49-62
Nagahara Y (1940) Experimentelle studien uber die histologischen Veranderungen des Geruchsorgans nach der Olfactoriusdurchschneidung. Beitrage zur Kenntnis des feineren Baus der Geruchsorgans. Jap J Med Sci Vet Pathol 5: 165-199
Nakashima T,Kimmelman CP, Snow JB Jr (1984) Structure of human fetal and adult olfactory neuroepithelium. Arch Otolaryngol 110: 641-646
Nishihata S (1984) Aging effect in nasal resistance. Nippon Jabiinkoka Gakkai Kaiho 87: 1654-1671
Pearson RCA, Esiri MM, Hiorns RW, Wilcock CK, Powell TPS (1985) Anatomical correlates of the distribution of the pathological changes in the neocortex in Alzhemier disease. Proc Natl Acad Sci USA 82: 4531-4534
Punter PH (1983) Measurement of human olfactory thresholds for several groups of structurally related compounds. Chem Sens 7: 215-235
Rehn B, Breipohl W, Mendoza AS, Apfelbach R (1986) Changes in granule cells of the ferret olfactory bulb associated with imprinting on prey odours. Brain Res 373: 114-125
Reyes PF, Golden GT, Fagel PL, Zalewska M, Fariello RG, Katz L, Carner E (1987) The prepiriform cortex in dementia of the Alzheimer type. Arch Neurol 44: 644-645
Sakakura Y, Ukai K, Majima Y, Murai S, Harada T, Miyoshi Y (1983) Nasal mucociliary clearance under various conditions. Acta Otolaryngol 96: 167-173
Schemper T, Voss S, Cain WS (1981) Odor identification in young and elderly persons: sensory and cognitive limiations. J Gerontol 36: 446-452
Schiffman S (1977) Food recognition by the elderly. J Gerontol 32: 586-592
Schiffman S (1979) Changes in taste and smell with age: Psychophysical aspects. In: Ordy JM, Brizze K (eds) Sensory systems and communication in the elderly. Raven Press, New York, p 227
Schiffman S, Pasternak M (1979) Decreased discrimination of food odors in the elderly. J Gerontol 34: 73-79
Smith CG (1942) Age incidence of atrophy of olfactory nerves in man. J Comp Neurol 77: 589-595
Somlyo AP, Somlyo AV (1968) Vascular smooth muscle. I. Normal structure, pathology, biochemistry, and biophysics. Pharmacol Rev 20: 197-272
Stevens JC, Cain WS (1985) Age-related deficiency in the perceived strength of six odorants. Chem Sens 10: 517-529

Stevens JC, Cain WS (1986) Smelling via the mouth: Effect of aging. Percept Psychophys 40: 142-146

Stevens JC, Cain WS, DeMarque A (1990) Memory and identification of simulated odors in elderly and young persons. Bull Psychon Soc 28: 293-296

Stevens JC, Cain WS, Schiet FT, Oatley MW (1989) Olfactory adaptation in old age. Perception 18: 265-275

Stevens JC, Plantinga A, Cain WS (1982) Reduction of odor and nasal pungency associated with aging. Neurobiol Aging 3: 125-132

Tomlinson BE, Blessed G, Roth M (1968) Observations on the brains of demented old people. J Neurol Sci 11: 205-242

Weiffenbach JM (1984) Taste and smell perception in aging. Gerodontology 3:137-146

Wysocki CJ, Gilbert AN (1989) National Geographic Smell Survey: Effects of age are heterogenous. Ann NY Acad Sci 561: 12-28

Yoshida M (1984) Correlation analysis of detection threshold data for "standard test" odors. Bull Facul Sci Eng Chuo Univ 27: 343-353

PART 4

BASIC CHARACTERISTICS OF HUMAN OLFACTION

10

OLFACTORY ADAPTATION

E P KÖSTER

RENE A de WIJK

I. INTRODUCTION

Sensory adaptation is the reduction of sensitivity following stimulation, and is common to all senses. The phenomenon is more striking in some senses (vision and olfaction) than in others (hearing). Adaptation is thought to be an important functional mechanism preventing overflow in the central parts of the nervous system by neural activity resulting from stimuli that are either too strong or of long duration. Thus, it helps the organism to remain alert for new information.

The extent of adaptation appears to be dependent on both the strength and the duration of the preceding stimulation and adaptation has a degree of specificity, so as not to block the influx of new information. Accordingly, cross-adaptation, which is the reduction of sensitivity to a stimulus caused by the previous presentation of a stimulus of another quality within the same modality, is almost always smaller than self-adaptation, the adaptation resulting from the same stimulus, provided the stimuli are of equal intensity and duration.

Adaptation is also a temporary phenomenon. As soon as exposure to a particular stimulus ceases, recovery of sensitivity to that stimulus occurs, and after a given time sensitivity is fully restored. The rate of recovery and the time needed to reach full recovery seems to be dependent on the extent of the previous adaptation and therefore on the intensity and the duration of the preceding adapting stimulus. This is also common to all senses, but there are considerable differences in the rate at which recovery takes place.

In olfaction, adaptation may be sufficient to lead to the cessation of the perception of the stimulus. This phenomenon is called 'complete adaptation' and the time required for the cessation of the sensation (ATCS) depends on the intensity of the stimulus.

Two other phenomena related to adaptation are facilitation, which is characterised by an increase of the response to a stimulus as a result of previous exposure to another stimulus, and habituation which is characterised not by a loss of sensitivity, but by a loss of interest in or reactivity to an unchanging monotonous stimulus. Novelty or contrast effects may well explain the instances of facilitation found by some authors.

The effects of adaptation can be measured using psychophysical methods to assess changes in olfactory threshold and perceived intensity, and by electrophysiological methods to monitor activity in receptor cells and secondary neurons. Basically, there are two types of measurement that should be distinguished, one that monitors the changes in the response during continuous exposure to the stimulus and another in which the effect of a preceding adapting stimulus on the response to a separate test stimulus is measured. A third type of measurement that is sometimes used, is not directly concerned with changes in the stimulus intensity, but uses time (response latencies, reaction times, ATCS) as an indication of adaptation. In olfaction each of these measurements has its own problems which will be discussed shortly before a survey of the results obtained with them is given.

II. EFFECTS OF ADAPTATION ON THE OLFACTORY THRESHOLD

A. Self-adaptation

Changes in the olfactory detection threshold during self-adaptation cannot be satisfactorily monitored on a continuous basis. The extent of adaptation and speed of recovery after adaptation to an odor, however, can be determined by presenting only one concentration of an odorant which lies somewhat above the detection threshold, and determining the detectability of this stimulus when it is presented a large number of times at different intervals. By varying the intervals in a systematic way and by using intervals of sufficient duration to let complete recovery occur, an estimate of the adapting effect of the stimulus can be obtained by comparing the percentage of positive responses to the stimulus after the longest and shortest interval. The rate of recovery is indicated by the percentages of positive detection at different intervals, and is referred to here as the Detection Recovery Method (DRM).

Vaschide (1902), Komuro (1922) and Zwaardemaker (1925) (Fig 1), reported on the influence of self-adaptation on the olfactory threshold.

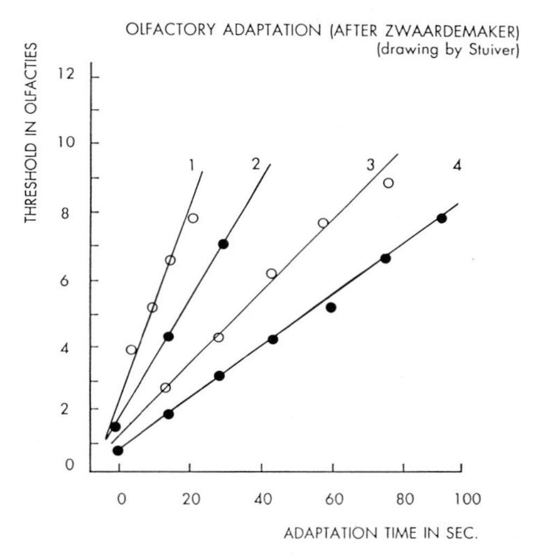

Fig 1. The results of Zwaardemaker (1925) on raising of the olfactory threshold during self-adaptation. Curves 1 and 2 represent the change in threshold for benzoin following adaptation to one of two different concentrations of benzoin (9 and 3.5 olfacties). Curves 3 and 4 represent the change in threshold for India rubber following adaptation to two different adapting concentrations of India rubber (14 and 10 olfacties). An 'olfactie' represents the threshold concentration.

As the curves of Figure 1 show, the threshold rises with an increase of both the duration and the intensity of the adapting stimulus, and adaptation rate may vary for different odors. Since the curves are linear, it seems reasonable to assume that the level of complete adaptation was not reached in these experiments. Later, Cheesman and Mayne (1953) and Cheesman and Townsend (1956), found that the logarithm of the threshold is a linear function of the logarithm of the adapting concentration. Stuiver (1958), using a sophisticated olfactometer, also studied the effect of self-adaptation on threshold, by varying the presentation time and intensity of the adapting stimulus (Fig 2).

The Figure shows that Stuiver's results do not agree with the linear relation between the raised threshold and adaptation time reported by Zwaardemaker.

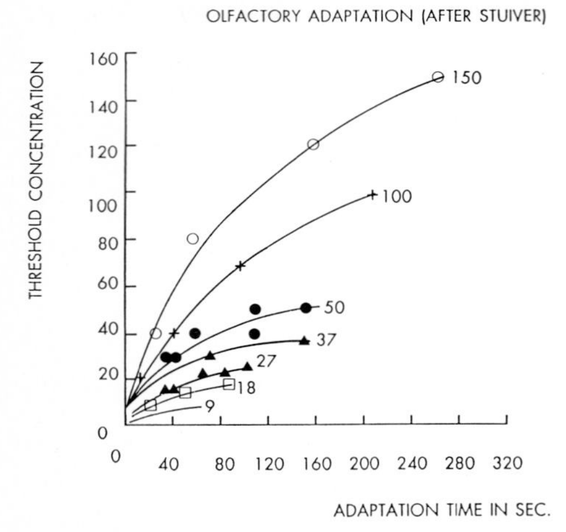

Fig 2. From Stuiver (1958), showing the change in threshold to d-octanol following adaptation to d-octanol. The numbers on the curves indicate the concentration of the adapting stimulus in terms of multiples of the unadapted threshold.

1. *Recovery after self-adaptation*

Stuiver (1958) investigated the decrease in threshold during the recovery period after adaptation to a number of different concentrations of an odorant. The results are given in Figure 3.

Recovery of olfactory sensitivity is rapid during the first few minutes and then slows down considerably. In particular, the decrease of threshold during the first 100 s is large. The results indicate that after a few minutes of recovery, the threshold is controlled by the total amount of odorous substance used for adaptation.

Köster (1965,1968,1971) using a DRM method, drew attention to the irregular shape of the recovery curves for some substances. The clearly biphasic appearance of these curves (Fig 4), led to the hypothesis that each of them show the super-imposed recovery curves of two separate types of receptors with different rates of recovery, in analogy with curves showing the recovery of the cones and rods in the retina after light adaptation.

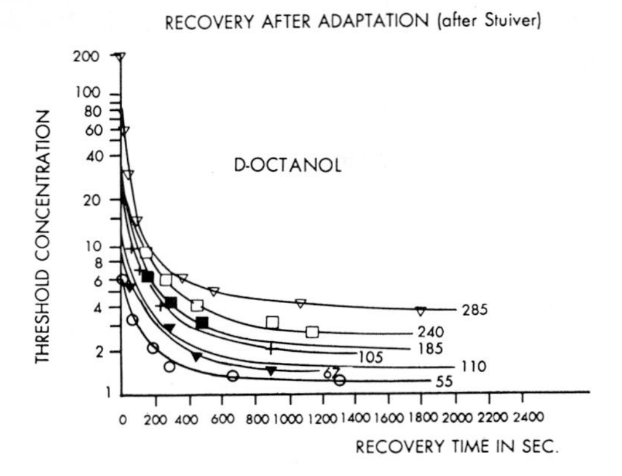

Fig 3. From Stuiver (1958), showing the relationship between the decrease of threshold and recovery time for various adaptation times and adapting intensities. The adapting intensities are equal to the threshold concentrations when recovery begins. The adaptation times are indicated on the right of the curves in seconds. Odorant: d-Octanol.

In an effort to localize the mechanisms involved in adaptation, Elsberg (1935) used coffee and citral as adapting stimuli and presented them at a constant concentration for 30, 60, 120 and 180 s. He also varied the rate at which the stimulus was injected into the nose of the subject and compared the effects of bilateral and unilateral injection. After cessation of the stimulus, the time necessary for complete recovery of the olfactory sensitivity was measured. Some of the results are shown in Figure 5.

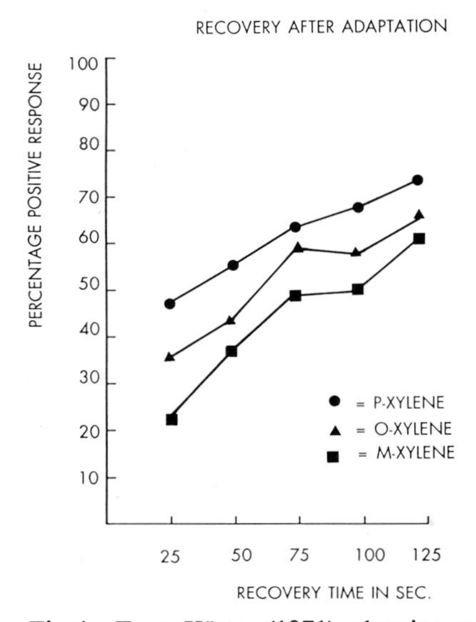

Fig 4. From Köster (1971), showing recovery after adaptation to three xylene isomers.

The Figure shows the increase in recovery time with increasing adaptation time at two rates of injection. The curves show the effect of birhinal adaptation on the monorhinally measured recovery time. With monorhinal injection, the effect appears to be strongest on the ipsilateral side of the nose. The existence of adaptation on the contralateral side, seems to indicate that adaptation is partly due to adaptation of central parts of the olfactory nerves.

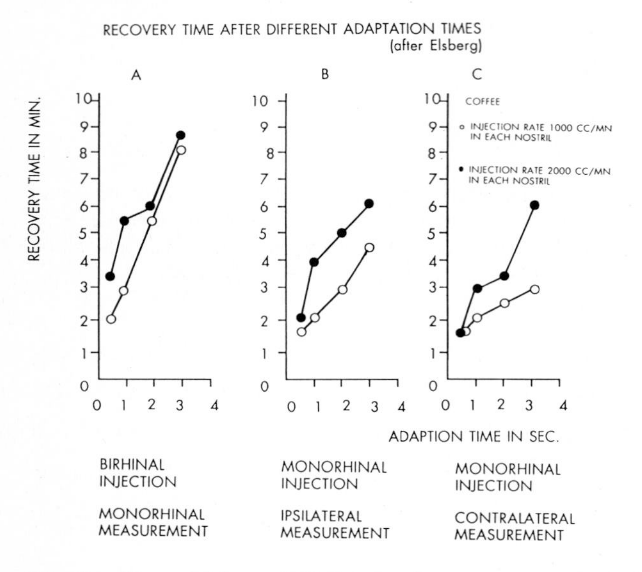

Fig 5. From Elsberg (1935), showing recovery after different adaptation times and different odor injection rates. Odorant substance: coffee.

Recovery time is not only dependent upon adaptation time, but also upon the rate at which the adapting stimulus is injected i.e., on the amount of odorous substance per unit time. Stuiver (1958) also investigated the recovery of the olfactory sensitivity of each of the two sides of the nose after adaptation of one side, and obtained a similar result, as shown in Figure 6.

In 1971, Köster used the DRM method to confirm these results and concluded that recovery from contralateral adaptation is fast (less than 60 s for a threshold stimulus), whereas the recovery from the much stronger ipsilateral adaptation effect caused by the same weak stimulus is slow (over 120 s). Thus, there are at least two mechanisms involved in adaptation, a fast recovering central process and a slow recovering peripheral one.

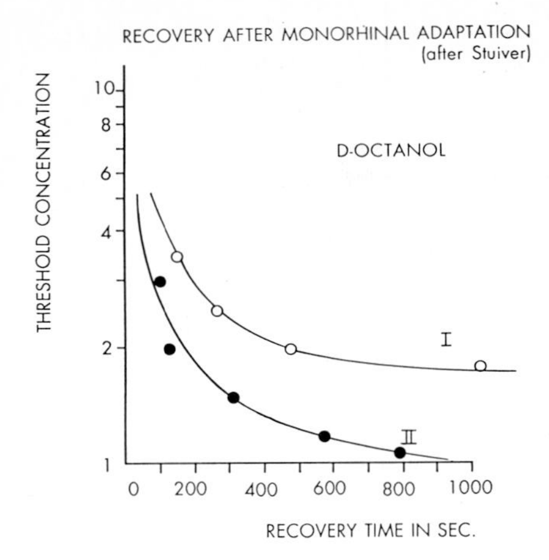

Fig 6. From Stuiver (1958), on the recovery of one side of the nose when the same side of the nose has been adapted (I) and the recovery of the contralateral side to which the odor had not been presented. (II) The adapting stimulus was d-octanol presented for 100s at a concentration 50 times higher than the absolute threshold level.

B. Cross-adaptation

Aronsohn (1886), who was the first investigator to devote attention to cross-adaptation in olfaction, adapted two subjects completely to one substance and tested their sensitivity to a number of others, with the intention of using the results as a basis for a scientific classification of odorous substances.

A number of authors (Vaschide 1902, Nagel 1903, Backman 1917, Komuro 1922, Ohma 1922, LeMagnen 1947, Cheesman and Mayne 1953, Mullins 1955, Stuiver 1958, Moncrieff 1966, Köster 1971) followed his example. On the basis of these results, Köster proposed a number of general rules. The most important of these rules were:

1. No adapting substance enhances the sensitivity to another substance.
2. No other substance reduces the sensitivity to a given odorous substance to a larger degree than that substance itself.
3. An odorous substance may have a larger adapting effect on the sensitivity to another substance than it has on the sensitivity to itself.

4. Most cross-adaptational relationships are non-reciprocal. One substance influences the sensitivity to another substance to a larger degree than the second one influences the sensitivity to the first one.
5. The sensitivity to an odorous substance which self-adapts rather strongly is usually also reduced strongly by other odorous substances.

However, he concluded that it is difficult to classify odors reliably and unequivocally on the basis of the data obtained. Recently, De Wijk (1989) investigated self- and cross-adaptation effects with two pairs of substances. The results were in good agreement with the five general rules, except that facilitation occurred with one of the three subjects during cross-adaptation. Facilitation has been reported by Engen and Bosack (1969) and Corbit and Engen (1971).

III. EFFECTS OF ADAPTATION ON PERCEIVED INTENSITY

The perceived intensity of an odor can be monitored during adaptation by direct scaling methods (magnitude estimation, category scaling), or by cross-modality matching procedures (finger span, auditory matching; see Chapter 5, this book). Time-intensity measurements are commonly achieved by having a subject estimate perceived intensity using a form of slide rule that is interfaced to a computer which displays the estimate on a screen. One of the major handicaps when assessing perceived intensity during adaptation, is that olfactory perception is linked to breathing. In order to overcome this problem a number of authors have used injection olfactometers to deliver stimuli continuously to the nose of subjects who have been instructed to breath exclusively through the mouth. Although the variability between the curves for various subjects is large, the shape of the curves for an individual subject is remarkably constant and reliable. Such curves give insight to response latency, rate of initial response growth, time at which maximal response is reached, shape and speed of the decline of the response under the influence of adaptation, and time at which complete adaptation is reached (ATCS).

Cain and Engen (1969) used magnitude estimation to investigate the effects of self- and cross-adaptation on the psychophysical functions (relationship between perceived intensity and stimulus concentration) of pentanol and propanol. The psychophysical functions for both odorants became steeper after adaptation. Largest adaptation effects were found with low concentrations of test stimuli and high concentrations of adapting stimuli, whilst with high concentrations of test stimuli adaptation had little effect. Cross-adaptation effects were smaller than self-adaptation effects, and the effects pentanol and propanol exerted on each other were asymmetrical. De Wijk (1989), investigated self- and cross-adaptation processes using geraniol, cineole, hexane and butanol as adapting and test stimuli at suprathreshold concentrations in all possible combinations, and confirmed most of these effects.

The duration of the adapting stimulus proved to be of minor importance for its adapting effect (Cain and Engen 1969, De Wijk 1989), however, the limited effects of duration may have been due to the adaptation durations used. Pryor et al. (1970), for example, varied the adaptation duration between 25 and 800 s and found that the psychophysical exponent changed with increasing adaptation duration. Berglund et al. (1978) found facilitation effects in cross-adaptation using hydrogen sulfide, dimethyldisulfide and pyridine. However, they used odorous substances that may have trigeminal effects. Facilitation of perceived intensity was also frequently observed by De Wijk (1989), but again only during cross-adaptation. Since facilitation was obtained only during perceived intensity measurements and not when reaction times were simultaneously recorded, and facilitation effects largely disappeared when the quality of the test or adapting stimulus was not perceived, it is possible that facilitation is primarily a contrast effect rather than an olfactory phenomenon. The contrast between the qualities of adapting and test stimuli, results in an overestimation of the perceived intensity of the test stimulus.

Monorhinal (single nares) and birhinal (both nares) odor delivery methods have been used to localize the mechanisms involved in the effects of adaptation on perceived intensity. Cain (1977), used a method in which an adapting stimulus of a fixed concentration of linalyl acetate was followed by a test stimulus of various concentrations of the same odorant. The results obtained were very similar to those reported by investigators who studied the effect of adaptation on threshold. In another study, De Wijk (1989) presented various combinations of geraniol, cineole and butanol as adapting and test stimuli, each at a single concentration. The results confirmed those of Stuiver (1958), Köster (1971) and Cain (1977), in relation to self-adaptation processes. Cross-adaptation was also investigated, but in this case, ipsi- and contralateral adaptation led to equal effects on the perceived intensity of the test stimulus. This could indicate that cross-adaptation is primarily a central effect. However, it was suggested that possible peripheral effects of cross-adaptation were obscured by the frequent occurrences of facilitation with some combinations of odorants.

The rate of change in the perceived intensity of hydrogen sulphide over time during exposure to an adapting concentration was reported by Ekman et al. (1967). Exposure to the adapting stimulus was not continuous but occurred during inspirations. At several fixed points in time, either every 30 or 60 s, a subject indicated the perceived intensity of the stimulus by a finger-span method. The results showed that variation in time-intensity measurements was large between subjects. Complete adaptation did not occur with most subjects and the rate of adaptation was not systematically related to concentration. Similar results were found by Cain (1974), who measured time-intensity functions of ozone, butyl acetate, eugenol and propanol. To overcome recovery during exhalations, De Wijk (1989) used a stimulus injection technique to present several concentrations of cineole, geraniol, hexane, camphor, eugenol, butanol or limonene continuously to subjects. The results indicated that the ascending part of the time-intensity function accurately reflects the temporal integration process, which dominates the adaptation process. The time to reach the limit of the temporal integration process, Tmax, is dependent on the odorant and subject, but does not vary with stimulus concentration. Unlike the results of other authors (Ekman et al. 1967, Cain 1974), complete adaptation was achieved in all cases. The different results are probably attributable to the methods of stimulation used.

Time-intensity measurements with binary mixtures composed of one relatively fast adapting odorant, cineole, and one relatively slow adapting odorant, geraniol, indicate that the rate of adaptation to a mixture is primarily determined by the rate of adaptation to the slowest adapting odorant (De Wijk 1989).

It is obvious that measurements of the adaptation time required for the cessation of smell (ATCS), can only be carried out if prolonged stimulation leads to complete adaptation, which is not always the case. The time-intensity measurements carried out by De Wijk (1989), indicate that cessation of smell can be achieved for a large number of odorants presented at moderately high concentrations, when a stimulus injection method is used.

Woodrow and Karpman(1917) and Stuiver(1958) found a positive relationship between concentration and ATCS. The longest ATCSs measured by Stuiver were about 500 s, and the curve for octanol was linear for ATCSs larger than 100 s. Continuous stimulation (De Wijk 1989) did not lead to shorter ATCSs, although recovery processes were minimal.

IV. EFFECTS OF ADAPTATION ON REACTION TIMES

Reaction time measurements have been reported for various sensory modalities. Most often when simple reaction times are measured, a subject responds as soon as the stimulus is detected. Simple reaction times range from 140 ms (sound), to 500-1000 ms (smell). Reaction times are influenced by adaptation. Since reaction times for detection primarily reflect peripheral processes, they may be particularly useful for the investigation of peripheral adaptation. A technical prerequisite for reaction time measurements, however, is that the commencement of the signal must be well defined and a square stimulus pulseform delivered. These requirements are not easily met in olfactory research.

In 1909 Hermanides used prolongation of reaction time as an indication of cross-adaptation and found many instances of non-reciprocity. Adaptation to iso-amylacetate, for instance, had no influence on the reaction time to nitrobenzene, but nitrobenzene increased the reaction time to iso-amylacetate. Skatole, however, did not affect the reaction times to the other odorants tested. De Wijk (1989), using an injection olfactometer, measured reaction times to investigate cross- and self-adaptation. Average olfactory reaction times for detection of moderately intense odorants without previous adaptation, ranged from one to two seconds, depending on the subject and the odorant. Stimuli presented at much higher concentrations led to reaction times as short as 500-600 ms. Reaction times to test stimuli were always longer after self- or cross-adaptation. The increase in reaction times after adaptation varied between 19 and 40% for self-adaptation, and between 1 and 18% for cross-adaptation, depending on the test stimulus. This result is in accordance with results found for experiments involving the measurement of perceived intensity or threshold. All of the rules for adaptation at threshold level which were described earlier in this Chapter, proved to be valid for reaction time measurements.

An increase in adaptation duration from three to five seconds, which failed to affect the perceived intensity of test stimuli, led to longer reaction times to the test stimuli. This result demonstrates that reaction time may be a more sensitive measure of olfactory adaptation than perceived intensity. An effort to localize the mechanisms of adaptation using reaction times failed, since ipsi- and contralateral adaptation affected reaction times to test stimuli in the same way.

V. PHYSIOLOGICAL MEASUREMENTS OF ADAPTATION

A. Effects of Adaptation on the Responses of Receptor Neurons

The responses of olfactory receptor cells during adaptation have been measured using the electro-olfactogram (EOG) and extracellular single unit recordings. The EOG is considered to be the integrated response of a large number of receptor cells. It is a slow negative response which is characterized by a response latency, a rising phase to a maximum response (Rmax), and a declining phase to a plateau level of response which is maintained as long as the stimulus is present. The response returns to zero only after cessation of the stimulus. Clearly, the EOG does not show complete adaptation. Usually, the rate of decline of the response and the ratio between Rmax and the plateau are used as measures of adaptation, but in a number of studies the influence of previous adapting stimuli on all aspects of the EOG have been studied. In extracellular recordings of olfactory receptor cells and secondary neurons in the bulb, response latencies and changes in response frequency are the most common measures of adaptation.

Ottoson (1956), came to the conclusion that the EOG was very resistant to adaptation, while Mozell (1962) showed that adaptation of the EOG in the frog was related to a decrease of integrated olfactory nerve activity. Van Boxtel and Köster (1978) studied recovery of the EOG in the frog following repeated stimulation with m-xylene, toluene and dioxan, and found that stimulus concentration and recovery rate were negatively related, and recovery was usually almost complete within 25 s. The latter results were in agreement with those of Ottoson. In a later study Van As et al. (1985) measured recovery of the EOG during the 15 s period immediately following the adapting stimulus. Using 12 different substances including three pairs of enantiomers, they concluded that the rate of recovery of the EOG was remarkably similar for all of these substances. An 80% reduction in the amplitude of the EOG occurred when the test stimulus followed the adapting stimulus by an interval of 1 s. After 15 s the magnitude of the EOG of the test stimulus had recovered to 80 % of its original value.

Getchell and Shepherd (1978) investigated adaptation of the olfactory receptor cells by performing extracellular unit recordings in the tiger salamander. By stimulating with step pulses of odor varying in duration between 1 and 10 s, they concluded that most olfactory receptor cells were slowly adapting with variable phasic responsiveness which was dependent on odor concentration and other factors. Some cells showed essentially no adaptation over the period after the phasic part of the response was completed, but most appeared to reflect a distinct component of adaptation during this period. In 1979 Baylin and Moulton used extracellular recordings of single receptor cells to measure self- and cross-adaptation in the same species. Seven odors were used, some with similar molecular properties (butanol and propanol) and some with similar types of odors (benzaldehyde and nitrobenzene). Two stimuli were presented during standardized pulses separated by an interval of 0, 5, 10 or 15 s between the adapting and the test stimulus. Self- and cross-adaptation were observed in some receptor cells but not in others, and cross-adaptation was non-reciprocal in all cells studied and could occur in cells which showed no self-adaptation. Kaissling et al. (1987) and Zack Strausfeld and Kaissling (1986) gave a clear account of the different forms of adaptation in the receptor cells of insects to pheromones and food odors. They concluded that the receptor potential and nerve impulse generators have different and very likely spatially separate adaptation mechanisms. From their results it is clear that there is a reduction of the phasic response to repeated stimuli separated by intervals of several seconds or even minutes. In a number of other studies (Voigt and Atema 1987, Atema et al. 1988, Borroni et al. 1988) adaptation processes in the receptor cells of the lobster have been investigated with special attention to the influence of the effects of background stimulation in which the stimuli were presented. It was demonstrated that the signal to noise ratio and not the absolute stimulus level predicted the responses.

B. Effects of Adaptation on Cells in the Olfactory Bulb

Although there is a large number of studies on the responses of single mitral cells in the olfactory bulb, specific studies of the effects of adaptation on the responses of bulbar neurons are rare. In 1976 Potter and Chorover reported experiments involving the measurement of EOGs and responses of mitral cells in the hamster. They concluded that during the course of a single continuous odor presentation, decrements in the EOG were seen which were mirrored in the responses of mitral cells. When odorless periods of 30-90 s were introduced between stimuli, complete recovery of the EOG occurred, as in the experiments mentioned above, but recovery of the mitral cell response took much longer (15-30 min). Potter and Chorover concluded that these long recovery periods indicate that the response decrements in the bulb cannot be explained by receptor adaptation and must therefore be the result of a central process called 'habituation'. Mair (1982) compared the effects of adaptation on the EOG and the responses of single mitral cells in the bulb of the rat using repeated stimulation, and found that exposure to an adapting stimulus did not qualitatively change the patterns of mitral cell activity evoked by an unadapted stimulus, but did affect the amount of evoked excitation. Neurons exhibited increased (enhanced), decreased (suppressed), or unchanged activity in response to test stimuli in the adapted state. No neuron showed both enhancement and suppression. Intervals between the adapting and test stimuli as large as 15 s resulted in a reduction of the adaptation effects. In a number of cross-adaptation experiments, instances of both symmetrical and asymmetrical adaptation effects were reported. Doving (1987) studied the responses of bulbar neurons in the rat using ramp stimulation and found adaptation on only 18% of the trials.

VI. DISCUSSION

Most of the psychophysical results described in this Chapter are coherent with the functional ideas about adaptation mentioned in the Introduction, but there seems to be a discrepancy between the results of psychophysical and electrophysiological experiments. The first point of disagreement concerns the speed of adaptation. Whereas most psychophysicists would consider the sense of smell to be a rapidly adapting sense with a rather slow recovery, electrophysiologists have demonstrated that both the receptor cells and bulbar cells show a slow decrease of their tonic response during prolonged stimulation. They admit, however, that the initial phasic response decreases more rapidly upon repeated stimulation. Since the minimal reaction times found for most odorous substances are between 600 and 800 ms (De Wijk 1989), the latency of the EOG is about 200 ms and the risetime of this same response is about 500-600 ms, it is clear that psychophysical reaction times are based on the phasic component of the neural responses, which also show a more rapid decrease after adaptation in electrophysiological measurements. The second point of disagreement, is that although psychophysical experiments show that complete adaptation is usually reached when continuous stimulation is applied (De Wijk 1989), most neurons (peripheral and bulbar) keep firing at their tonic level. A possible cause for this disagreement is given by the results of Chaput and Panhuber (1982), who measured mitral cell activity during prolonged odor stimulation in awake rabbits. Their results did not show large adaptation effects, but a respiration-related synchronisation of firing activity was observed which resulted in activation during inhalation and inhibition during exhalation. The respiration related patterning of the firing activity might prevent complete adaptation. This would explain why complete adaptation is observed only in psychophysical experiments in which the stimulus is presented continuously. Thus, it becomes clear that the second discrepancy can be resolved easily.

REFERENCES

Aronsohn E (1886) Experimentelle Untersuchungen zur Physiologie des Geruchs. Archiv Anat und Physiol, Physiol Abteilung: 321-357

Atema J, Borroni PF, Johnson BR, Voigt R, Handrich LS (1989) Adaptation and mixture interactions in chemoreceptor cells: mechanisms for diversity and contrast enhancement. In: Laing DG, Cain WS, McBride RL, Ache BW (eds) Perception of complex smells and tastes. Academic Press, Sydney, p 83

Backmann EK (1917) Experimentella undersokingar aft luktsinnets fysiologi Uppsala. Lak-Foren Forh 22: 319-464

Baylin F, Moulton DG (1979) Adaptation and cross-adaptation to odor stimulation of olfactory receptors in the tiger salamander. J Gen Physiol 74: 37-55

Berglund B, Berglund U, Lindvall T (1978) Separate and joint scaling of odor intensity of n-butanol and hydrogen sulfide. Percept Psychophys 23: 313-320

Borroni PF, Atema J (1988) Adaptation in chemoreceptor cells. J Comp Physiol 164: 67-74
Cain WS (1974) Perception of odor intensity and the time-course of olfactory adaptation. ASHRAE Trans 80: 53-75
Cain WS (1977) Bilateral interaction in olfaction. Nature 268: 50-52
Cain WS, Engen T (1969) Olfactory adaptation and the scaling of odor intensity. In: Pfaffmann C (ed) Olfaction and taste, vol III. Rockefeller University Press, New York, p 127
Chaput MA, Panhuber H (1982) Effects of long duration odor exposure on the unit activity in awake rabbits. Brain Res 250: 41-52
Cheesman GH, Mayne S (1953) The influence of adaptation on absolute threshold measurements for olfactory stimuli. Q J Exp Psychol 5: 22-25
Cheesman GH, Townsend MJ (1956) Further experiments on the olfactory thresholds of pure chemical substances, using the 'sniff-bottle method'. Q J Exp Psychol 8: 8-14
Corbit TE, Engen T (1971) Facilitation of olfactory detectors. Percept Psychophys 10: 436-443
De Wijk RA (1989) Temporal factors in human olfactory perception. Unpublished doctoral dissertation, University of Utrecht
Doving KB (1987) Response properties of neurones in the rat olfactory bulb to various parameters of odour stimulation. Acta Physiol Scand 130: 285-298
Ekman G, Berglund B, Berglund U, Lindvall T (1967) Perceived intensity of odor as a function of time of adaptation. Scand J Psychol 8: 177-186
Elsberg CA (1935) Olfactory fatigue. Bull Neurol Inst 4: 479-495
Engen T, Bosack TN (1969) Facilitation in olfactory detection. J Comp Physiol Psychol 68: 320-326
Getchell TV, Shepherd GM (1978). Adaptive properties of olfactory receptors analyzed with odour pulses of varying durations. J Physiol 282: 541-560
Hermanides J (1909) Uber die Konstanzen der in der Olfaktologie gebrauchlichen neun Standardgeruche. Unpublished doctoral thesis, University of Utrecht
Kaissling KE, Zack Strausfeld C, Rumbo ER (1987) Adaptation processes in insect olfactory receptors: mechanisms and behavioral significance. In: Atema J, Roper S (eds) Olfaction and taste, vol IX. Ann NY Acad Sci 510: 692-694
Komuro K (1922) L'Olfactometrie dans l'air parfume. Arch Neerl de Physiol 6: 58-76
Köster EP (1965) Adaptation, recovery and specificity of olfactory receptors. Rev de Laryngologie Octobre Suppl 86: 880-894
Köster EP (1968) Recovery of olfactory sensitivity after adaptation. In: Tanyolac N (ed) Odor theories and odor measurement. Robert Gordon College Research Centre, Istanbul, p 307
Köster EP (1971) Adaptation and cross-adaptation in olfaction. Unpublished doctoral dissertation, University of Utrecht
LeMagnen J (1947) Etude d'une methode d'analyse qualitative de l'olfaction. Annee Psychologique 43-44: 249-264
Mair RG (1982) Adaptation of rat olfactory bulb neurones. J Physiol 326: 361-369
Marks LE (1974) Sensory processes: The new psychophysics. Academic Press, New York
Moncrieff RW (1966) Odour preferences. Leonard Hill, London
Mozell MM (1962) Olfactory mucosal and neural responses in the frog. Am J Physiol 203: 353-358
Mullins LJ (1955) Olfaction. Ann NY Acad Sci 62: 247-276

Nagel W (1903) Handbuch der Physiologie des Menschen. Braunschweig
Ohma S (1922) La classification des odeurs aromatiques en sous classes. Arch Neerl de Physiol 6: 567-590
Ottoson D (1956) Analysis of the electrical activity of the olfactory epithelium. Acta Physiol Scand 35 Suppl 122:1-83
Potter H,Chorover SL (1976) Response plasticity in hamster olfactory bulb: peripheral and central processes. Brain Res 116: 417-429
Pryor GT, Steinmetz G, Stone H (1970) Changes in absolute detection threshold and subjective intensity of supra-threshold stimuli during olfactory adaptation and recovery. Percept Psychophys 8: 331-335
Stuiver M (1958) Biophysics of the sense of smell. Excelsior, The Hague
Van As W, Menco BPhM, Köster EP (1985) Quantitative aspects of the electro-olfactogram in the tiger salamander. Chem Sens 10: 1-21
Van Boxtel A, Köster EP (1978) Adaptation of the electro-olfactogram in the frog. Chem Sens Flav 3: 39-44
Vaschide N (1902) Recherches experimentales sur la fatigue olfactive. J de l'Anat et de la Physiol 38: 85-103.
Voigt R, Atema J (1987) Signal-to-noise ratios and cumulative self-adaptation in chemoreceptor cells. In: Atema J, Roper S (eds) Olfaction and taste, vol IX. Ann NY Acad Sci 510: 692-694
Woodrow H, Karpman B (1917) A new olfactometric technique and some results. J Exp Psychol 2: 431-447
Zack Strausfeld C, Kaissling KE (1986) Localized adaptation processes in olfactory sensilla of Saturniid moths. Chem Sens 11: 499-512
Zwaardemaker H (1925) L'Odorat. Librairie Octave Doin, Paris

11

MEMORY FOR ODORS

FRANK R SCHAB

WILLIAM S CAIN

I. INTRODUCTION

In this Chapter, we seek to integrate current knowledge on odor memory with what might be called mainstream knowledge on memory. The coverage is divided into five main topics: The commonly mentioned ability of odors to cue episodic memories, odor imagery, varieties of memory functioning, odor recognition, and odor identification.

The coverage begins with the observation that people sometimes find odors very memorable but yet in some circumstances seem to find it difficult to learn arbitrary responses to odors. Although little is known about why this happens, speculation is made about whether the personal meaning of the odor cue or sluggishness in the speed of odor processing may resolve the disparity. Within the context of how well people can learn about odors, the question "Does odor imagery exist?" is addressed. Some data support its existence, but the case is incomplete.

Essentially all published research on odor memory has addressed only explicit memory for odors, where the subject deliberately attempts to remember a previous encounter with an odor e.g., odor recognition. Nevertheless, implicit tasks, where prior exposure to a stimulus modifies subsequent behavior without conscious retrieval of the stimulus from memory, has revealed much about memory for other sensory stimuli and verbal material, and should prove useful to understand odor memory.

In the explicit tasks of short-term and long-term recognition, odor memory seems relatively durable, though important parametric investigations of the issue remain to be done. Studies of the type of interference that will impair odor memory and studies that bridge the gap between short-term and long-term processing should have high priority. Although recognition memory can in principle circumvent verbal processing, such processing has apparently played a large role in many odor recognition memory experiments. In brief, people may have remembered what they could identify. The need now is to study the workings of episodic odor memory without the influence of such verbal processing.

Despite the presumed role of semantic (verbal) processing in existing studies of episodic odor memory, odor identification with verbal labels often seems poor. Specifically, most people fail to identify large numbers of odors with precision, without some sort of assistance. In this Chapter it is argued that both sensory (viz., discriminative) and cognitive factors underlie the limitations. In the past, sensory factors have received no attention. Finally, group differences in semantic memory for odors are addressed. Differences between males and females, the young and the old, and the blind and the sighted may prove productive in the search for mechanisms of odor memory.

II. ODOR-EVOKED MEMORY

A survey of individuals from all walks of life would almost certainly reveal that odors appear to evoke past experiences more vividly than do other sensations. Such folk wisdom concerning the power with which odors can awaken memories of distant events, however, stands somewhat in contrast to experiments that show either no memory advantage or an actual disadvantage for odor cues over other stimuli e.g., Eich (1978).

In paired-associate (PA) learning experiments, for example, where subjects learned to associate numerals with odors or visual stimuli (electrical symbols), the latter led to more rapid learning (Fig 1) (Davis 1975). The same held true in a comparison of odors to visual free forms, simple and relatively meaningless line-drawings that people might commonly call "blobs" (Davis 1975, 1977). Free forms seemed a fair comparison to odors in their simplicity. Nevertheless, the semantically meaningless forms proved to be better memory stimuli even when judged to be less familiar and no more discriminable than the odors. Davis speculated that some hidden perceptual limitation in olfaction made odors more difficult to encode than the forms.

The conditions under which an odor triggers the memory of a past experience in everyday life e.g., ambient odor and a personally significant event, and those of the PA experiment (many, personally irrelevant odors presented in small containers) differ enough for us to question the actual relevance of the PA experiment to the personal observations of our hypothetical respondents. Nevertheless, the results of Davis raise important questions: What constitutes the inherent limitation on odor encoding? Will it lend itself to analysis? Could the poor results in PA learning derive from such characteristics as poor or nonexistent olfactory imagery, or weakness in the formation of olfactory-verbal associations?

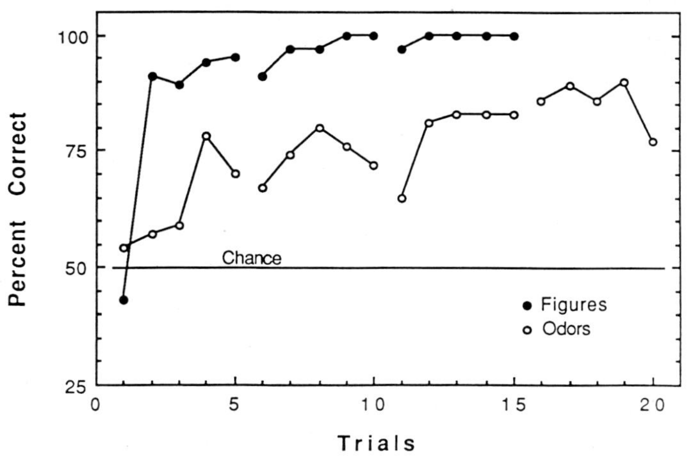

Fig 1. Paired-associate learning of numerals to four odorants and four visual stimuli over four blocks of trials. Trials 1 to 5 and 6 to 10 took place on one day and trials 11 to 15 and 16 to 20 took place on the next day. From Davis (1975).

Free forms may afford little opportunity for verbal encoding, but do afford imaginal encoding. In contrast, odors may afford little imaginal encoding since, in the view of some, odor imagery may fail to exist at all or may exist only in rudimentary form (Engen 1982). The absence of such encoding would preclude a potent form of rehearsal. Odors may accordingly need an alternate form of encoding, such as labels of some sort e.g., "musty basement smell", for good encoding. At the very least, the assignment of accurate labels such as calling a lemon odor "lemon" rather than "fruit", greatly enhances the encoding of odors and hence memory for them (Rabin and Cain 1984, Lyman and McDaniel 1986). Within that context, the many seconds it may take to generate a label for an odor may prove limiting (Lawless and Engen 1977). The relative sluggishness of olfactory processing may force subjects to spend inordinate resources on stimulus processing, leaving fewer resources for associating the stimuli with their respective responses. Perhaps then, the PA task sets up a temporal mismatch whereby too many odors appear too fast.

In another investigation into odors as cues to retrieval, Schab (1990) found that an ambient odor present during incidental learning i.e., learning without instruction to learn, enhanced free recall of a list of words over retention intervals of up to 48 hr. The facilitation went beyond any engendered by labels assigned to the odors. Subjects recalled about 50% better in the context of the label and the odor e.g., chocolate smell and being told to think of chocolate, than in the context of the label alone (being told to think of chocolate) (Fig 2).

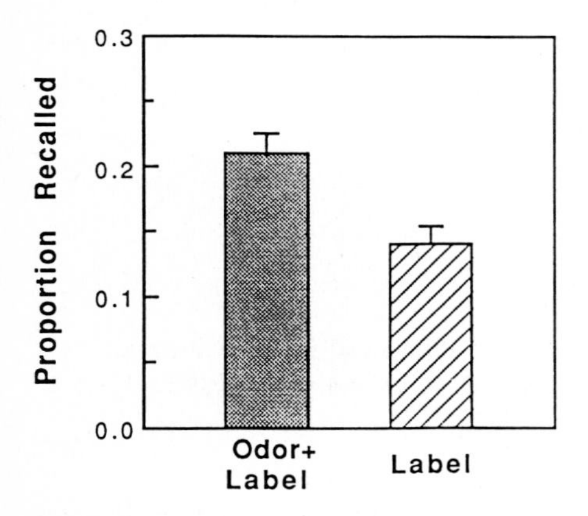

Fig 2. Proportion of words recalled after 24 h as a function of whether an ambient chocolate odor was present at learning and testing. From Schab (1990).

When an odor triggers memory in everyday experience, the odor occurs within a context that often has personal significance and an emotional component for the perceiver. Consider a case where a subject is recruited under the pretence of a questionnaire study and arrives at a room which has a characteristic smell or a characteristic sound. While the subject completes the bogus questionnaire, the experimenter leaves the room ostensibly to take a phone call, and a collaborator of the experimenter insults and argues with the subject. Later, in the presence of the ambient odor or sound, the subject seeks to recount the details of the experience to the experimenter. After the recall task, independent observers could rate the accuracy of the subject's recall relative to that captured by covert video-taping of the original incident. Such an experiment might support the popular expectation regarding odor-evoked retrieval simply because it may simulate the environmentally realistic event more faithfully.

III. ODOR IMAGERY

When asked whether or not they experience odor imagery, about four out of 10 people will say yes (Brower 1947, Lindauer 1969). Nevertheless, few people remark on a role of imagery in odor rehearsal or memory. Experiments could presumably decide the issue. Lyman and McDaniel (in press), for example, showed that odor recognition improved when subjects sought to image the smell associated with an object over when they sought to image the object visually. The outcome suggests some odor imagery.

By the common definition of imagery e.g., Finke and Shepard (1986), imaging a stimulus should prove similar to perceiving it physically and images should reveal themselves by their influences on actual perception. When Perky (1910) directed subjects to relate their visual images of common objects during actual presentation of low intensity copies, the ability of the subjects to detect faint stimuli depended upon whether they had received imagery instructions. Under such instructions, subjects reported no objects actually seen, only internal images (see also Segal and Fusella 1970). Under non-imagery instructions, the subjects always saw the projected images. Other investigators, however, have shown facilitatory effects of imagery instructions on the perception of modality-congruent stimuli (Farah 1989, Peterson and Graham 1974; see Farah and Smith 1983, for a possible explanation for the discrepancy). In either case, odor imagery might reveal itself by facilitatory or inhibitory influences on odor perception. One might, for instance, measure the detectability of an odorant as a function of whether a subject sought to image that odor, a different odor, a visual referent for the odor, or nothing.

In similar fashion, one might employ a memory task in which subjects seek, during a retention interval, to form the image of the test odor, a different odor, or no odor. Forming the image of the to-be-remembered odor might improve memory performance relative to the no-imagery control insofar as imagery can constitute a form of rehearsal. Forming the image of a different odor would then presumably interfere with memory for the target odor.

IV. DIFFERENT FORMS OF MEMORY

Memory may manifest itself variously, as, for example, in the difference between episodic memory and semantic memory (Tulving and Thompson 1973). Episodic memory refers a subject's experience back to a specific previous encounter with a stimulus. In the typical test of odor recognition memory, a subject is exposed to odors on one occasion and asked later to recognize which items had appeared previously. Semantic memory refers the subject's experience back to more general encounters with a stimulus. Odor identification, which requires a subject to report the name of an item, exemplifies semantic memory. Insofar as episodic odor memory behaves differently from semantic odor memory, it is possible that the two tasks tap different systems. If so, what processes might underlie them and how do they interact? As will be shown below, researchers in odor memory have barely begun to address such issues. First, however, mention of some other distinctions relevant to the study of odor memory will be made.

Over the last decade, considerable research on verbal memory has focused on implicit memory tasks. Odor recognition and odor identification, as described above, constitute explicit tasks in which subjects consciously attempt to retrieve the record of prior experience. An implicit task, on the other hand, measures retention of a previous event in the absence of deliberate recollection. Implicit memory may reveal itself in facilitated performance in sensory-motor as well as in more cognitive i.e., episodic and semantic tasks. Facilitation in cognitive tasks, often now called priming, emerges in certain word completion and lexical decision tasks. In word completion, for example, retention for the word "stealth" may show itself by an increase in the probability that a subject will complete the word-stem "ste" with "stealth" rather than with, say, "stealing".

Implicit memory tasks take on theoretical importance when, as in verbal memory, experimental manipulations influence implicit and explicit measures differently (Blaxton in press, Jacoby and Dallas 1981, Warrington and Weiskrantz 1970). Changes in what are called stimulus surface features e.g., modality of presentation at learning and testing, affect implicit though not explicit memory, whereas the reverse holds true of changes in elaboration at encoding e.g., graphemic vs semantic orienting tasks. Such dissociations suggest the operation of more than one memory system (Cohen 1984, Squire 1987).

Data on amnesics, who generally exhibit severe memory deficits in explicit memory, but who frequently appear unimpaired in implicit tasks, bolster the case for more than one memory system. In severe amnesia, performance on a perceptual-motor task e.g., rotary-pursuit, improves continuously with practice even though the subject has no recollection of having performed the task (Milner 1962). The ability of an amnesic person to assemble a jigsaw puzzle similarly improves with training, despite the continuing subjective novelty of the task (Brooks and Baddeley 1976). When amnesics and normals receive a list of words to remember and later seek to recall those words when cued with the first three letters of each, amnesics perform very poorly. If, however, subjects receive the same cues with instructions to complete each stem with the first word that comes to mind, amnesics and normals complete the fragments equally well (Graf et al. 1984). Hence, exposure to the material of interest led to learning that would have been missed by an explicit measure.

Whereas the distinction between implicit and explicit memory has spawned intense research in verbal and visual memory, it has spawned almost none on odor memory. Our own results, however, show that priming can facilitate odor identification. Previous exposure to just the name of an odor enhanced identification and shortened its latency, but previous exposure to both the name and the odor enhanced performance even more.

Additional priming effects in odor memory remain unexplored. For example, would prior experience with a tomato odor lead to a greater probability, and perhaps shorter latency, of identifying pizza odor? Would the magnitude of the effect depend on the ability of the subject to identify pizza odor as such? Would tomato odor still facilitate the identification of pizza if the subject interpreted and labeled the tomato odor as squash?

V. EPISODIC ODOR MEMORY

The ongoing issue of whether more than one memory system exists, has roots in findings of differences in verbal memory performance over short retention intervals (less than one minute) vs longer intervals (minutes and longer) (Baddeley 1986, Crowder 1982, James 1890, Melton 1963, Waugh and Norman 1965, Zechmeister and Nyberg 1982). For example, subjects encode verbal information primarily phonologically in short-term tasks and semantically in long-term tasks (Baddeley and Dale 1966, Dale and Baddeley 1969, Conrad 1964). The recency effect in free recall i.e., better recall of the last few words of a sequentially presented list, dissipates with a delay in recall. An increase in the presentation rate of the original items has no influence on the recency effect, but does influence the phenomenon known as the primacy effect i.e., better recall of the first few words of the list (Glanzer 1972). The verbal memory literature contains many examples of differences in short-term vs long-term memory. No study has yet compared short-term to long-term memory for odors, although several investigators have examined each separately.

A. Short-term Odor Memory

Engen et al. (1973) found odor recognition memory to decay little over intervals of 3 to 30 s (Fig 3), where one might have expected a much steeper decline. Correct recognition equalled a surprisingly poor 80% at 3 s, indicating a rather severe encoding limitation, but then increased to about 90% at 12 s, with a decline back to 80% at 30 s. The relatively flat forgetting function suggested something unusual about the process of short-term odor memory.

Although short-term odor memory may indeed prove special, it should be noted that the matter of rehearsal during the retention interval requires more attention in order to decide just how special. The subjects in the experiment of Engen and colleagues counted backwards during the retention interval. Although the investigators believed that such counting prevented rehearsal, the matter actually remains open. The well known result that a visual distractor will interfere more with the memory for a visual stimulus than with that for a verbal stimulus prompts the concern.

The tendency for performance in the short-term experiment to peak after an interval of 12 s also merits further attention. Perhaps the finding represents an important time-constant in the processing of olfactory information. Within this context, we should seek to learn whether the loss of odor information from memory observed between the 12- and 30-s retention intervals would continue with longer retention intervals.

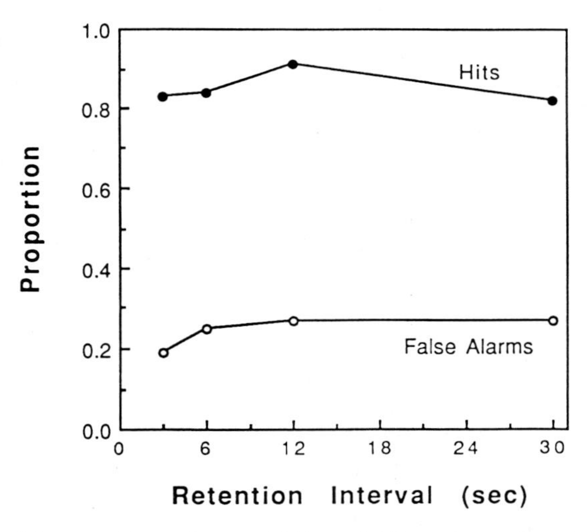

Fig 3. Proportions of hits and false alarms in odor recognition memory at retention intervals of 3 to 30 s. From Engen et al. (1973).

Walk and Johns (1984) concluded that olfactory-specific interference impaired short-term odor memory, but that verbal interference did not. Those investigators tested recognition memory for common food odors over a 26 s interval during which subjects 1) verbalized associations to the name of the target odor, 2) verbalized associations to the name of an odor in the same category as the target e.g., vegetables, fruits, 3) smelled and verbalized associations to a categorically related odor, or 4) did nothing. Memory proved best when subjects verbalized associations to the label of the target odor and worst when they smelled and verbalized associations to a categorically related odor (Fig 4). Despite their conclusion regarding modality-specific interference, however, the pattern of Walk and John's results left open the possibility of a semantic basis for the interference. Although subjects performed worse after smelling and verbalizing associations to a categorically related odor than after merely verbalizing associations to the name of such an odor, the difference failed to achieve statistical significance. Perhaps the use of odor distractors without the need to verbalize associations would yield a different estimate of perceptual interference. Moreover, perhaps easily identifiable distractor odors would cause more interference than less identifiable ones because of the richer semantic processing they would engender.

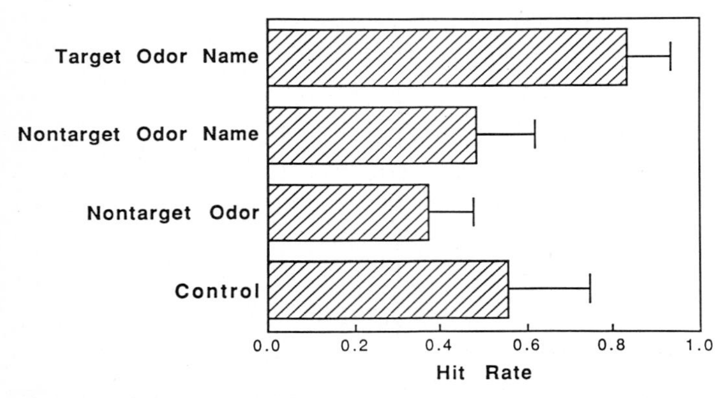

Fig 4. Hit rates and standard deviations for odor recognition memory scores where subjects performed one of the following distractor tasks in a 26 s retention interval: verbalized associations to the name of the target odor (target odor name), verbalized associations to the name of another odor (nontarget odor name), smelled a nontarget odor and verbalized associations to it (nontarget odor), or did whatever they pleased (control). From Walk and Johns (1984).

B. Long-term Odor Memory

Studies of long-term memory have shown that successfully encoded odors show relatively slow forgetting. Engen and Ross (1973), for example, found that subjects who inspected a set of common and uncommon odors could recognize only 67% on an immediate test, but could still recognize more than 65% after one month, and continued to recognize the stimuli with about the same accuracy even after one year (Fig 5). The relatively flat forgetting function for odors led Engen and Ross to suggest that subjects encode odors as unitary perceptual events. Lawless and Cain (1975) confirmed the shallow decline of memory over a one-month period. Neither study, however, uncovered an influence of rated odor familiarity or pleasantness on forgetting.

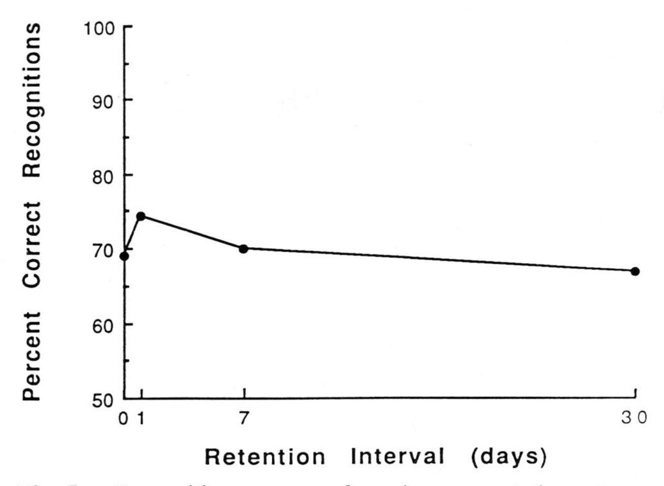

Fig 5. Recognition memory for odors presented up to one month after inspection. Chance performance in the forced-choice task equalled 50%. From Engen and Ross (1973).

In a direct comparison of odors with visual stimuli, Lawless (1978) found relatively flat, parallel forgetting functions for odors and ameboid forms and a much steeper function for pictures (Fig 6). He speculated that odors, like the forms, may be encoded as relatively featureless stimuli, an explanation consistent with both poor initial encoding and slow forgetting. More complex stimuli, such as pictures, probably allow redundant encoding because of their many features. Insofar as they share features with other stimuli, however, they may allow more interference over time than the simpler, though more unique, odors and forms. The overlap of features of items held in memory can impair accurate retrieval because the presentation of one feature as a retrieval cue could activate memory of all items that possess that feature.

If the difference between memory for pictures and memory for odors lies in the number of encodable features, then would experts such as perfumers and culinary experts, who may encode odors as feature-rich stimuli rather than holistic stimuli, remember more odors initially but forget them faster than laymen? This outcome would flow from the

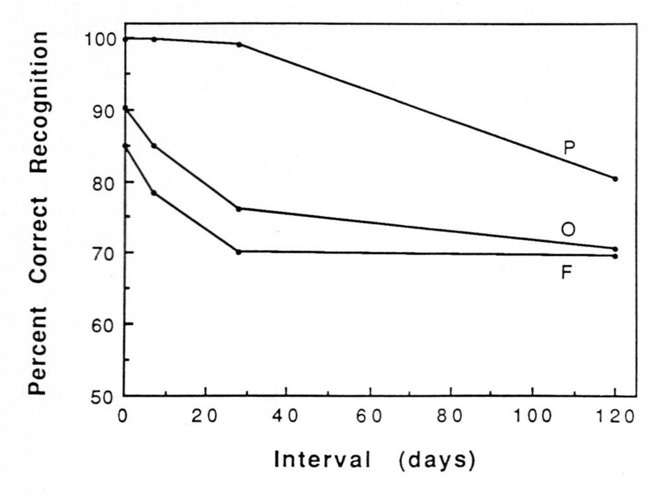

Fig 6. Recognition memory for odors (O), pictures from travel magazines (P), and free forms (F). Chance performance equalled 50%. From Lawless (1978).

holistic-encoding hypothesis for odors if indeed experts analyze odors into component features and nonexperts do not.

Although verbal encoding improves memory for auditory and visual stimuli (Bartlett 1977, Freedman and Haber 1974), some early studies (Engen and Ross 1973, Lawless and Cain 1975) found no association between verbal encoding and odor recognition memory. Rabin and Cain (1984), however, concluded that the previous investigations failed to detect the relationship because of large individual differences in the way people encode odors. Rabin and Cain found that the more familiar and identifiable an odor was to a given subject, the more likely that subject was to remember it (Fig 7). Congruent with the previous findings, though, Rabin and Cain demonstrated a flat forgetting curve for odors over retention intervals up to seven days.

In another study that supported the importance of labeling, Lyman and McDaniel (1986) directed subjects to inspect odors during one of four orienting tasks; 1) visual imagery of the referent for the odor, 2) generation of the name of the odor, 3) description of a life

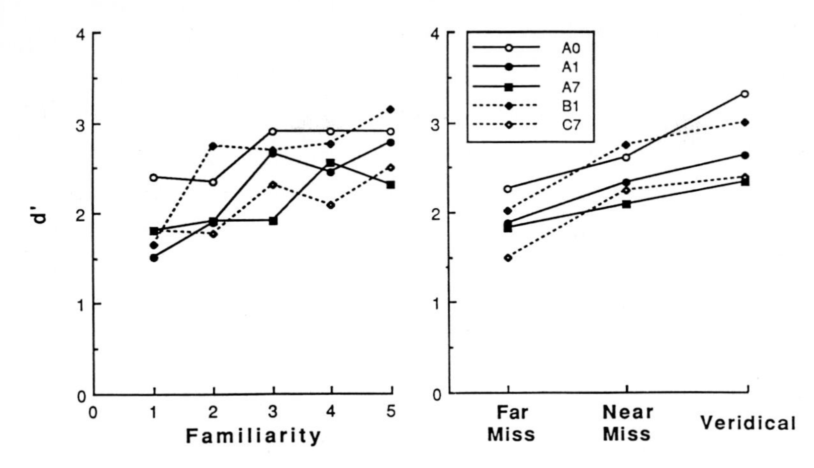

Fig 7. Left: Recognition memory (d') vs rated familiarity of odors. Letters refer to different groups of subjects and numbers refer to the number of days between inspection and the recognition test. Right: Recognition memory vs the quality of label assigned to an odor at inspection. From Rabin and Cain (1984).

episode that involved the odor, or 4) nothing. Subjects who identified the odors verbally, and those who related life episodes, recognized the odors better after a seven-day retention interval. The visual imagery group, in contrast, failed to perform differently than the control group. Subjects directed to identify the odors verbally performed best when they attached veridical labels and less well when they attached less accurate labels.

Although odor recognition memory may yield a relatively flat forgetting function because of holistic encoding, other possibilities exist. Under conditions of intentional learning, subjects will characteristically use any available strategy to improve performance. Merely naming an odor improves its encoding (Rabin and Cain 1984). It may therefore be assumed that subjects will attempt identification during odor inspection. Such behavior could lead investigators to confuse memory for the previous occurrence of the odor with the previous occurrence of its label. Hence, do subjects remember odors or do they remember the labels they attached to the odors? Future investigations on odor recognition could attempt to determine the relative contributions of perceptual and verbal information to the odor memory code by, for example, comparing recognition performance for odors, odor names, and both combined, as a function of whether odors only, odor labels only, or both odors and their labels were presented during learning.

The flat forgetting function for odor recognition memory could also occur because of a lack of transfer-appropriate processing of odors encountered outside the experiment (Morris et al. 1977, Bransford et al. 1979). If subjects in the recognition experiment attempt verbal identification of odors in order to enhance encoding, but avoid semantic encoding of odors otherwise, then the difference in mode of processing, rather than the nature of odor memory itself, might account for a shallow forgetting function. An experiment that compared odor recognition where distractor odors that intervened between learning and testing were processed either similarly to or dissimilarly from the targets might uncover a role of transfer-appropriate processing.

VI. ODOR IDENTIFICATION

Early studies implied that even with considerable practice and feedback, people could identify only about 16 odors at a time without error (Engen and Pfaffman 1960, Jones 1968). More recent evidence indicates that with everyday odorants people can identify many more e.g., Desor and Beauchamp (1974). Cain (1979) argued that *in the long run* only the inherent confusability among a set of stimuli should limit identification. With highly discriminable stimuli, identification could easily go into the hundreds. The term *in the long run* requires comment: Since paired-associate learning of odors proceeds slowly, a person's ability to identify odors will reveal itself best for stimuli overlearned through years of everyday experience. One cannot, therefore, choose just any set of discriminable stimuli and expect that identification will reach high levels in a few practice sessions. Presumably because of factors already discussed e.g., limitations on odor rehearsal via imagery, the possible need for a verbal code, a person might require hundreds of trials of learning. Such practice might prove easy to give in a training session with visual stimuli, where presentation rate could equal many per minute and the need for practice is lower to begin with, but would prove prohibitively time-consuming with odors, where presentation rate could equal a couple per minute.

Few people make conscious attempts to add odors to their repertoire of remembered information. Most odor learning seems to occur incidentally. Some school teachers instruct young children about smells once in a while, but society remains generally passive about the need for intentional learning of odors. Professionals, such as perfumers and flavorists, who try to learn about odors, intentionally develop their expertise slowly. One cannot master the repertoire in mere months or a year. The time-scale typically spans many years, but with the appropriate investment of time the expert can achieve an amazing superiority over the ordinary person.

Even with the generally overlearned odors of everyday life, exact identification often eludes a person. Given a stimulus like chocolate, a person may misname it as coffee, or may draw what seems like a temporary blank. Everyone has experienced the tip-of-the-tongue phenomenon or, as Lawless and Engen (1977) called it, the tip-of-the-nose phenomenon. When first given everyday items to smell, subjects can usually name only about half of them with precision (Cain 1977, 1979, Desor and Beauchamp 1974, Engen 1987, Lawless and Engen 1977, Jones 1968, Sumner 1962). For the other half, subjects give names that range from reasonably good approximations e.g., cherry for strawberry, to surprisingly poor choices, such as soy sauce for strawberry. Although it may seem a minor error to call strawberry odor cherry, such an error to the sight of a strawberry would seem large indeed. Interestingly, even when subjects make relatively large errors, they often feel that they could do better on another occasion. Whether they could or not remains an issue, but subjects at least feel something temporary about their failings. The data tend to justify those feelings up to the degree that small manipulations may increase identification dramatically. Cain (1979), for example, found that simple feedback with the veridical names of 80 stimuli improved performance from approximately 45% correct to close to more than 90% in three trials per item.

The rapid learning reported by Cain would never have occurred with new stimuli, and it is of considerable interest to determine the basis of the differences between the initial slow learning and the fast subsequent re-learning. The mere finding that a person might not always find the correct word for an odor suggests to some workers that a weak connection between odors and language causes the apparently temporary blockage of retrieval of well-learned material (Engen 1987). The finding that people can remember odors well once they have encoded them and that people have little difficulty retrieving the names of odoriferous objects by sight, lends credence to the notion of weakness in odor-name associations.

Some take the absence of a unique vocabulary for odors as other evidence for a tenuous link between odors and words (Richardson and Zucco 1989). Since odors come from things, however, we may ask, "Why do we need new names for the odor of an object when the name of the object itself suffices?". Special names for odors would appear necessary only insofar as the same odor arises from many different objects, as say sourness comes from many different products. Clearly, an odor categorization scheme that cut across categories of objects would make sense. In general, though, a banana smell comes only from a banana, a fish smell from fish, a rose smell from roses, and so on. With the exception of certain hedonic terms e.g., putrid, stinking, of specific olfactory origin, the odor word follows the object word with admirable parsimony.

Yet further evidence invoked regarding a vagueness in the connection between odors and words is the malleability of subjects when led to expect a certain odor (Engen 1987). For example, a person may perceive a red piece of candy flavored with apricot flavor as having a cherry flavor (Cain 1980, DuBose et al. 1980). Such examples appear in every sense domain of course, but people seem particularly gullible with respect to odors and flavors.

Whereas difficulty in retrieving the association between an odor and its lexical representation appears to impede odor identification, no unambiguous data suggest it as the only factor. Noise in memory-based discrimination undoubtedly serves as another important impediment. An important question, so far not addressed, regarding the limitations on odor identification, and hence the limitations on semantic memory for odors, is whether misidentification implies misperception. When a person confidently calls the apricot flavor cherry was the flavor actually perceived as cherry? If so, then the limitation would appear fundamentally perceptual and might lead us to question the discriminative power of olfaction. Curiously, the discriminative power of olfaction usually goes unchallenged. One can say, probably accurately, that olfaction can discriminate among thousands or tens of thousands of stimuli. This would seem to indicate that limitations of discrimination could not possibly coincide with the limitations of identification. Nevertheless, olfactory discrimination is usually thought of as a more or less simultaneous comparison of items placed side-by-side. Surprisingly little research has addressed, no less confirmed, such discriminative capacity. When Eskenazi et al. (1983, 1986) gave categorically different odorants using the triangle test method (see Chapter 5 this book), for example, almost all subjects made errors. These results suggest it is time to learn more about discriminative limitations of olfaction, particularly outside the domain of side-by-side comparisons. This, of course, is in part what the short-term memory experiment taps, but with a different focus than that needed to study discrimination.

Whereas we may normally view quality discrimination as the underlying capability that permits all other operations of olfaction, certain kinds of experience can sharpen discrimination itself. Rabin (1988) showed that training subjects to label odor stimuli enhanced discrimination of quality in side-by-side comparisons. Labeling sharpened perceptual boundaries in a way that other kinds of prior experience with the stimuli did not. In view of the clear association between what might seem strictly a sensory task (discrimination) versus a cognitive task (odor identification), we can hardly afford to assume that errors of identification fail to arise from failures of discrimination. Before we invoke linguistic limitations on olfaction, therefore, we need to understand the veridicality of olfactory perception.

The approximations to correct labels that subjects give in odor identification experiments largely seem consistent with a discriminative limitation. When a subject calls the smell of a peach by the name fruit, a not uncommon kind of error, rather than by the name of a specific fruit, it seems unlikely that the subject simply cannot think of the word peach. Instead, it seems that the subject has processed the stimulus information well enough only to get into a general category, in this case the category fruit, but not well enough to reach the sharpness required for specific identification. The mere possibility that subjects can make partial or semi-specific identifications raises interesting questions about how the olfactory system processes information. To some degree, it apparently extracts some of the chemical commonality e.g., the presence of various esters, among the complex stimuli that make up fruit, though it apparently also responds to the total profile across chemicals in such a way that exact identification can take place. Analogous processes of extraction occur everywhere in sensory and perceptual experience and in a sense allow an enormous range of hierarchical and non-hierarchical categorization. It is pointed out here that olfaction participates in such experiences, but that its perceptual limitations like those of other modalities will define the kind of experiences possible. Experiments on free odor identification, though in some respects difficult to score because of the variety of possible responses, probably give more insight into the richness of olfactory experience than any other.

Semantic memory for odors can also be investigated using multiple-choice recognition tests. Not surprisingly, subjects perform better on such tasks than in free odor identification. The usual reason given for the superiority concerns the reduction of the cognitive challenge imposed when subjects seek to retrieve an odor name by free recall (Cain and Gent 1986). It seems just as likely, however, that the advantage could derive from the reduction in the discriminative burden. If the four or so labels presented to a person after smelling an item represent the only parts of the odor space of interest then the demand for precision shrinks markedly over that in the free identification mode.

It remains unclear why odor identification often fails. Only further investigation can decide the relative roles of inherently weak odor-language associations, poor memorial odor discrimination, and other factors.

A. Males vs Females

When asked how well they thought the typical person could identify 80 different everyday odors, men and women agreed very well (r=0.93). When asked whether men or women would identify the odors better, the two sexes agreed well again (r=0.95), and both predicted superiority for women. Women, however, predicted female superiority more strongly than did men (Cain 1982). In an actual test, women did indeed outperform men (Fig 8) (Cain 1982, see also Cowart 1989). Using common odors and a multiple-choice technique, Doty et al. (1984 a,b) confirmed the superiority of women. Furthermore, they found the difference to hold across four different cultural groups (Doty et al. 1985).

Whence comes the superiority of women? Does it come from superior discriminative ability, superior verbal ability, more intentional learning, better memory? In experiments where men and women sought to recognize the odors of human hands or of t-shirts worn by themselves or others, women also performed better (Wallace 1977, Schleidt 1980). Such data imply something other than strictly a verbal advantage, but isolate it no further.

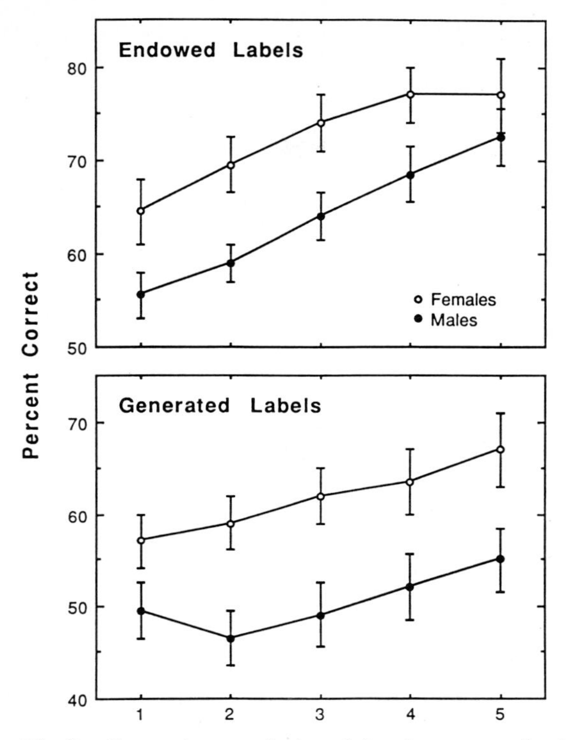

Fig 8. Shows the superiority of females over males in odor identification. The top panel shows how well subjects performed when the experimenter uttered the names of the odors (40 everyday substances) during an inspection period and then gave corrective feedback with those names during five test sessions. The lower panel shows performance in the case where subjects generated their own labels in the inspection period and received feedback with those labels during the five test sessions. From Cain (1982).

B. Blind vs Sighted Persons

In connection with the use of object names for odors, the present authors argued that in a categorical fashion items essentially smell as different as they look. When a person identifies a smell, the object from which it emanates is also identified. Hence if deprived of sight, a person could rely upon smell. Insofar as success at odor identification depends upon one's investment, the blind should outperform the sighted at the task. Murphy and Cain (1986) found this to be true. In their particular sample, the blind actually had poorer measured olfactory sensitivity than their sighted controls, but nevertheless outperformed them at identification.

C. Aging

Ability to identify odors declines rather markedly and progressively with age. The age where performance first begins to fall-off varies with the ease of the task and has led some to conclude that aging begins to show itself only near old age, above about 60 years. In a multiple-choice task where young people obtain correct scores at or near 100%, the effects of age do seem blunted, presumably only because of the eased burden of discrimination discussed above. On the other hand, in a more challenging 80-item free-identification task, aging influenced performance of people under 40 (Murphy 1987). By the time people reach their 70's, their ability to identify odors has typically become profoundly impaired, and their olfactory sensitivity has also suffered (Stevens and Cain 1987). Although the two functions of identification and sensitivity show some association, it appears that something more than sensitivity drives the loss of identification. Schemper et al.(1981) implicated the process of spontaneous verbal mediation in age-related losses in identification. Young people would spontaneously re-code an odor when the opportunity presented itself ("Oh yes, now I realize that that's the smell of lemon, I'll remember that for next time".), whereas old people did not. When directed to re-code through veridical feedback, the elderly showed an ability to do so. Despite their problems in semantic memory, which deals essentially with old memories, the elderly seem also to have serious problems with new memories, as measured via episodic memories. Cain and Murphy (1987), using pictures of faces, schematic symbols, and odors, found that episodic odor recognition, though, comparable in young and old in an immediate recognition test, fell to chance after two weeks for the elderly, but remained above chance for the young over six months.

VII. CONCLUSIONS

Odor memory has been discussed in this Chapter in a way that should make the fundamental issues obvious to readers irrespective of their background. The literature on odor memory has been reviewed within the framework of mainstream memory research and future efforts based on current perspectives and attitudes have been suggested. Research on odor memory has remained largely descriptive rather than analytic. The focus should probably now shift to facilitate incorporation of memory for odors into the larger body of research on verbal and sensory memory.

A theoretical structure, whether borrowed or created strictly from the data on odor memory itself, will undoubtedly aid progress especially insofar as it generates testable hypotheses. There even exists the likelihood that data on memory for odors will reveal the need to alter general theories of memory derived from visual or auditory data. Indeed, this has already begun. Visual and auditory experiments implied that the psychophysical function for remembered intensity would generally have a shallower slope than that for directly perceived intensity. The results led to two competing theories, one based on the uncertainty of memorial estimates of intensity and one based on what can be called a re-perception mechanism. Both theories predicted that the psychophysical function for memorial odor intensity would prove flatter than that for directly perceived intensity. In fact, the functions for the two modes of judgment had the same slope (Algom and Cain 1990). Such data seem to reinforce notions of the fidelity of odor memory and, if substantiated by further experimentation, would seem to require reformulation of the existing theories.

ACKNOWLEDGEMENT

Preparation supported in part by grant DC 00284.

REFERENCES

Algom D, Cain WS (1990) Chemosensory representation in perception and memory. In: Gescheider GA, Bolanowski SJ (eds) Ratio scaling of psychological magnitude. Erlbaum, Hillsdale, NJ, p183

Baddeley AD (1986) Working memory. Oxford University Press, Oxford

Baddeley AD, Dale HCA (1966) The effect of semantic similarity on retroactive interference in long- and short-term memory. J Verb Learn Verb Behav 5: 417-20

Bartlett JC (1977) Remembering environmental sounds: the role of verbalization at input. Mem and Cognit 5: 404-14

Blaxton TA (in press) Investigating dissociations among memory measures: Support for a transfer appropriate processing framework. J Exp Psychol: Learn Mem Cognit

Bransford JD, Franks JJ, Morris CD, Stein BS (1979) Some general constraints on learning and memory research. In: Cermak LS, Craik FIM (eds) Levels of processing in human memory. Erlbaum, Hillsdale N J, p 331

Brooks DN, Baddeley AD (1976) What can amnesic patients learn? Neuropsychologia 14: 111-22

Brower D (1947) The experimental study of imagery: II. The relative predominance of various imagery modalities. J Gen Psychol 37: 199-200

Cain WS (1977) Physical and cognitive limitations on olfactory processing in human beings. In: Mueller-Schwarze D, Mozell M M (eds) Chemical signals in vertebrates. Plenum, New York, p 287

Cain WS (1979) To know with the nose: Keys to odor identification. Science 203: 467-470

Cain WS (1980) Chemosensation and cognition. In: van der Starre H (ed) Olfaction and taste, vol VII. IRL, London, p 347

Cain WS (1982) Odor identification by males and females: Predictions and performance. Chem Sens 7: 129-141

Cain WS, Gent JF (1986) Use of odor identification in clinical testing of olfaction. In: Meiselman H L, Rivlin R S (eds) Clinical measurement of taste and smell. Macmillan, New York, p 170

Cain WS, Murphy CL (1987) Influence of aging on recognition memory for odors and graphic stimuli. Annal NY Acad Sci 510: 212-215

Cohen NJ (1984) Preserved learning capacity in amnesia: Evidence for multiple memory systems. In: Squire LR, Butters N (eds) Neuropsychology of memory. Guilford Press, New York, p 83

Conrad R (1964) Acoustic confusions in immediate memory. Brit J Psychol 55: 75-84

Cowart BJ (1989) Relationships between taste and smell across the adult life span. Annal NY Acad Sci 561: 39-55

Crowder RG (1982) The demise of short-term memory. Acta Psychologica 50: 291-323

Dale HCA, Baddeley A (1969) Acoustic similarity in long-term paired-associate learning. Psychon Sci 16: 209-11

Davis RG (1975) Acquisition of verbal associations to olfactory stimuli of varying familiarity and to abstract visual stimuli. J Exp Psychol: Human Learn Mem 104: 134-42

Davis RG (1977) Acquisition and retention of verbal associations to olfactory and abstract visual stimuli of varying similarity. J Exp Psychol: Human Learn Mem 3: 37-51

Desor JA, Beauchamp GK (1974) The human capacity to transmit olfactory information. Percept Psychophys 16: 551-6

Doty RL, Applebaum S, Zusho H, Settle RG (1985) Sex differences in odor identification ability: A cross-cultural analysis. Neuropsychologia 23: 667-72

Doty RL, Shaman P, Applebaum SL, Giberson R, Sikorski L, Rosenberg L (1984a) Smell identification ability: Changes with age. Science 226: 1441-1443

Doty RL, Shaman P, Dann M (1984b) Development of the University of Pennsylvania Smell Identification Test: A standardized microencapsulated test of olfactory functioning. Physiol Behav 32: 489-502

DuBose CN, Cardello AV, Maller O (1980) Effects of colorants and flavorants on identification, perceived flavor intensity, and hedonic quality of fruit-flavor beverages. J Food Sci 45: 1393-1400

Eich JE (1978) Fragrances as cues for remembering words. J Verb Learn Verb Behav 17: 103-111

Engen T (1982) The perception of odors. Academic Press, New York

Engen T (1987) Remembering odors and their names. Am Sci 75: 497-503

Engen T, Kuisma JE, Eimas PD (1973) Short-term memory of odors. J Exp Psychol 99: 222-5

Engen T, Pfaffmann C (1960) Absolute judgments of odor quality. J Exp Psychol 59: 214-9

Engen T, Ross BM (1973) Long-term memory of odors with and without verbal descriptions. J Exp Psychol 100: 221-7

Eskenazi B, Cain WS, Friend K (1986) Exploration of olfactory aptitude. Bull Psychon Soc 24: 203-6

Eskenazi B, Cain WS, Novelly RA, Friend KB (1983) Olfactory functioning in temporal lobectomy patients. Neuropsychologia 21: 365-74

Farah MJ (1989) Mechanisms of imagery-perception interaction. J Exp Psychol: Human Percept Perform 15: 203-11

Farah MJ, Smith AF (1983) Perceptual interference and facilitation with auditory imagery. Percept Psychophys 33: 475-8

Finke RA, Shepard RN (1986) Visual functions of mental imagery. In: Roff K, Kaufman L, Thomas JP (eds) Handbook of perception and human performance, vol II. Cognitive processes and performance. Wiley, New York, p 37

Freedman J, Haber RN (1974) One reason why we rarely forget a face. Bull Psychon Soc 3: 107-9

Glanzer M (1972) Storage mechanisms in recall. In: Bower GH, Spence JT (eds) The psychology of learning and motivation, vol 5. Academic Press, New York

Graf P, Squire LR, Mandler G (1984) The information that amnesic patients do not forget. J Exp Psychol: Learn Mem Cognit 10: 164-78

Jacoby LL, Dallas M (1981) On the relationship between autobiographical memory and perceptual learning. J Exp Psychol: Gen 110: 306-40

James W (1890) Principles of psychology. Henry Holt, New York

Jones FN (1968) The informational content of olfactory quality. In: Tanyolac N N (ed) Theories of odors and odor measurement. Robert College Research Center, Istanbul, p 133

Lawless HT (1978) Recognition of common odors, pictures, and simple shapes. Percept Psychophys 24: 493-5

Lawless HT, Cain WS (1975) Recognition memory for odors. Chem Sens Flav 1: 331-7

Lawless HT, Engen T (1977) Associations to odors: Interference, memories, and verbal labeling. J Exp Psychol: Human Learn Mem 3: 52-9

Lindauer MS (1969) Imagery and sensory modality. Perceptual Motor Skill 29: 203-215

Lyman BJ, McDaniel MA (1986) Effects of encoding strategy on long-term memory for odours. Q J Exp Psychol 38: 753-65

Lyman BJ, McDaniel MA (in press) Memory for odors and odor names: Modalities of elaboration and imagery. J Exp Psychol: Learn Mem Cognition

Melton AW (1963) Implications of short-term memory for a general theory of memory. J Verb Learn Verb Behav 2: 1-21

Milner B (1962) Les troubles de la memoire accompagnant des lesions hippocampiques bilaterales. In: Physiologie de l'hippocampe. Centre National de la Recherche Scientifique, Paris, p 257

Morris CD, Bransford JD, Franks JS (1977) Levels of processing versus transfer appropriate processing. J Verb Learn Verb Behav 16: 519-33

Murphy C (1987) Olfactory psychophysics. In: Finger TE, Silver WL (eds) Neurobiology of taste and smell. Wiley, New York, p 251

Murphy C, Cain WS (1986) Odor identification: The blind are better. Physiol Behav 37: 177-180

Perky CW (1910) An experimental study of imagination. Am J Psychol 21: 422-52

Peterson MJ, Graham SE (1974) Visual detection and visual imagery. J Exp Psychol 103: 509-14

Rabin MD (1988) Experience facilitates olfactory quality discrimination. Percept Psychophys 44: 532-540

Rabin MD, Cain WS (1984) Odor recognition: Familiarity, identifiability, and encoding consistency. J Exp Psychol: Learn Mem Cognit 10: 316-25

Richardson JTE, Zucco GM (1989) Cognition and olfaction: A review. Psychol Bull 105: 352-360

Schab FR (1990). Odors and the remembrance of things past. J Exp Psychol: Learn Mem Cognit. 16: 648-655

Schemper T, Voss S, Cain WS (1981) Odor identification in young and elderly persons: sensory and cognitive limitations. J Gerontol 36: 446-52

Schleidt M (1980) Personal odor and nonverbal communication. Ethol Sociobiol 1: 225-231

Segal SJ, Fusella V (1970) Influence of imaged pictures and sounds on the detection of visual and auditory signals. J Exp Psychol 83: 458-64

Squire LR (1987) Memory and brain. Oxford University Press, Oxford

Stevens JC, Cain WS (1987) Old-age deficits in the sense of smell gauged by thresholds, magnitude matching, and odor identification. Psychol Aging 2: 36-42

Sumner D (1962) On testing the sense of smell. Lancet II: 895-7

Tulving E, Thompson DM (1973) Encoding specificity and retrieval processes in episodic memory. Psychol Rev, 80: 352-73

Walk HA, Johns EE (1984) Interference and facilitation in short-term memory for odors. Percept Psychophys 36: 508-14

Wallace P (1977) Individual discrimination of humans by odor. Physiol Behav 19: 577-579

Warrington EK, Weiskrantz L (1970) Amnesia: Consolidation or retrieval? Nature 228: 628-30

Waugh NC, Norman DA (1965) Primary memory. Psychol Rev 72: 89-104

Zechmeister EB, Nyberg SE (1982) Human memory: An introduction to research and theory. Brooks/Cole, Monterey, California

12

CHARACTERISTICS OF THE HUMAN SENSE OF SMELL WHEN PROCESSING ODOR MIXTURES

DAVID G LAING

I. INTRODUCTION

Odors emanate from many sources, and can be pleasant or unpleasant. Aromas from bakeries and coffee shops, perfumes from chic passers-by and fragrances from flowers, are almost all pleasant. In contrast, sewage plants, chemical factories and automobile exhausts produce odors that are unpleasant. All of these frequently encountered odors, however, have one feature in common, they are complex mixtures of odorants.

Since it is difficult to find two single odors that cannot be discriminated, our sense of smell appears to be very well equipped to discriminate the dozens, often hundreds of odorants present in commonly encountered odors.

But how well can we discriminate the constituents of mixtures? General observations suggest we can discriminate very few constituents and that the remainder blend to form a homogeneous background which is difficult to describe. That is, we may perceive mixtures like we do colors. For example, if we blend red and yellow we see only one homogeneous color, orange. Although it is possible to describe orange as so much of red and so much of yellow, in reality when we see an orange object it is registered as a single homogeneous percept.

Although it could be argued that semantics limit our descriptions of complex odors, many including chocolate, rose and bread contain constituents that have very distinctive odors which, when presented alone, can be readily described. Why, then, can't we continue to describe them as readily when they are present in mixtures? Clearly, something changes our perception of the constituents once they are presented en masse to the nose.

The mechanisms underlying the perception of odorants in mixtures are essentially unknown. Thus the question remains, how many constituents can we smell in a mixture? If we cannot smell them all, what are the factors that determine which ones can be perceived? Are the factors physiological, biochemical or psychological, or a combination of all three?

Another property of odor mixtures that has particular relevance to the environment, is the prediction of their perceived intensity. In studies of air pollutants, for example, it is important to be able to predict not only the type of odor that will be produced by an industrial process, but also the perceived odor strength outside the factory in nearby communities. Mathematical models for calculating the dispersal of plumes of pollutants emanating from factories depend in part on information that relates the concentration of odorants to perceived odor intensity. At present the information available about the types of constituents of many complex pollutants is scarce and even less is known about the perceived intensity of their mixtures. Predicting the strength and impact of pollutants on nearby communities is therefore difficult and inaccurate.

In this Chapter studies of the perception of odor mixtures by humans will be reviewed and the possible underlying factors that influence perception of the constituents will be discussed.

II. PHENOMENA OBSERVED WITH ODOR MIXTURES

When the vapors of two odors are mixed and sniffed there can be a number of outcomes. According to Zwaardemaker (1900) the general rule is that a mixture does not smell as strong as the sum of its unmixed components. However, more recently Cain and Drexler (1974) defined a more detailed set of outcomes (Fig 1). They indicated that the perceived strength of a mixture may smell 1) as strong as the sum of the perceived intensities of the unmixed components, exemplifying complete addition; 2) more intense than the sum of its components, exemplifying hyperaddition, or 3) less intense than the sum of its components, exemplifying hypoaddition. As shown in Figure 1, there can be three types of hypoaddition: Partial addition, where the mixture smells more intense than the stronger component smelled alone, but less intense than the sum of the components; compromise, when the mixture smells more intense than one component smelled alone but less intense than the other, and compensation, when the mixture smells weaker than both the stronger and weaker components. Which outcome occurs depends on the type of odorants in a mixture and their concentrations (Cain and Drexler 1974, Laing et al. 1984).

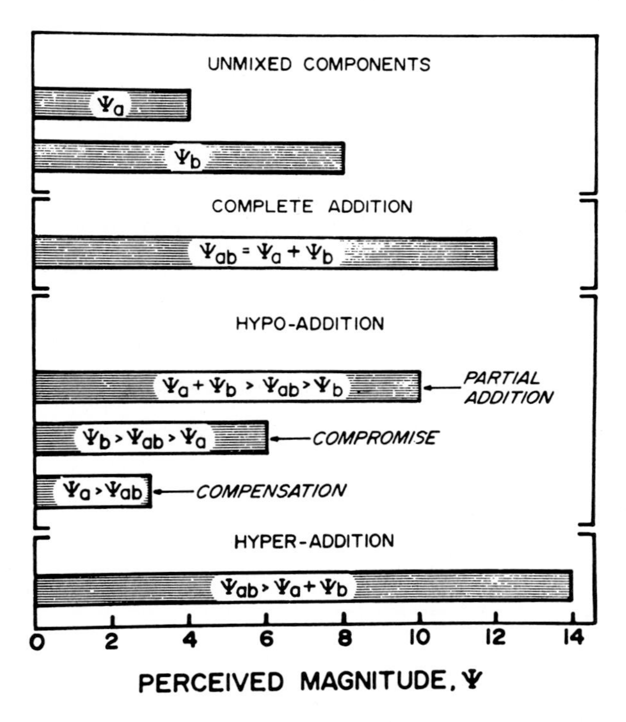

Fig 1. The perceived magnitude (ψ_{ab}) of a two-component mixture may be equal to, less than, or greater than the perceived magnitudes (ψ_a, ψ_b) of its components. Various possibilities are shown here. In this illustration the perceived magnitude of component *a* smelled alone is 4, and the perceived magnitude of component *b* smelled alone is 8. From Cain and Drexler (1974) with permission.

These effects, however, refer only to the overall perceived intensity of a mixture compared to the intensities of the components smelled alone, and do not indicate changes in the perceived intensity of the individual constituents in the mixture.

Studies of the effects of mixing on the intensity and identification of the individual constituents indicate a number of outcomes are possible (Zwaardemaker 1900, Laing et al. 1984). For example, the effect that one odorant has on the perception of another is not reciprocal. This is clearly demonstrated with the citrus smelling substance (+) -limonene, which strongly reduces or suppresses the perceived intensity of the vinegar odor of propionic acid. In contrast, the acid has little effect on the perception of limonene (Bell et al. 1987a). In turn, the odor of limonene is reduced in binary mixtures with (-)-carvone (spearmint), but there is little or no reduction in the smell of carvone (Laing 1988). Although reciprocal interactions depend largely on the type of odorants present in a mixture, whether one or both odorants are perceived also depends on their concentrations, hence (unmixed) perceived intensities. Commonly, with mixtures consisting of two odorants of equal perceived intensity (unmixed), both are perceived; however, one may be perceived to be considerably weaker than the other (Laing et al. 1984). With binary mixtures that contain odorants of unequal (unmixed) intensity, asymmetric interaction is often exaggerated (Table 1). The stronger component will often reduce the perceived level of the other to a far greater extent than itself is reduced. Such effects indicate that the odor qualities perceived in mixtures can change quite sharply with relatively small changes in the concentrations of the constituents. The above effects, however, refer only to binary mixtures and comparable systematic studies have not been reported with more complex mixtures. Furthermore, the results give few clues to the existence of a relationship between the chemical structure of odorants and their interactions in mixtures, or to the mechanisms underlying the effects observed with mixtures.

TABLE 1 : Interactions between odorants of unequal perceived intensity.

Odorants[a]	Intensities (Unmixed)		Intensities (Mixed)	
B_5C_4	88	56	82	10
P_5B_4	89	56	77	24
C_5P_4	84	69	81	16
P_5E_4	77	62	77	26
E_5P_4	73	53	63	26

a The odorants were benzaldehyde (B), (-)-carvone (C), eugenol (E), propionic acid (P). The subscripts indicate ascending order of concentration. Adapted from Laing et al. (1984).

III. WHAT DO WE SMELL IN MIXTURES?

Few studies have addressed this question in a systematic way, or in a way which would indicate how many constituents can be distinguished.

A common approach to characterising the smell of a mixture is to use the technique of odor profiling. It is used widely in the food, flavor and fragrance industries. In brief, subjects are provided with a large list of descriptors from which they select those that describe the odorous characteristics of a mixture. In industry, odor profiles are developed for individual products and often a group of people are trained to recognise key features, both desirable and undesirable. The features described, however, may not fully represent the smell of a constituent. A single odorant, for example, may be described by several descriptors e.g., sweet, green, fatty and fruity. However, when mixed with other odors only the green feature may be discernible, the others being masked by the other smells in the mixture. Laing and Willcox (1983) showed that features of one odor were gradually eliminated as its concentration was decreased. Whether subjects still recognised an odorant devoid of many of its features was not established and this is a question that remains to be answered. Thus, because subjects do not provide an absolute yes or no to the presence of an odorant when profiling, the technique cannot be used to establish the number of odorants humans perceive in mixtures.

The fact that only a limited number of descriptors are used to describe natural oils consisting of many hundreds of odorants (Table 2), suggests humans may have a very limited capacity to discriminate the constituents of mixtures. This suggestion was the subject recently of four experiments in this laboratory where the aim was to determine the capacity of humans to identify odors in mixtures (Laing and Francis 1989, Livermore and Laing (unpublished),Glemarec and Laing (unpublished)). In the first experiment, 123 untrained subjects were given the task of identifying the constituents of stimuli consisting of 1-5 common odorants. The results showed that humans have great difficulty in identifying odors even with the simplest of mixtures (Fig 2). Only 12% of judgements of the constituents of binary mixtures were absolutely correct i.e., only the correct two odors were chosen. The result is in general agreement with another recent finding that subjects could not distinguish between single odorants and binary mixtures (Schiet and Frijters 1988).

TABLE 2 : Odor profiles of natural oils[a].

Substance	Descriptors
Mandarin Oil	citrus,fresh,fruity,vegetable, green, tart, earthy, aldehydic, floral, sweet
Lemon Oil (Italian)	citrus, fresh, fruity, tart, green, aldehydic, sweet
Sandalwood Oil (East Indian)	woody, balsamic, amber, powdery, sweet, vegetable, earthy, spicy

a Adapted from 'Fragrance profile survey. Published by Naarden International Fragrance Division, Holland.

In the second experiment, the effect of training on identification was investigated. Subjects were trained over one week to identify the same single odors used in the first experiment. Once it had been established that all subjects had achieved this goal, they were given the same task as the untrained subjects, but in this case were tested daily for two weeks. Although these subjects achieved higher correct scores than the untrained group, few subjects identified more than three of the components in 4- and 5-component mixtures. The effects of training and experience on mixture perception were investigated further with a group of eight flavorists and perfumers in a third experiment. In this study each subject was given the opportunity to sniff the single odorants each day for a week before commencing the test. The results indicate that although the performance of these experts was substantially better than untrained subjects, in only 3% of the trials with 5-component mixtures did they correctly identify all five components. The fourth experiment involved the use of a selective attention task which has been shown in studies of vision and audition to improve discrimination. Again, however, subjects were limited to identifying three components. Taken with the results of the other three experiments, these findings suggest that humans have a very limited capacity to discriminate odors in mixtures, with three or four odors being about the limit.

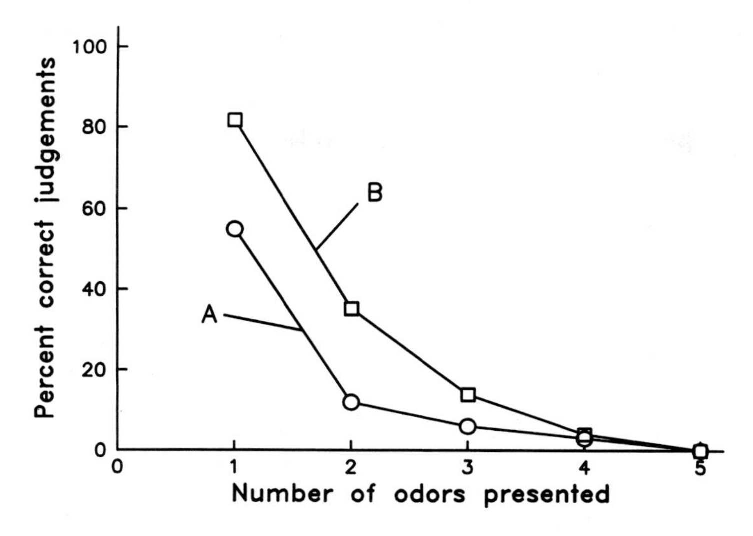

Fig 2. Percentage of judgements correctly identifying the components of stimuli consisting of 1-5 odorants. Function A indicates the percentage of times that the correct odor(s) and no other was selected. Function B shows the percentage of times that the correct odor(s) was selected but others were incorrectly selected.

This suggestion is reinforced by the finding, in all four experiments, that when 3-, 4- or 5-component mixtures were presented, the majority of responses indicated that only 2 or 3 components were present regardless of whether the response was correct or incorrect. Since the total number of odorants selected provides a measure of the perceived complexity of the stimulus, it appears that subjects found 4- and 5-component mixtures to be no more complex than mixtures with three components. The perceived complexity of mixtures, therefore, appears to have a ceiling at three components, again supporting the view that humans have a limited capacity to distinguish components of mixtures.

The above results generally support the finding by Moskowitz and Barbe (1977) and Jellinek and Köster (1979) that mixtures have a small range of perceived complexity. In the former study subjects rated the complexity of stimuli containing up to five components. The small range recorded, suggests that with increasing numbers of components perceived complexity stabilizes at an asymptote reflecting the degree to which olfaction is a synthetic sense. Furthermore, the study showed that perceived complexity is not additive. For example, the complexity estimate for camphor was low. Combining camphor with isobutylbutyrate produced the highest complexity rating of all of the ten binary mixtures studied. Adding a third component diminished the complexity.

Overall, these several studies show that:

1. perceived complexity of a single odorant can be equal to or greater than a multi-component mixture;
2. perceived complexity may reach a constant value with relatively low numbers of odorants;
3. there is little or no correlation between perceived complexity and chemical complexity;
4. perceived complexity of odorants is not additive.

The evidence from studies of the absolute identification of mixture components and perceived complexity, therefore, strongly indicate that humans have a very limited capacity to distinguish odors in mixtures and that characteristic features of odors are totally or partially lost in mixtures containing 4, 5, or more components.

IV. ROLE OF COGNITIVE FACTORS IN MIXTURE PERCEPTION

In the study of mixture perception described above (Laing and Francis 1989), the task involved the absolute identification of mixture constituents. Although common odors and labels were used to facilitate identification, e.g., almond, spearmint, there was no attempt to choose odorants that were highly familiar to each subject. Furthermore, the task of absolute identification is more difficult than one involving relative discrimination where a subject has only to indicate whether two stimuli are different. Thus, it is possible that the results obtained by Laing and Francis (1989) do not accurately reflect the discriminative ability of the sense of smell.

Recent studies (Rabin 1989, Rabin and Cain 1989) suggest the latter may be the case. Using a task that involved discrimination, Rabin and Cain showed that the familiarity and pleasantness of mixture constituents influenced perception of the constituents. Briefly, the task involved discrimination of a binary mixture and a single constituent of the mixture. At each trial each target odor was paired with itself or with a transform of itself. Each transform consisted of the target odor plus a less intense familiar or unfamiliar component. The results showed that a familiar target odor is more likely to be discriminated from a contaminated version of itself than is an unfamiliar target odor. Furthermore, a familiar contaminant is more likely to be apparent to an observer than an unfamiliar one. One explanation given for the influence of familiarity was that participants possessed a more finely tuned perceptual representation of a familiar stimulus and were better able to judge whether or not another stimulus, a transform of the familiar one, violated the perceptual boundaries of the first (Rabin and Cain 1989). Accurate labelling may therefore have indicated the presence of a unique perceptual category congruous with discriminative capacity.

Pleasantness of odors was also found to influence detection of the minor component in binary mixtures, with unpleasant stimuli more detectable than pleasant ones (Rabin and Cain 1989). As noted by the authors, this finding is not too surprising given the common experience of detecting impurities or spoilage in food products.

Overall, the experiments on the effects of familiarity and pleasantness of odors on their detection in mixtures indicate that these are important factors that need to be taken into consideration when assessing the ability of humans to perceive the constituents of mixtures. Since earlier studies of odor mixtures demonstrated that the type of odorant and stimulus intensity are also important (Cain and Drexler 1974, Laing et al. 1984, Laing 1989), it is clear that both experiential and stimulus factors as well as the test method need careful choice in future studies aimed at determining the ability of humans to perceive the constituents of mixtures.

V. PREDICTING THE ODOR INTENSITY OF MIXTURES

As mentioned earlier in this Chapter, derivation of methods for calculating the perceived intensity of mixtures has very useful practical applications. Being able to estimate the strength of odors in a plume at different distances from an odor source such as a factory or sewage treatment plant can provide a basis for siting such plants at locations that will cause least public nuisance. Furthermore, being able to predict the intensity of mixtures and their constituents, may provide insight into the relevant physico-chemical properties of odor molecules that influence perceived intensity and mixture interactions.

One of the earliest attempts to measure the effect of mixing on the perceived intensity of odors was that of Zwaardemaker (1900). He found that when two odors are mixed the perceived intensity of the mixture was always less than the sum of the perceived intensities of the component odors. Jones and Woskow (1964) reported later that the perceived intensity of a mixture, although less than the sum of the component intensities, was more than a simple average of the two. In essence, almost all subsequent studies of mixture intensity have supported this finding indicating that the sense of smell compresses information about odor intensity. The mutual weakening of the perceived intensity of components of mixtures was visualized by Zwaardemaker (1930) as "The two sensations can be imagined as two vectors representing two forces counterbalancing each other in our intellect." In an attempt to formalise the interaction between two odorants, Berglund et al. (1973) developed a mathematical model which incorporated the application of vector addition to odor mixtures and termed it the vector model. Thus they proposed that the perception of odor mixtures should be treated with the same rule as that used to add vectors, like a parallelogram of forces (Fig 3). Under these conditions a unique number, cos α, characterises a pair of odorants whatever their concentrations and levels of intensity. The concept of an angle characterising a given pair of odorants was previously applied to describe odor similarity by Ekman et al. (1964). As shown in Figure 3, two odors in a mixture are represented by the vectors i and j and the lengths of these ψ_i , ψ_j, denote their odor intensities. The length of the resultant vector ψ_{ij} represents the odor intensity of the mixture ij, as perceived by an observer. The angle α_{ij} between any two vectors i and j is constant for each pair of odorants, is independent of odor concentration, and is assumed to represent a qualitative perceptual relationship between odors. A mathematical formulation of the vector model for binary mixtures is

$$\psi_{ij} = (\psi_i^2+\psi_j^2+2\psi_i\psi_j\cos\alpha_{ij})^{1/2}$$

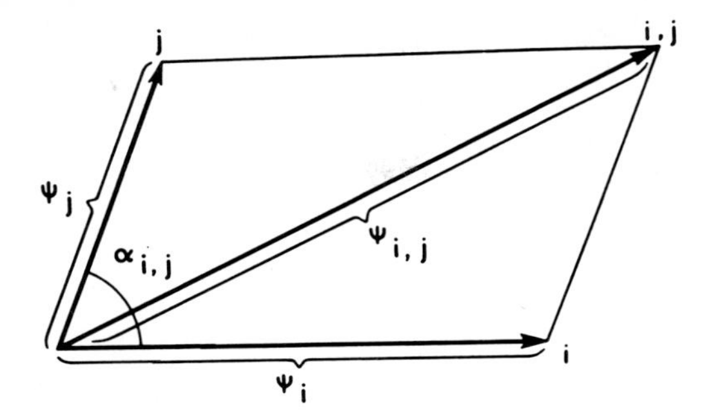

Fig 3. The vector model of perceptual odor interaction illustrated for two odor components. From Berglund et al. (1973) with permission.

Thus the authors proposed that α_{ij} may be used as an index of the common perceptual qualitative content between two odors. So far the value of α for all odor pairs studied has been between 102° (Berglund 1974) and 115° (Laing et al. 1984).

A shortcoming of the vector model is that it assumes the effects of components are symmetrical. Clearly, the results described earlier in this Chapter (Laing et al. 1984, Laing 1989) indicate that many odor interactions are asymmetrical. Furthermore, the model cannot handle synergism. Nevertheless, in tests of the model with binary mixtures (Berglund 1974), high values for correlation coefficients for predicted versus experimental intensity estimates were obtained with a mean value of $r = 0.94$ recorded with the five odor pairs. Application of an expanded version of the model to 3-, 4- and 5-component mixtures of the same odorants (Fig 4) gave reasonable agreement between the predicted and experimental values.

Following suggestions by Cain (1975) and Moskowitz and Barbe (1977) that cos α reflects not only sensory interaction but may also indicate the direct influence of the psychophysical power law, Patte and Laffort (1979) and Laffort and Dravnieks (1982) developed the 'U' and 'UPL' models respectively, with the latter model reflecting the influence of the power law on the constituents. Combining these two models mathematically allowed derivation of an index Γ which provides a means of defining olfactory interaction in binary mixtures by a number (Laffort 1989, Laffort et al. 1989). For

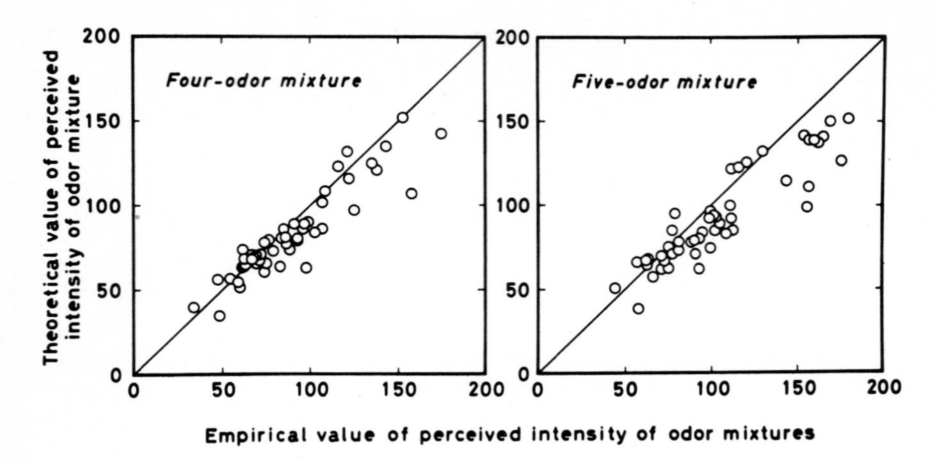

Fig 4. Results of mixing four and five odors varying in perceived odor intensity. Empirical scale values of perceived odor strength plotted against theoretical values computed according to the vector model. Left hand diagram is for a mixture of hydrogen sulfide, dimethyldisulfide, pyridine and dimethylsulfide. Right hand side is for a mixture of the latter four odors and methyl mercaptan. Reprinted with permission from Berglund (1974).

example, a value of $\Gamma > 1$ or < 1 signifies true synergy or inhibition respectively. A value of $\Gamma = 1$ signifies lack of interaction. So far values of Γ have proven to be a reasonable indication of the extent and type of interaction between two odorants. Clearly, however, there is some way to go before accurate predictions can be made of the perceived odor intensity of mixtures containing more than two components and of the perceived intensity of the individual constituents.

VI. MECHANISMS OF MIXTURE PERCEPTION

Physiological studies of the mechanisms underlying how odor mixtures are perceived by insects (O'Connell 1975), and by aquatic (Derby et al. 1985) and terrestrial animals (Bell et al. 1987a, 1987b), indicate that both peripheral and central mechanisms can be involved. Possible peripheral mechanisms involving interaction at the receptor cell are; 1) suppression of one odor by another through competition for receptor sites on a receptor cell; 2) allosteric effects where occupation of a site by one odorant adjacent to the receptor site of another affects binding of the latter odor, or 3) via intracellular effects where reception at a distant site on the same cell surface triggers the release of internal substances e.g., Ca++, which may inhibit the transduction process (Shirley et al. 1986) and prevent stimulation of a cell by another odorant. This particular method of suppression could occur if one transduction process involved adenylate cyclase, an enzyme which is believed to be involved in the transduction process used by many odorants (Sklar et al. 1986), and which is inhibited by Ca++.

Perireceptor effects, that is, effects that occur in the mucus before contact of odor molecules with receptor cells, could also affect the perception of components of odor mixtures. It is possible, for example, that during stimulation with a mixture, the concentration of a particular odorant in mucus could affect the quantity of other odors partitioning between inspired air and mucus, and thereby alter the number of molecules reaching receptor cells. The subsequent reduction in cell response could be wrongly interpreted as resulting, for example, from competition of odor molecules for sites on a receptor cell. Another perireceptor effect which could alter responses of receptor cells is the interaction of odorants with the so-called odorant-binding protein (Bignetti et al. 1985, Pevsner et al. 1985). This protein which differs slightly in structure in different species (Bignetti et al. 1988) and is similar to a family of transport proteins e.g., the retinal-binding protein, has been found in olfactory mucus. Since odorants differ widely in their ability to bind to the protein and there is a broad correlation between odor threshold and binding constants, low thresholds correlating with high binding efficiencies (Topazzini et al.1985), it is possible that selective binding and transport of odorants to the receptors by the protein could alter the ratio of odorants interacting with the receptor cells and subsequently the responses of the cells. Currently, there is no evidence to support or reject either of these possible perireceptor effects.

Modulation of peripheral responses to odors could also occur by the action of chemicals released in the vicinity of receptor cells e.g., Substance P (Papka and Matulionis 1983), or by the trigeminal nerve which innervates the olfactory epithelium and tends to suppress olfactory activity (Stone 1969). Thus, the presence of an odorant in a mixture at a concentration high enough to stimulate the trigeminal nerve could affect perception of the other odorants.

From a series of psychophysical experiments with humans (Laing et al. 1984, Bell et al. 1987a, Laing 1988), Laing (1989) proposed that it is now possible to construct a scenario of what may be happening at the receptor cells in the nose when odor mixtures are perceived. Essentially it was proposed, on the basis of the many asymmetric interactions observed, that when one odor has no effect on the perception of another, that this provides evidence that cells responsive to the latter odorant have no sites than can accommodate the former odorant. For example, the odorants limonene and propionic acid have little effect on the perception of carvone, whilst the latter odorant suppresses perception of the former pair. This suggests that cells which are responsive to carvone have no sites to accommodate limonene or propionic acid. In contrast, limonene-sensitive, and propionic acid-sensitive cells have sites that allow carvone to bind, and by unknown mechanisms, inhibit responses to limonene and the acid. Thus, from psychophysical studies of binary mixtures, it should be possible to produce an information base on cell types that exist in the olfactory epithelium of humans and animals and to test these predictions in animals using single-cell recording techniques.

Suppression of one odor by another, however, need not signify a peripheral mechanism. Should two odors have no cells in common that involve stimulation or inhibition, then any suppression or interaction observed must arise from central effects. The olfactory bulb, for example, is well set up for modifying input from the receptor cells with lateral inhibition likely between glomeruli via periglomerular cells, or between mitral cells (Shepherd 1974).

Another psychophysical approach that has been used to investigate the peripheral versus central question, involved comparison of the effects observed when two odorants were delivered simultaneously but separately to each nostril (dichorhinic stimulation), or simultaneously as a mixture to both nostrils, as occurs during natural sniffing (birhinal stimulation). With dichorhinic stimulation no interaction can occur between the two odorants at the receptor cells since the nostrils are essentially isolated from each other. Thus any interaction observed must be of central origin. A weakness with the theory behind this approach is that the constituents of mixtures are not perceived in this way and interhemispherical interaction would be minimal with the natural presentation procedure. With dichorhinic presentation, interaction can only arise through interhemispherical interactions. Although Cain (1977) showed there were similarities between the extent of suppression using both techniques, Laing and Willcox (1987) clearly demonstrated that in many instances where strong suppression of one odor by another occurred during birhinal stimulation, little or no suppression occurred with dichorhinic stimulation. Furthermore, in almost all instances the extent of suppression with birhinal stimulation was greater than with dichorhinic stimulation. Overall, the evidence from the study by Laing and Willcox (1987) indicates that the suppression observed during dichorhinic presentation of stimuli does not represent the suppression observed during the natural perception of mixtures.

The mechanisms underlying the apparent limited capacity of humans to perceive very few constituents in mixtures (Laing and Francis 1989) may have both peripheral and central origins. Loss of information about constituents could occur if the activity pattern that characterises stimulation by an odorant in the receptor epithelium (Bell et al. 1987b) and bulb (Stewart et al. 1979, Royet et al. 1987, Bell et al. 1987a) was combined with those of other odorants in a mixture into a new pattern at another olfactory centre. Loss of information through changes in activity patterns in the central olfactory system has its origins in the arrangement of anatomical projections that connect the bulb to the olfactory cortex. For example, it is well known that there are ordered topographical projections from the nose to the bulb (Astic and Saucier 1986) and that neural responses to odorants are characterised by discrete patterns of responsive receptor (Bell et al. 1987b) and bulbar cells (Bell et al. 1987a, Royet et al. 1987, Stewart et al. 1979). However, this order is lost once axons leave the bulb, with projections from the bulb to the cortex exhibiting a very low degree of order (Haberley and Bower 1988). Small areas in the bulb project to large regions of the cortex and small regions in the cortex sample broad areas of the bulb. This anatomical arrangement suggests that processing at the cortex and beyond could involve a significant degree of combinatorial operations in which axons from scattered bulbar sites converge on a target cell. Such convergence would result in loss of any topographical information that characterises an odor (Cattarelli et al. 1988) at the periphery and bulb, and produce a representation of a complex odor that provides little information about the constituent odors (Staubli et al. 1987).

Although studies with humans (Laing and Francis 1989) suggest that blending and subsequent loss of information about mixture constituents occurs with mixtures containing three or more odors, it is clear that some of the constituents can be discriminated. A possible explanation is that odors may be partly discriminated on a temporal basis. Neurophysiological studies have shown that different odors differ greatly in the times taken to stimulate receptor cells, with differences in the order of hundreds of milliseconds being recorded (Getchell et al. 1984). If these temporal differences are maintained at central olfactory structures such as the olfactory cortex, then if only two to three odors are presented in a mixture it may be possible for them to be distinguished and identified. However, when more than three odors are presented the time intervals separating the arrival of neural activity arising from stimulation by these odorants at the olfactory cortex may be too small to allow differentiation and only the very fast and slow odorants, for example, may be discriminated. Identification of the constituents of mixtures may therefore be limited by the convergence of neural input at the olfactory cortex but be aided to some extent by temporal separation of input from each odorant.

Finally, it may be that identification of odors is limited to the 'magical number seven plus or minus two' (Millar 1956), as has been found for tones. In other words the sense of smell may not be capable of processing information about more than say 5 or 6 odors when they are presented simultaneously. Multiple tones, for example, can be identified accurately up to a total of seven but above this number the error rate increases substantially. As discussed above, there are anatomical and physiological reasons that could explain why the olfactory system may comply with the 'magic number'.

VII. SUMMARY

How we perceive odor mixtures remains a mystery. However, piece by piece, information from diverse sources including physiological and biochemical studies with aquatic and terrestrial animals, and psychophysical studies with humans is providing a framework upon which we can explain and understand the phenomena we observe when we sniff complex odors.

REFERENCES

Astic L, Saucier D (1986) Anatomical mapping of the neuroepithelial projection to the olfactory bulb. Brain Res Bull 16: 445-454

Bell GA, Laing DG, Panhuber H (1987a) Odour mixture suppression: evidence for a peripheral mechanism in human and rat. Brain Res 426: 8-18

Bell GA, Laing DG, Panhuber H (1987b) Early-stage processing of odor mixtures. Ann NY Acad Sci 510: 176-177

Berglund B (1974) Quantitative and qualitative analysis of industrial odors with human observers. Ann NY Acad Sci 237: 35-51

Berglund B, Berglund U, Lindvall T, Svensson LT (1973) A quantitative principle of perceived intensity summation in odor mixtures. J Exp Psychol 100: 29-38

Bignetti E, Cattaneo P, Cavaggioni A, Damiani G, Tirindelli R (1988) The pyrazine-binding protein and olfaction. Comp Biochem Physiol 90B: 1-5

Bignetti E, Cavaggioni A, Pelosi P, Persuad KC, Sorbi RT, Tirindelli R (1985) Purification and characterisation of an odorant-binding protein from cow nasal tissue. Eur J Biochem 149: 227-231

Cain WS (1975) Odor intensity: mixtures and masking. Chem Sens Flav 1: 339-352

Cain WS (1977) Bilateral interaction in olfaction. Nature 268: 50-52

Cain WS, Drexler M (1974) Scope and evaluation of odor counteraction and masking. Ann NY Acad Sci 237: 427-439

Cattarelli M, Astic L, Kauer JS (1988) Metabolic mapping of 2-deoxyglucose uptake in the rat piriform cortex using computerized image processing. Brain Res 442: 180-184

Derby CD, Ache BW, Kennel EW (1985) Mixture suppression in olfaction: electrophysiological evaluation of the contribution of peripheral and central neural components. Chem Sens 10: 301-316

Ekman G, Engen T, Künnapas T, Lindman R (1964) A quantitative principle of qualitative similarity. J Exp Psychol 68: 530-536

Getchell TV, Margolis FL, Getchell ML (1984) Perireceptor and receptor events in vertebrate olfaction. Prog Neurobiol 23: 317-345

Haberley LB, Bower JM (1988) Olfactory cortex: model circuit for study of associative memory. TINS 12: 258-64

Jellinek JS, Köster EP (1979) Perceived fragrance complexity and its relation to familiarity and pleasantness. J Soc Cosmet Chem 30: 253-262

Jones FN, Woskow MH (1964) On the intensity of odor mixtures. Annal NY Acad Sci 116: 484-494

Laffort P (1989) Models for describing intensity interactions in odor mixtures: a reappraisal. In: Laing DG, Cain WS, McBride RL, Ache BW (eds) Perception of complex smells and tastes. Academic Press, Sydney, p 205

Laffort P, Dravnieks A (1982) Several models of suprathreshold quantitative olfactory interaction in humans applied to binary, ternary and quaternary mixtures. Chem Sens 7: 153-174

Laffort P, Etcheto M, Patte F, Marfaing P (1989) Implications of power law exponent in synergy and inhibition of olfactory mixtures. Chem Sens 14: 11-23

Laing DG (1988) Relationship between the differential adsorption of odorants by the olfactory mucus and their perception in mixtures. Chem Sens 13: 463-471

Laing DG (1989) The role of physicochemical and neural factors in the perception of odor mixtures. In: Laing DG, Cain WS, McBride RL, Ache BW (eds) Perception of complex smells and tastes. Academic Press, Sydney, p 189

Laing DG, Francis GW (1989) The capacity of humans to identify odors in mixtures. Physiol Behav 46: 809-814

Laing DG, Panhuber H, Willcox ME, Pittman EA (1984) Quality and intensity of binary odor mixtures. Physiol Behav 33: 309-319

Laing DG, Willcox ME (1983) Perception of components in binary odor mixtures. Chem Sens 7: 249-264

Laing DG, Willcox ME (1987) An investigation of the mechanisms of odor suppression using physical and dichorhinic mixtures. Behav Brain Res 26: 79-87

Millar GW (1956) The magical number seven, plus or minus two: Some limits in our capacity for processing information. Psychol Rev 63: 81-97

Moskowitz HR, Barbe CD (1977) Profiling of odor components and their mixtures. Sensory Processes 1: 212-226

O'Connell RJ (1975) Olfactory receptor responses to sex pheromone components in the redbanded leafroller moth. J Gen Physiol 65: 179-205

Papka RE, Matulionis DH (1983) Association of substance-p-immunoreactive nerves with the murine olfactory mucosa. Cell Tissue Res 230: 517-525

Patte F, Laffort P (1979) An alternative model of olfactory quantitative interaction in binary mixtures. Chem Sens Flav 4: 267-274

Pevsner J, Trifiletti RR, Strittmatter SM, Snyder SH (1985) Isolation and characterisation of an olfactory receptor protein for odorant pyrazines. Proc Natl Acad Sci 82: 3050-3054

Rabin MD (1989) Experience facilitates olfactory quality discrimination. Percept Psychophys 44: 532-540

Rabin MD, Cain WS (1989) Attention and learning in the perception of odor mixtures. In: Laing DG, Cain WS, McBride RL, Ache BW (eds) Perception of complex smells and tastes. Academic Press, Sydney, p 173

Royet JP, Sicard G, Souchier C, Jourdan F (1987) Specificity of spatial patterns of glomerular activation in the mouse olfactory bulb: computer-assisted image analysis of 2-deoxyglucose autoradiographs. Brain Res 417: 1-11

Schiet FT, Frijters JER (1988) An investigation of the equiratio-mixture model in olfactory psychophysics: A case study. Percept Psychophys 44: 304-308

Shepherd GM (1974) The synaptic organization of the brain. Oxford University Press, New York London Toronto

Shirley SG, Robinson CJ, Dickinson K, Aujla R, Dodd GH (1986) Olfactory adenylate cyclase of the rat. Biochem J 240: 605-607

Sklar PB, Anholt RRH, Snyder SH (1986) The odorant sensitive adenylate cyclase of olfactory receptor cells. J Biol Chem 261: 15538-15543

Staubli U, Fraser D, Faraday R, Lynch G (1987) Olfaction and the "Data" memory system in rats. Behav Neurosci 101: 757-765

Stewart WB, Kauer JS, Shepherd GM (1979) Functional organization of rat olfactory bulb analysed by the 2-deoxyglucose method. J Comp Neurol 185: 715-34

Stone H (1969) Effect of ethmoidal nerve stimulation on olfactory bulbar electrical activity. In: Pfaffman C (ed) Olfaction and taste, vol III. Rockefeller University Press, New York, p 216

Topazzini A, Pelosi P, Pasqualetto PL, Baldaccini NE (1985) Specificity of a pyrazine binding protein from cow olfactory mucus. Chem Sens 10: 45-49

Zwaardemaker HC (1900) Die compensation von Geruchsempfindungen. Arch Physiol Leipzig pp 423-432 as translated in Perf Ess Oil Rec (1959) 50: 217-221
Zwaardemaker HC (1930) An intellectual history of a physiologist with psychophysical aspirations. In Carl Murchison (ed) 'A History of Psychology in Autobiography'. vol 1. Clark University Press, Worcester, Massachusetts, p 491

13

COMPARISON OF ODOR PERCEPTION IN HUMANS AND ANIMALS

JAMES C WALKER

ROGER A JENNINGS

I. INTRODUCTION

In this Chapter the abilities of humans and animals to detect odorants and to perceive the strengths of suprathreshold concentrations of odorants are compared and an effort is made to provide conceptual and practical reasons for such comparisons. In addition, current obstacles to more profitable use of psychophysical data for determining the neural mechanisms that underlie human odor perception are discussed. Results from odor psychophysical experiments in humans and animals are then summarized and general conclusions and recommendations for future research are offered.

There are at least two reasons for comparing odor perception in humans and animals. First, it provides scientific evidence to support or refute hypotheses about the relative olfactory prowess of humans and animals. Second, integration of human psychophysical results with those from physiological studies of the olfactory systems of animals may reveal the neural mechanisms that mediate the processing of odor information in humans.

Over the past three decades there have been a number of reviews of olfactory psychophysics with humans (Stone 1966, Hyman 1977) and animals (Passe and Walker 1985) and these have highlighted and outlined the reasons for much of the variability in results from different laboratories. These include inadequate control of stimulus concentration and purity, psychological "impurity" (the perception of irritation as well as odor) and procedures which bias subjects to either over-report or under-report odor stimulation.

Significant progress has been made during the past two decades in solving some of the problems noted in these reviews. Cain (1977) showed that when care is taken to control and measure odorant concentration, the olfactory system exhibits precision comparable to that seen with other sensory systems. Laing (1982, 1983, 1986) has determined some of the minimum requirements for air dilution olfactometers used in human research. His studies indicate that such olfactometers should either deliver a final volume flow rate of at least 30 l/min or allow a volume of least 500 ml in the sampling port.

The issue of psychological "impurity" has also been investigated (Walker et al. 1990a, Walker et al. 1990b). The nasal irritation thresholds of acetic acid, propionic acid and amyl acetate were 30, 50 and 200 times higher, respectively, in four anosmics than in 16 normal human subjects. This finding is consistent with the idea that stimulation of the trigeminal nerve is sufficient but not necessary for the perception of irritation.

Researchers studying the visual, auditory or somatosensory systems readily agree on the physical continua along which stimuli can be varied. In contrast, those studying olfaction make decisions about stimulus selection without benefit of a widely accepted theory concerning the important physico-chemical determinants of odor sensation. As a result, a wide variety of odorants have been used in olfactory psychophysical studies but systematic comparisons have been hindered since most stimuli are common to only a few studies. In Figure 1, the total number of odorants that have been employed in odor psychophysical studies in humans, non-human mammals and each of the remaining vertebrate classes are shown. Since the number of odorants used in animal studies is so much smaller than that for human studies, the majority of odorants tested in humans have not been used in animal studies.

A second indication of the extent of the research effort with humans and animals is given in Figure 2, which shows that the number of studies of odor thresholds with animals is roughly equivalent to that in humans. Perhaps due to the longer training and testing times required for animal studies, investigators have tended to use a smaller number of odorants per experiment in animal investigations. Further, there is greater overlap in odorant selection among laboratories studying odor perception in animals.

Although the set of odorants that has been used in studies involving both humans and animals is small, it is instructive to summarize the tentative comparisons that can be

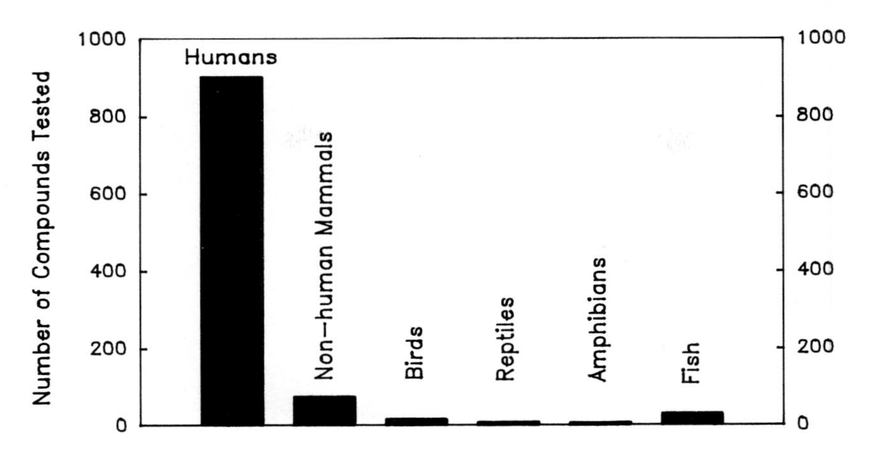

Fig 1. Total number of compounds used in odor psychophysical studies with humans, non-human mammals and four other vertebrate classes. Information for this Figure was based on published literature obtained through computerized searches, from the period 1965 to the present, on EMBASE, BIOSIS, and Medline.

made. In this way, it may be possible to determine which experimental approaches may best advance our understanding of olfactory perception. In this Chapter, comparisons are limited to studies of detection thresholds, or those in which intensity aspects of suprathreshold stimuli were measured. Studies of odor quality are excluded because the psychophysical tasks used to study odor quality in humans (cross-adaptation, perception of mixtures, odor identification) have not been used, with few exceptions, in animals. Conversely, virtually no data on simple odor discrimination exist for humans although this has been the predominant tool in animal research. An excellent review of odor discrimination performance in animals was published recently (Slotnick 1990).

II. OLFACTORY SENSITIVITY

Over the past 40-50 years there has been a consistent effort to measure the responses of animals and humans to different concentrations of a variety of odorants. The most commonly studied measure is the absolute threshold. Although there are a number of definitions of this term (Engen 1971), it is generally taken to mean the lowest concentration of odorant that can be detected. Concentrations higher than this are required for odor identification.

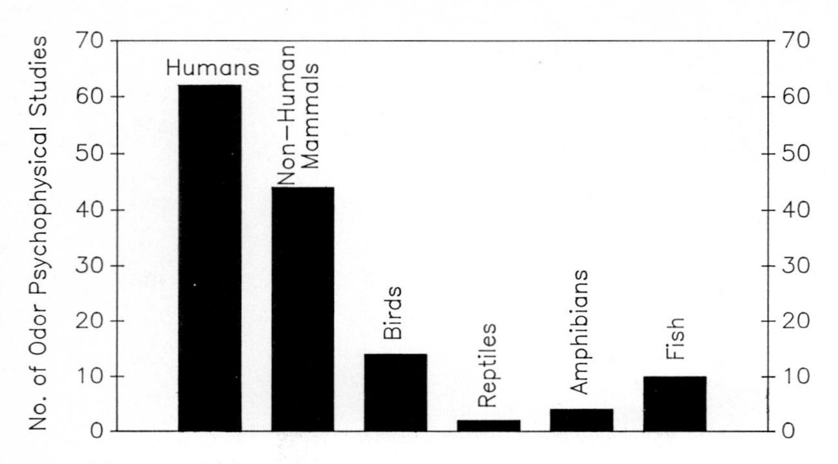

Fig 2. Total number of odor psychophysical studies with humans, non-human mammals, and the remaining four vertebrate classes. Information for this Figure was based on the same sources as those used for Figure 1.

Procedures for measuring thresholds in humans are discussed in Chapter 5. The major difference between tests with humans and animals is that, with animals, conditioning techniques must be used to establish different behaviors in the presence and absence of odor. Once this training is complete, the odor concentration is lowered until detection no longer occurs. A concentration-dependent decrease in differential responding to odor and clean air supports the interpretation that the animal is relying solely on odor stimuli. More practically, threshold measurements in a given species provide an estimate of the minimum concentration required in order for a particular odorant to influence behavior in the natural environment.

The most direct method of comparing the odor sensitivity of humans and animals is to determine their thresholds for the same compounds. However, as shown in Figure 3, there is little overlap in the odorants tested. In this Figure, the area of each circle is directly proportional to the number of compounds tested in a given group and the area of overlap shows the number of compounds tested in both humans and a particular species. For example, 18 odorants have been used in studies with birds and, of these, 10 have been tested with humans.

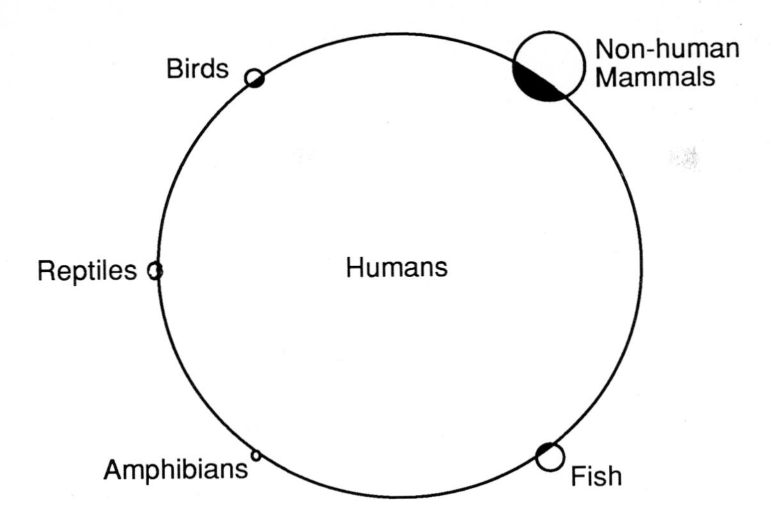

Fig 3. Degree of overlap between human and animal psychophysical studies in terms of odorants used. Data for this Figure were obtained from sources used to generate Figures 1 and 2.

One of the more commonly held views of human and animal olfaction is that dogs are able to detect much lower concentrations of odorants than humans. Currently, there are five odorants for which two or more threshold values have been reported in both species and these data are summarized in Figure 4. Since thresholds in the Figure are depicted in terms of the negative logarithm of molarity, values near the bottom of the Figure represent maximum sensitivity. Each symbol represents a published threshold. This compilation is consistent with the idea that dogs are more sensitive to odors. However, the considerable range of values within several of the subject-by-odorant combinations raises questions about the magnitude of the dog's superiority and whether the advantage is similar for different odorants. For example, the evidence for a difference in sensitivity of humans and dogs to propionic or butyric acid is much more convincing than that for amyl acetate.

The notion that the olfactory sensitivity of the dog is better than that of humans is supported by reports where dogs and humans were compared by the same researchers. Krestel et al. (1984), for example, reported that dogs (beagles) are about 300 times more

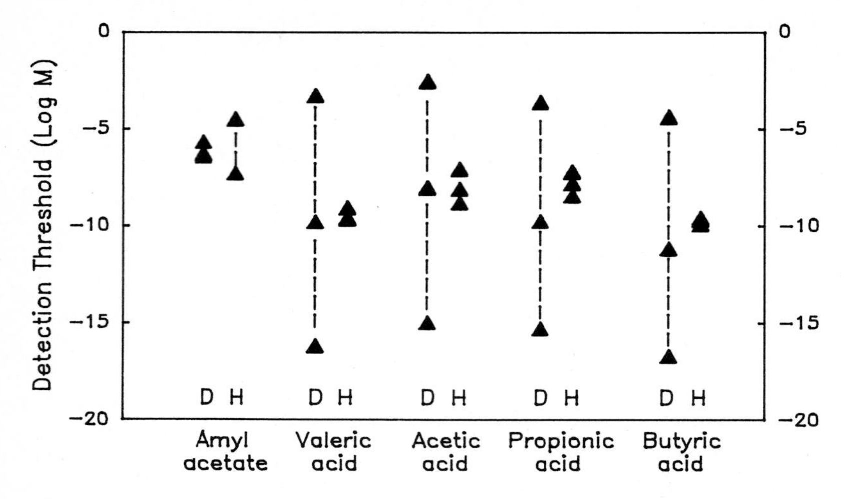

Fig 4. Comparison of odor detection threshold concentrations for several odorants in dogs (D) and humans (H). Each symbol represents a published threshold value cited in Krestel et al.(1984), Walker et al. (1990a), Moulton et al.(1960), Ashton et al.(1957), Neuhaus (1979), Punter (1983), or Naus (1982).

sensitive than humans to amyl acetate, and Marshall and Moulton (1981) found that dogs (German shepherds) are 1000-10,000 times more sensitive than humans to alpha-ionone. Similarly, an intra-laboratory comparison of rats and humans (Laing, 1975) provides convincing evidence of greater sensitivity of the rat. The superiority of the rat for the five compounds tested was between 8 and 50-fold: n-propanol(x8); benzaldehyde(x10); cyclohexanone(x10); isobutyl-n-butyrate(x30); n-heptanol(x50).

Another method of gaining insight into human-animal differences is to compare thresholds for compounds that have been tested in a large variety of subjects. Figure 5 summarizes the results for amyl acetate, an odorant that has been used in numerous studies. The data suggest that the sensitivity of humans for this odorant is poorer than that for all of the animals that have been tested, although the evidence for pigeon superiority over humans is not as strong as that for mammals. Conclusions about the relative sensitivity of different species of animals cannot be drawn due to the large variation in the rat data and the limited sampling of other mammals.

It is possible that the variation in threshold values within species is due primarily to methodological differences among laboratories, as discussed in the Introduction, rather than to variation in detection ability. Since the detection threshold is probably the simplest means for characterizing a sensory system, less variation among individuals of the same species than between species is expected.

With improvements in psychophysical test methods and a greater degree of overlap among published reports in terms of odorants and types of subjects, it should be possible to use threshold results to address the following important issues:

1. In relation to the stimulus, ascertain the physico-chemical determinants of odor sensitivity?
2. Determine whether there are differences between species in odor sensitivity and, if so, are these differences general or specific to certain kinds of odorants?
3. Determine the effects of drugs, hormones and long term exposure to inhaled toxins or odorants on odor sensitivity?
4. Determine how nasal structure and respiratory behavior affect odor detection?

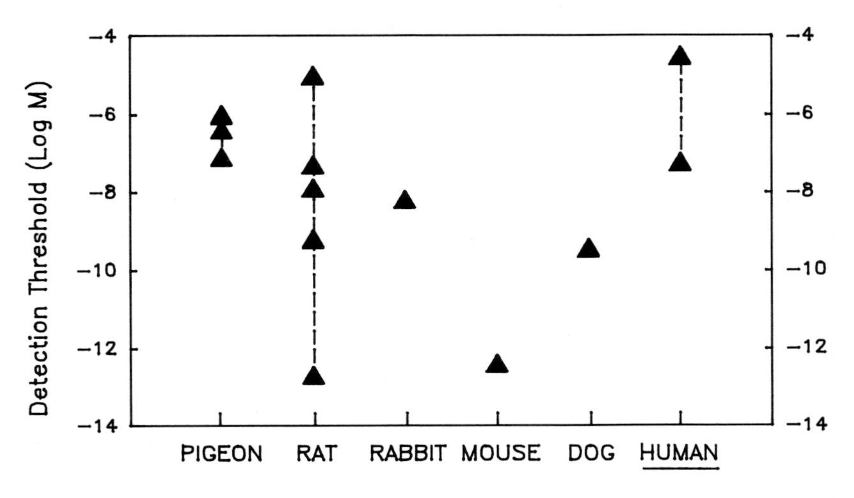

Fig 5. Comparison of human odor detection thresholds for amyl acetate with those for different animal groups. Data for this Figure were gathered from a review paper (Passe and Walker 1985) and from Walker et al. (1990a) and Krestel et al. (1984).

III. SUPRATHRESHOLD INTENSITY

Suprathreshold odor intensity in humans is studied either by quantifying the growth in the magnitude of odor sensation with increases in odor concentration (suprathreshold scaling), or by measuring the ability to detect small changes in odor concentration (differential sensitivity). Except for the responses to suprathreshold stimuli recorded during detection threshold training and testing, and what might be estimated from performance during intensity difference testing, there are no data on the relationship between odor intensity and odorant concentration in animals. Accordingly, the present discussion is limited to measures of differential sensitivity.

In humans, differential sensitivity is measured by determining the ability of a subject to discriminate between two concentrations of a given odorant. Whereas subjects in detection threshold tests attempt to discriminate between a reference stimulus of clean air and various concentrations of an odorant, those in intensity difference tests attempt to discriminate between a suprathreshold reference stimulus and comparison stimuli which differ from this reference by small increments in concentration. Until the 1960's, studies of differential sensitivity in humans employed either a Zwaardemaker type of olfactometer (Gamble 1898, Zigler and Holway 1935) or used the blast injection technique described by Elsberg et al. (1935).

In the sixties, Stone and colleagues (Ough and Stone 1961, Stone et al. 1962) used an air dilution olfactometer to determine intensity difference thresholds in humans. In three groups of subjects, Stone (1963) determined difference thresholds to 2-octanone, n-heptyl alcohol and ethyl-n-valerate. With each odorant, a set of three reference concentrations was used. However, since the highest reference used was less than 7 times higher than the detection threshold of the group, it is unlikely that trigeminal stimulation occurred. When the difference threshold was calculated using the formula $\Delta I/I$, the Weber fractions for 2-octanone, n-heptyl alcohol and ethyl-n-valerate were 0.24, 0.27 and 0.43 respectively. However, when a correction factor was incorporated into the formula $\Delta I/I+IT$, where IT is the detection threshold for the group of subjects, difference thresholds were lowered to 0.20, 0.21 and 0.31 and there was less variability among the different reference concentrations. Stone's use of the threshold concentration as a correction factor was based, in part, on work in audition (Miller and Garner 1944), and took into account the expectation that it is increasingly difficult for subjects to judge intensity as the reference concentration is lowered.

Difference thresholds of about 0.2 and 0.5 were seen with heptanone and acetic acid, respectively, when the same approach was used in a later study (Stone 1964). Since the highest reference concentration of acetic acid (12.6 ppm) was just below the recently reported threshold of anosmic humans for this compound (Walker et al. 1990b), the rather high Weber fraction for acetic acid is not likely to have been influenced by trigeminal stimulation.

The importance of precise stimulus control in intensity difference testing was emphasized in work by Cain (1977), where subjects inhaled through one nostril the odors from vessels containing cotton balls soaked with different concentrations of n-butyl alcohol, n-amyl alcohol, or ethyl-n-butyrate. Since the vapor phase concentrations of the sample stimuli were measured by gas chromatography, the variation in odor intensity discrimination could be partitioned into two components, variation in delivered concentration of odor, and that of the olfactory system itself. "Noise" in the stimulus accounted for 87, 88, and 36% of the variance in performance for n-butyl alcohol, ethyl n-butyrate, and n-amyl alcohol, respectively. When this variance was "factored out", the mean value of the Weber fractions for these odorants was roughly halved. Measurement of the Weber fraction for n-butyl alcohol using an air dilution olfactometer yielded a value of 0.04 which is close to the value of 0.05 predicted if variation in the stimulus concentration is removed. These data demonstrate that the size of Weber fractions is strongly dependent on the apparatus used to deliver odorants and illustrate the necessity of using air dilution olfactometry in studies of differential sensitivity.

As regards the differential sensitivity of animals, only two studies with pigeons and rats, have been reported. Shumake et al. (1969) used electric shocks to suppress key pecking in pigeons in the presence of a reference stimulus of amyl acetate at 7% of vapor saturation. Intensity discrimination was lost when the test stimulus, with which no shocks were paired, was raised to 2 or 3% of vapor saturation. Using the convention whereby the lower concentration serves as the denominator, Weber fractions were obtained which ranged from 1.33 to 2.50. Although these data appear to show that the pigeon is extremely poor at discriminating odor intensities, several cautions are in order. First, a great deal of stimulus generalization from the reference stimulus to the lower comparison stimuli is expected with the conditioned suppression procedure because it relies on aversive stimulation. Secondly, since the pigeon's head was not held within the breathing chamber, it is likely that there was considerable variation in the concentration of odorant delivered to the nasal cavity. Third, it should be noted that the concentrations of amyl acetate employed were rather high. The 7% reference stimulus was 1.5 to 2.0 log units above the detection threshold reported for pigeons (Henton 1969) and was close to the trigeminal threshold for this odorant (Walker et al. 1979, Walker et al. 1986). Accordingly, it would be of interest to measure differential sensitivity in a lower concentration range using a technique in which the pigeon could obtain reinforcement in the presence of either odorant intensity. One such technique is that developed by Walker (1983) in which pigeons, restrained so that the head was held within the breathing chamber, were reinforced with food for pecking at either of two locations, depending on whether or not odor was presented.

Slotnick and Ptak (1977) measured the differential odor sensitivity of rats and humans to amyl acetate. From the reported data and calculations based on Raoult's Law, the two reference stimuli are estimated as $10^{-9.1}$ and $10^{-10.1}$ M. As shown in Figure 5, published thresholds for the rat for amyl acetate range from $10^{-12.7}$ M (Davis 1973) to $10^{-5.0}$ M (Moulton 1960). Rats were tested with both of these reference stimuli, but humans were tested with only the lower one. Rats were reinforced for bar presses in the presence of a reference stimulus but not in the presence of any of a number of weaker comparison stimuli. Humans were instructed to press a single button only when the reference stimulus was presented. No feedback was given for failure to report the presence of the reference stimulus, or for pressing the button when a weaker comparison stimulus was presented. Slotnick and Ptak found that, when two rats were tested with the higher reference stimulus, Weber fractions of 0.185 and 0.283 were obtained, whilst with the lower reference stimulus, values of 0.056 and 0.036 were recorded. When four humans were tested using only the lower reference concentration, Weber fractions ranged from 0.23 to 0.40. Direct comparison of the differential sensitivity of humans and rats to this odorant, using concentrations which spanned the perceptual range of each species would be valuable. This would determine whether there is an overall difference in the differential sensitivity of humans and rats and would reveal the dependence of differential sensitivity on the subjective intensity range in which testing is conducted. The finding of lower difference thresholds with the lower reference stimulus is somewhat surprising given Stone's (1964) finding that differential sensitivity declines (Weber fractions increase) as the reference stimulus concentration is lowered.

Given the limitations in the data described above there is perhaps even less quantitative evidence for comparing the ability of humans and animals to discriminate differences between two odor intensities than is the case with thresholds. In future studies several steps should be taken to improve this situation. First, the technique of using a range of concentrations of reference stimuli with each odorant (Stone 1963,1964) should be adopted. With both humans and animals, concentrations of reference and comparison stimuli should be above the normal detection threshold of each species, but below the trigeminal threshold. Second, it may be useful to consider the role of short-term memory in tests of differential sensitivity in both humans and animals. It is possible, for example, that one organism may appear to have a greater differential sensitivity than another simply because its short-term memory for odor intensity is better. Third, the minimum volume, flow rate and duration of odorant stimulation required for odor intensity judgments by different animals should be determined. This would allow a degree of standardization in animal studies comparable to that made possible by Laing's (1982, 1983, 1986) work on such requirements for humans. Finally, it may also be wise to determine the optimum intertrial interval for different psychophysical tasks when studying olfaction to avoid the problem of olfactory adaptation.

IV. ODOR QUALITY PERCEPTION

An approach to investigating the neural bases of odor quality perception in humans is to relate psychophysical performance in humans and/or animals to neural responses from structures in the olfactory pathways of animals. The likelihood of success with this approach is increased by the several findings discussed above, which suggest that the sense of smell of humans and animals operate by similar mechanisms. Additionally, such comparisons may indicate the ability of humans and animals to extract qualitative information from environmental chemicals. Large dogs, for example, have significantly more receptor cells than humans, which may provide them with a superior ability to detect and discriminate between odors.

The kinds of psychophysical tasks that have been used with humans and animals to study odor quality perception can be grouped into two general areas. The first area includes odor discrimination and odor recognition. In the former, the subject is simply required to respond differently in the presence of each of two odorants. In odor recognition tests, the subject must demonstrate recognition of each of several odorants by making a unique response. The second area involves cross-adaptation and/or mixture perception, in which a subject's ability to perceive an odorant after or during exposure to at least one other odorant is measured. In the first area, most animal studies have required the discrimination of two odors, while many human studies have involved verbal identification of sets of 10-20 odorants. Since the ability to discriminate odorants is necessary but not sufficient to ensure performance in the more cognitive task of recognition, comparative statements about the discriminative abilities of humans and animals must be made with caution. Comparisons of the performances of different species during cross-adaptation and perception of odor mixtures can be made more easily.

A. Quality Discrimination and Odor Recognition

Slotnick and colleagues (Slotnick and Katz 1974, Nigrosh et al. 1975, Slotnick 1985) have conducted a number of experiments in which rats were required to discriminate between two odorants. The stimuli used were ethyl, propyl and iso-amyl acetate. With each discrimination, responses were reinforced with water in the presence of one odorant (S+), but not the other (S-). In each case, the discrimination was achieved rapidly and the rats maintained a high degree of correct responding. When the assignment of odorants as S+ and S- was reversed, the rats were able to adjust rapidly to the new conditions and to again achieve a high number of correct responses. Assuming that the performance of these rats was not based on differences in subjective intensity within stimulus pairs, the ability of the rat to discriminate between odors is quite impressive. Unfortunately, there are apparently no data on the ability of humans to discriminate or identify these three odorants. Performance in such reversal tasks is, in part, a measure of the animal's cognitive abilities for the olfactory modality and not simply a gauge of the ability to perceive different odor qualities.

A study by Laing et al.(1974), suggests that some odorant pairs may be quite difficult for the rat to discriminate. In that study benzaldehyde, benzonitrile or isobutylbutyrate were delivered directly into the nasal cavity of the rat via a chronically implanted nasal cannula. Concentrations of the odorants were adjusted until they had similar intensities, as judged by a human panel. Rats demonstrated their ability to discriminate between the odorants by registering different response rates on two levers. Rats were much poorer at discriminating between the odors of benzaldehyde and benzonitrile, both of which are described by humans as having an "almond" odor, than between either of these odors and the fruity odor of isobutylbutyrate. These results are consistent with the view that odor similarity data from humans and animals may be comparable.

Braun and Marcus (1969) also reported data suggesting that humans and rats perceive odor quality similarly. They trained rats to respond differentially in the presence of pairs of odorants from the camphoraceous, ethereal and floral classes, as defined by Amoore's (1952,1962) stereochemical theory. In animals rewarded for nose presses in the presence of a single floral odorant, but not in the presence of a single ethereal odorant, responses to new stimuli from each of these two classes were recorded. Similar tests were conducted in animals trained to discriminate between cineole and propanol or between camphor and benzyl acetate. In almost all instances the degree of similarity in response rates between the new stimulus and the two training odorants was predicted by stereochemical class. For example, in rats rewarded for responses to the floral odor of geraniol, but not to that of acetone (ethereal), higher response rates were recorded with butyl acetate (floral) than with methylene chloride (ethereal). These data are in general agreement with the human psychophysical data reported by Amoore and Venstrom (1967), and like those of Laing et al. (1974), suggest that mechanisms of odor quality perception may be similar in humans and rats. Clearly one means of improving comparative information on the odor discrimination capabilities of humans and animals would be to conduct two-odor discrimination experiments with humans.

A second approach to studying odor recognition by animals was reported recently by Youngentob et al. (1990). These investigators constructed a device consisting of a set of five response bays and a single conditioning tunnel. Rats were trained to associate each of five test odorants with one of the response bays. Once trained, they were required to indicate their recognition of which of the five odorants was being presented by entering the appropriate response bay to receive a water reward. This approach is directly applicable to work with odorant confusion matrices in humans (Wright 1987, see Chapter 5 this book) and it provides important information about the degree of similarity between different pairs of odorants.

The task of recognising an odor by humans and animals places greater demands on memory than discriminating between two odorants. Any human-animal comparison should consider the odor memory capacities of humans and the particular species chosen. Recent investigations have shown that rats, for example, have a considerable capacity to remember the perceptual qualities of long "lists" of odorants (Staubli et al. 1987, Kufera et al.1988). In these studies, rats successfully remembered the response-reinforcement contingencies for at least 30 or so two-odor discrimination tasks. Unfortunately, similar experiments of human odor memory have not been conducted.

B. Adaptation and Odor Mixtures

Another approach to understanding odor quality perception by humans and animals involves investigation of the interactive effects of odorants. This assumes that the degree to which one odorant can affect the perception of another is a measure of the degree of qualitative similarity between the two stimuli. Two variants of this approach have been used: 1) In cross-adaptation studies the effect of exposure to one odorant on the intensity of a subsequently presented different odorant is measured. 2) In odor mixture perception, the ability to perceive one or more components of a mixture, as a function of the identity, number or concentration of other components is determined.

Studies of cross-adaptation in humans were conducted in the 1960's and 1970's to define the degree of similarities among pairs of odorants. Much of this work involved testing cross-adaptation between alcohols within a homologous series. For example, Cain (1970) found that the intensity of propanol is lowered more by adaptation to pentanol than is the intensity of pentanol by pre-exposure to propanol. Engen (1963) found asymmetrical cross-adaptation effects among a homologous series of alcohols, and similar results were reported by Corbit and Engen (1971) and Rovee (1972).

A reasonably direct comparison between these kinds of results with human subjects and the rat is possible since Laing and Mackay-Sim (1975) tested the effect of pre-exposure to heptanol on the perception of propanol. Unlike many of the experiments on odor quality in animals, this study included measurements of the detection thresholds to the adapting and test stimuli. Interestingly, concentrations of heptanol that were well above threshold either decreased sensitivity to propanol or had no effect. Subthreshold adapting concentrations of heptanol actually increased detectability of propanol. Human psychophysical studies with alcohols are consistent with the pattern of results reported by Laing and Mackay-Sim (1975) in the rat. With both groups, high concentrations of the adapting stimulus inhibited detection of the test stimulus (Engen 1963), whilst detection of the test stimulus was enhanced by sub-threshold adapting stimuli (Corbit and Engen 1971).

An insight into the neural mechanisms underlying cross-adaptation effects was obtained from the study by Bennett (1968) who showed that the degree of cross-adaptation could be greatly reduced by damage to the anterior limb of the anterior commissure. He also provided evidence that interbulbar connections allow input from the two olfactory mucosae to summate, resulting in increased sensitivity. This finding of a strong CNS component of cross-adaptation appears to be a drawback to attempts to use this task to generate odor classification schemes which reveal the degree to which odorants share receptor sites in the olfactory epithelium.

Several recent studies have investigated the lowering of the intensity of one component of a mixture as a function of the intensity or identity of the other components. This effect is referred to as mixture suppression and work by Laing and colleagues suggests that similar processes occur in humans and rats. In the first study to examine mixture suppression with animals, Laing et al. (1989) trained rats to respond to propionic acid and then examined the effect of carvone, limonene, and acetic acid on the detectability of propionic acid. The suppressive effect of acetic acid, which shares the vinegar-like smell of propionic acid, was much greater than that of the other two odorants. These data are in agreement with those reported by Laing and Willcox (1987) whose data with humans showed that both limonene and pinene suppress perception of propionic acid. Interestingly, propionic acid had little effect on the orange smell of (+)-limonene or the camphoraceous smell of pinene in that study. It would be of interest to determine if non-reciprocal mixture suppression effects are seen in animals as well. If it can be shown that the rules that govern odor mixture effects in humans and animals are similar (Bell et al. 1987), research into the physiological mechanisms underlying these effects could be conducted profitably in animals.

V. SUMMARY AND CONCLUSIONS

This comparative review of the olfactory abilities of humans and animals supports the idea that mammals such as the dog and rat are more sensitive to odorants than humans and appear to be better at discriminating between different concentrations of an odorant. Additional conclusions must await further research and will depend, in part, on improvements and new developments in psychophysical techniques and increased coordination between investigators studying human and animal responses to odors.

Methods that yield much greater inter-laboratory agreement between olfactory detection thresholds will facilitate direct comparisons of sensitivity to different odorants and will allow the testing of hypotheses relating detectability to physico-chemical properties (Laffort 1968). With the development of methods for the scaling of suprathreshold intensities in animals that are comparable to those used with humans, it will be possible to better investigate the neural correlates of odor intensity in animals, compare the growth of odor intensity with concentration increases in humans and animals, and control for subjective intensity as a cue in odor discrimination experiments in animals.

Measurement of the differential sensitivity of humans and animals, using precision olfactometry and a common set of odorants, may answer the following important questions:

1. Do humans differ from animals in their ability to judge intensity differences?
2. For both groups, how is this ability altered by odorant identity, subjective intensity of the reference concentration and sampling behavior?
3. To what degree can differential sensitivity data, from humans or animals, be used to predict the outcome of suprathreshold scaling?

Most data on odor quality perception from animal studies have been collected from the rat, and these results support the tentative conclusion that similar neural processes underlie odor quality perception in humans and animals. Increased emphasis on systematic comparisons of odor quality perception in humans and animals could provide a firmer basis for the use of physiological data from animals to improve our understanding of the neural bases of odor quality perception in humans. In addition to the use of other species and the inclusion of common stimuli for different subject groups, two approaches could be taken to further this effort. First, experiments with humans could be conducted to quantify the inherent discriminability of different odorants in ways that take into account the subjective intensities of the stimuli and place minimal demands on memory. Second, systematic intra-laboratory investigations involving odor identification, cross-adaptation and mixture perception, using the same sets of human and animal subjects and the same odorants, should be pursued. These experiments could greatly facilitate comparisons of human and animal processing of odor quality information, reveal the relationships between discrimination, cross-adaptation and mixture data, and determine the relative importance of central and peripheral processes in different measures of odor quality perception.

Collectively, these future developments should provide much of the information necessary to understand the encoding of odor intensity and quality by the olfactory system.

REFERENCES

Amoore JE (1952) The sterochemical specificities of human olfactory receptors. Perf Ess Oil Rec 43: 321-323

Amoore JE (1962) The sterochemical theory of olfaction. 2. Elucidation of the stereochemical properties of the olfactory receptor sites. Proc Sci Sec Toilet Goods Assoc Suppl 37: 13-23

Amoore JE, Venstrom D (1967) Correlations between stereochemical assessments and organoleptic analysis of odorous compounds. In: Hayashi T (ed) Olfaction and taste, vol II. Pergamon Press, Oxford, p 3
Ashton EH, Eayrs JT, Moulton DG (1957) Olfactory acuity in the dog. Nature 179: 1069-1070
Bell GA, Laing DG, Panhuber H (1987) Odour mixture suppression: evidence for a peripheral mechanism in human and rat. Brain Res 426: 8-18
Bennett MH (1968) The role of the anterior limb of the anterior commissure in olfaction. Physiol Behav 3: 507-515
Braun JJ, Marcus J (1969) Stimulus generalization among odorants by rats. Physiol Behav 4: 245-248
Cain WS (1970) Odor intensity after self-adaptation and cross-adaptation. Percept Psychophys 7: 271-275
Cain WS (1977) Differential sensitivity for smell: "Noise" at the nose. Science 195: 796-798
Corbit TE, Engen T (1971) Facilitation of olfactory detection. Percept Psychophys 10: 433-436
Davis RG, (1973) Olfactory psychophysical parameters in man, rat, dog, and pigeon. J Comp Physiol Psychol 85: 221-232
Elsberg CA, Brewer ED, Levy I (1935) The sense of smell. V. The relative importance of volume and pressure of the impulse for the sensation of smell and the nature of the olfactory stimulation process. Bull Neurol Inst NY 4: 264-269
Engen T (1963) Cross-adaptation to the aliphatic alcohols. Am J Psychol 76:96-102
Engen T (1971) Olfactory psychophysics. In: Beidler LM (ed) Handbook of sensory physiology. Springer-Verlag, New York, p 216
Gamble EAM (1898) The application of Weber's law to smell. Am J Psychol 10: 82-142
Henton WW (1969) Conditioned supression to odorous stimuli in pigeons. J Exp Anal Behav 12: 175-185
Hyman AM (1977) Factors influencing the psychophysical function for odor intensity. Sensory Processes 1: 273-291
Krestel D, Passe D, Smith JC, Jonsson L (1984) Behavioral determination of olfactory thresholds to amyl acetate in dogs. Neurosci Biobehav Rev 8: 169-174
Kufera AM, Slotnick BM, Risser JM (1988) Rats learn to label lots of odors: A remarkable demonstration of learning-set and odor memory in the rat. Abstr. Tenth Ann Meet Assoc Chemoreception Sci, Sarosota, Florida.
Laffort P (1968) Some new data on the physico-chemical determinants of the relative effectiveness of odorants. In: Tanyolac NN (ed) Theories of odors and odor measurement. Spartan Books, New York, p 247
Laing DG (1975) A comparative study of the olfactory sensitivity of humans and rats. Chem Sens Flav 1: 257-269
Laing DG (1982) Characterization of human behavior during odour perception. Perception 11: 221-230
Laing DG (1983) Natural sniffing gives optimum odor perception for humans. Perception 12: 99-117
Laing DG (1986) Identification of single dissimilar odors is achieved with a single sniff. Physiol Behav 37: 163-170
Laing DG, Mackay-Sim A (1975) Olfactory adaptation in the rat. Denton DA, Coghlan JP (eds) Olfaction and taste, vol V. Academic Press, New York, p 291

Laing DG, Murray KE, King MG, Cairncross KD (1974) A study of olfactory discrimination in the rat with the aid of a new odor delivery technique. Chem Sens Flav 1: 197-212

Laing DG, Panhuber H, Slotnick BM (1989) Odor masking in the rat. Physiol Behav 45: 689-694

Laing DG, Willcox ME (1987) An investigation of the mechanisms of odor suppression using physical and dichorhinic mixtures. Behav Brain Res 26: 79-87

Marshall DA, Moulton DG (1981) Olfactory sensitivity to alpha-ionone in humans and dogs. Chem Sens 6: 53-61

Miller GA, Garner WR (1944) Effect of random presentation on the psychophysical function: Implications for a quantal theory of discrimination. Am J Psychol 57: 451-467

Moulton DG (1960) Studies in olfactory acuity. III. Relative detectability of n-aliphatic acetates by the rat. Q J Exp Psychol 12: 203-213

Moulton DG, Ashton EH, Eayrs JT (1960) Studies in olfactory acuity. 4. Relative detectability of n-aliphatic acids by the dog. Anim Behav 8: 117-128

Naus A (1982) Olfactory threshold of industrial substances. Prac Lek 34: 217-218

Neuhaus W (1979) Uber die riechscharfe des hundes fur fettsauren. Z Vergl Physiol 35: 527-552

Nigrosh BJ, Slotnick BM, Nevin JA (1975) Olfactory discrimination, reversal learning, and stimulus control in rats. J Comp Physiol Psychol 89:285-294

Ough CS, Stone H (1961) An olfactometer for rapid and critical odor measurement. J Food Sci 26: 574-578

Passe DH, Walker JC (1985) Odor psychophysics in vertebrates. Neurosci Biobehav Rev 9: 431-467

Punter PH (1983) Measurement of human olfactory thresholds for several groups of structurally related compounds. Chem Sens 7: 215-235

Rovee CK (1972) Olfactory cross-adaptation and facilitation in human neonates. J Exp Child Psychol 13: 368-381

Shumake SA, Smith JC, Tucker D (1969) Odor intensity-difference thresholds in the pigeon. J Comp Physiol Psychol 67: 64-69

Slotnick BM (1985) Olfactory discrimination in rats with anterior amygdala lesions. Behav Neurosci 99: 956-963

Slotnick BM (1990) Olfactory perception. In: Berkley MA, Stebbins WC (eds) Comparative perception, vol I. Basic mechanisms. John Wiley and Sons, New York, p 407

Slotnick BM, Katz H (1974) Olfactory learning-set formation in rats. Science 185: 796-798

Slotnick BM, Ptak JE (1977) Olfactory intensity-difference thresholds in rats and humans. Physiol Behav 19: 795-802

Staubli U, Fraser D, Faraday R, Lynch G (1987) Olfaction and the "data" memory system in rats. Behav Neurosci 101: 757-765

Stone H (1963) Determination of odor difference limens for three compounds. J Exp Psychol 66: 466-473

Stone H (1964) Behavioral aspects of absolute and differential olfactory sensitivity. Ann NY Acad Sci 116: 527-534

Stone H (1966) Factors influencing behavioral responses to odor discrimination - A review. J Food Sci 31: 784-790

Stone H, Ough CS, Pangborn RM (1962) Determination of odor difference thresholds. J Food Sci 27: 197-202

Walker JC (1983) An operant procedure for testing olfactory capacities in restrained pigeons. Physiol Behav 30: 165-168

Walker JC, Reynolds JH, Warren DW, Sidman JD (1990a) Responses of normal and anosmic subjects to odorants. In: Green BG, Mason JR, Kare MR, Chemical Senses vol 2. Irritation. Marcel Dekker, New York, p 95

Walker JC, Tucker D, Smith JC (1979) Odor sensitivity mediated by the trigeminal nerve in the pigeon. Chem Sens Flav 4: 107-116

Walker JC, Walker DB, Tambiah CR, Gilmore KS (1986) Olfactory and non-olfactory odor detection in pigeons: elucidation by a cardiac acceleration paradigm. Physiol Behav 38: 575-580

Walker JC, Warren DW, Jennings RA, Reynolds JH (1990b) Psychophysical and respiratory responses of anosmic humans to odorants. In: Døving KB (ed) Olfaction and taste, vol X. GCS A/S, Oslo

Wright HN (1987) Characterization of olfactory dysfunction. Arch Otolaryngol 113: 163-168

Youngentob SL, Markert LM, Mozell MM, Hornung DE (1990) A method for establishing a five odorant identification confusion matrix task in rats. Physiol Behav 47: 1-7

Zigler MJ, Holway AH (1935) Differential sensitivity as determined by amount of olfactory substance. J Gen Psychol 12: 372-382

PART 5

CLINICAL AND HEALTH ASPECTS OF OLFACTION

14

OLFACTORY DYSFUNCTION

DAVID V SMITH

ALLEN M SEIDEN

I. INTRODUCTION

It has been estimated that approximately two million Americans suffer from a disorder of taste or smell[1], but this is probably an underestimation. Although there are over 200,000 visits to physicians each year for chemosensory disorders, many of these problems tend to be dismissed by patients as well as by physicians. It is quite apparent, however, that individuals with an olfactory loss may suffer a significant impairment to their quality of life, both aesthetically and emotionally. Patients presenting with olfactory complaints are difficult to diagnose and treat; often they have been to a number of health professionals in an unsuccessful attempt to get help for their problem. Much of the difficulty in dealing with these patients lies in a general lack of knowledge about the sense of smell and its disorders. Although an olfactory dysfunction can occur as a secondary process in a number of disease states, often a reduced ability to smell or a distorted olfactory experience is a patient's primary complaint. Several recent reviews have addressed these clinical problems (Doty 1979, Doty and Kimmelman 1986, Doty and Snow 1987, Feldman et al. 1986, Kimmelman 1986, Leopold 1986, Schiffman 1983a, 1983b) and others have dealt with clinical olfactory assessment (Cain et al. 1983, 1988, Smith 1988).

---

[1] Report of the Panel on Communicative Disorders to the National Advisory Neurological and Communicative Disorders and Stroke Council. Washington, D.C.: Public Health Service, 1979: 319. (NIH publication no. 79-1914).

II. EVALUATION OF SMELL COMPLAINTS

The evaluation of patients with olfactory disturbance must involve a careful medical history, with particular attention to antecedent events that might be related to the onset of olfactory loss, such as an upper respiratory infection or a traumatic injury (Doty 1979, Goodspeed et al. 1987, Leopold 1986). In addition, the nature of the smell symptoms can be a clue to the most likely cause of the problem (Gent et al. 1987). For example, if a patient with an olfactory loss reports that smell sensitivity fluctuates, it is probable that nasal or sinus disease is involved. A direct question to a patient about smell loss has high predictive value. That is, nearly all patients with a diminished sense of smell will verbally report a decrease in olfactory function (Gent et al. 1987, Smith et al. 1987). On the other hand, most patients with an olfactory loss also report that they have lost their sense of taste. Patients tend to exaggerate taste problems because of their confusion between taste and flavor. The perception of flavor involves olfactory, tactile, and thermal sensations in addition to taste, so that a loss of any of these sensory modalities is usually reported as a taste loss, yielding little predictive value in such a question. Direct questions to the patient about the four qualities of taste (salty, sour, sweet and bitter) are much more revealing of an actual taste dysfunction (Gent et al. 1987).

Dysfunctions of the sense of smell can take the form of complete or partial loss of sensitivity, increased sensitivity, or a distortion of olfactory perception. Although numerous terms have been used to classify these disorders, there is beginning to be some agreement about their usage (Doty 1979, Kimmelman 1986). The terms used for olfactory dysfunction include: general (or total) anosmia, or more simply, anosmia, which is an inability to detect any qualitative odor sensation; partial anosmia, an ability to detect some, but not all, qualitative odor sensations; hyposmia, a decreased sensitivity to some or all odorants; hyperosmia, an increased sensitivity to some or all odorants; dysosmia a distortion in the perception of a smell, such as the perception of an unpleasant odor when there is no odor present (phantosmia), or the perception of an atypical odor in response to a particular stimulus (parosmia); agnosia, an inability to classify, contrast, or identify an odor sensation verbally, even though the ability to distinguish between odorants or to recognize them may be normal. Although several other terms have been used to describe olfactory dysfunction e.g., Henkin (1967a), Estrum and Renner (1987), these terms have found the most common usage.

III. PSYCHOPHYSICAL MEASUREMENT IN A CLINICAL SETTING

Much of the difficulty in dealing clinically with smell complaints is in obtaining good quantitative measurements of the patient's sensory function. This difficulty is multi-faceted; the sensory assessment must involve well-designed and standardized sensory tests, careful stimulus control, and must be done in a clinical setting. No aspect of this problem is simple and it has attracted a great deal of attention in recent years (Cain and Gent 1986, Cain et al. 1983, 1988, Doty 1979, Doty et al. 1984a, Gent et al. 1986, Mozell et al. 1986a, Wright 1987). As a result, several tests have been developed and standardized on normal populations for use in the clinical testing of smell (Cain et al. 1983, 1988, Doty et al. 1984a, Smith 1988).

The method of presentation of olfactory stimuli for the assessment of smell dysfunction poses no small problem for the clinician. For example, controlling the concentration of an odorant is obviously an important consideration, but it is difficult to achieve. Several approaches have been taken to this problem, including the use of sniff bottles, blast injection into the nostrils, and air dilution olfactometers (see Doty 1979, for a thorough discussion of stimulus delivery techniques). One of the major considerations for clinical testing is that stimulus delivery needs to be relatively easy and somewhat portable. Recent tests that have been developed for clinical olfactory testing employ the use of sniff bottles and microencapsulated ("scratch and sniff") odorants. The major goal of any kind of sensory testing for smell is to assess the degree of dysfunction. Thus, many different approaches could be, and have been, employed. However, a proper assessment of chemosensory dysfunction requires the adoption of well-designed and psychophysically sound testing procedures.

IV. FREQUENT CAUSES OF OLFACTORY DYSFUNCTION

A. General

Olfactory dysfunction has been associated with a number of systemic diseases and metabolic disorders (Doty 1979, Doty and Kimmelman 1986, Doty and Snow 1987, Estrum and Renner 1987, Feldman et al. 1986, Henkin 1981, Kimmelman 1986, Leopold 1986, Schiffman 1983a, 1983b). However, most patients presenting with a primary smell complaint fall into only a few diagnostic categories. Of 750 patients presenting to the University of Pennsylvania Smell and Taste Center, most (78.1%) complained of a reduced ability to smell and more than half (66.4%) of a reduced ability to taste (Deems et al. in press). Dysosmias, either parosmias or phantosmias, were reported by 32.1% of these patients. Measurement of chemosensory capacity in these patients revealed that 70.9% had olfactory deficits, but less than 3% had a measurable taste deficit. Thus, patients quite accurately report a smell loss, but are not so accurate in their reports of taste dysfunction (Gent et al. 1987). This discrepancy reflects the common confusion between taste and odor. Similar results have been reported on a smaller number of patients at the Connecticut Chemosensory Clinical Research Center (Goodspeed et al. 1987) and at the University of Cincinnati Taste and Smell Center (Smith et al. 1987).

Patients with a primary olfactory complaint most likely fall into one of four etiologic categories: 1) Nasal and/or sinus disease i.e., nasal polyposis, chronic sinusitis, allergic rhinitis, etc.; 2) prior upper respiratory infection i.e., a history of a viral-like upper respiratory illness just prior to the onset of the olfactory loss; 3) idiopathic, or 4) head trauma. These four categories account for 81% (Deems et al. in press), 83% (Goodspeed et al. 1987), 72% (Henkin 1981), and 85% (Smith et al. 1987) of olfactory deficits in patients presenting with a primary chemosensory complaint at four clinics. The terminology in the literature is somewhat different, with Henkin's (1981) post-influenza hyposmia and hypogeusia (PIHH) and allergic rhinitis categories corresponding to what is referred to here as prior upper respiratory infection and nasal and/or sinus disease, respectively. In the three largest studies, the most frequent cause of olfactory dysfunction was either nasal and/or sinus disease (Goodspeed et al. 1987), or prior upper respiratory infection (Deems et al. in press, Henkin 1981). This difference probably reflects the nature of the referral patterns to these Clinics. Since the Connecticut group i.e., Goodspeed et al. (1987), sees mostly self-referred patients, those with nasal disease are much more likely to appear than if they were referred by other physicians. Nevertheless, the great majority of patients with olfactory complaints falls into one of these four etiologic categories. Smaller numbers of patients have olfactory impairment due to exposure to toxic chemicals, craniofacial surgery, seizure disorders, cerebral vascular accidents, endocrine disorders, a congenital anomaly, or other causes (Feldman et al. 1986, Goodspeed et al. 1987, Henkin 1981, Smith et al. 1987). Relatively few patients present with a primary chemosensory complaint resulting from the many widespread systemic or metabolic diseases e.g., see Schiffman (1983a), (1983b), claimed to be responsible for taste or smell dysfunction (Goodspeed et al. 1987).

B. Nasal and/or Sinus Disease

Olfactory disorders can be characterized as 1) conductive disorders, arising from interference with the access of odorants to the olfactory receptors, or 2) sensorineural disorders, resulting from damage to the receptor cells or olfactory pathways (Doty and Snow 1987). Unfortunately, olfactory testing alone is not capable at present of distinguishing between these possibilities. Of the four major etiologic categories, only nasal and/or sinus disease typically involves a conductive loss, resulting from the restriction of upper airway patency. This obstruction may arise from intranasal polyposis, chronic sinusitis, or allergic rhinitis. The loss of smell resulting from nasal and/or sinus disease is usually quite severe, with most patients characterized as anosmic (about 75%) rather than hyposmic (Goodspeed et al. 1986a, 1987, Smith et al. 1987). Even a moderate degree of ostiomeatal disease without accompanying polyps can result in significant olfactory impairment (Seiden and Smith 1988). In one study, about a third of patients with IgE-mediated nasal allergy were shown to have a measurable olfactory impairment (Seiden et al. 1989), with the deficit being worse in those patients with nasal polyps. That a conductive loss is often involved in this etiology is demonstrated by the fact that patients with nasal and/or sinus disease are much more likely to verbally report fluctuations in smell sensitivity than those with loss due to a prior upper respiratory illness (Gent et al. 1987). That is, these patients sometimes have a temporary return in their sense of smell, often associated with exercise or with medications. Anosmic patients with allergic rhinitis and/or nasal polyposis have been shown to recover olfactory function following administration of systemic corticosteroids (Fein et al. 1966, Goodspeed et al.

1986b, Hotchkiss 1956), or after careful intranasal administration of topical corticosteroids (Scott et al. 1988, 1989). Although chronic administration of systemic steroids is not a recommended treatment for anosmia, a short course of steroid therapy may help in diagnosing nasal disease as the cause of an olfactory loss (Leopold 1986). Surgical procedures that reduce nasal obstruction have also been shown to produce marked improvement in olfactory sensitivity (Ghorbanian et al. 1983, Ophir et al. 1986). Endoscopic ethmoidectomy to alleviate allergic disease in the ostiomeatal complex has been shown to restore olfactory sensitivity, with some patients maintaining this improvement for as long as two years, even without concomitant steroids or immunotherapy (Seiden and Smith 1988). Thus, olfactory losses resulting from an interference with odor access to the receptors i.e., conductive losses, are often amenable to treatment (Doty and Snow 1987).

C. Prior Upper Respiratory Infection

Olfactory losses with sensorineural involvement are presently more difficult to manage than those resulting from conductive disorders. Anosmia or hyposmia following a viral-like upper respiratory infection (Henkin et al. 1975) may be of the sensorineural type. The diagnosis of post-viral anosmia or hyposmia rests primarily on the coincidence of a viral illness or cold just prior to the onset of olfactory loss along with an absence of other etiologic factors (Goodspeed et al. 1987). An olfactory biopsy study suggested that the olfactory epithelium in post-viral anosmia may be damaged, with extensive scarring and replacement of the olfactory epithelium with respiratory epithelium (Douek et al. 1975). A more recent ultrastructural study of the human olfactory epithelium demonstrated a loss of olfactory receptor cells and an absence of olfactory cilia on the remaining receptor cells in patients with viral-induced smell loss (Jafek and Eller 1989). Influenza viruses are known to affect ciliary activity in the respiratory epithelium and to produce necrosis of pseudostratified ciliated columnar epithelium (Snow 1969); thus similar processes may be acting on the olfactory neuroepithelium (Doty 1979). In addition, the olfactory epithelium has been shown to be a route of viral invasion of the central nervous system (Monath et al. 1983). Thus, the long-lasting effects of a viral illness may be the result of damage to the olfactory epithelium or could even involve central olfactory pathways.

In comparison to patients with nasal and/or sinus disease, those with smell dysfunction following an upper respiratory infection are more likely to be women, to be older, to be hyposmic rather than anosmic, and to also show some hypogeusia (Goodspeed et al. 1987). Patients with these etiologies (nasal disease or post-viral) often report parosmias or phantosmias, almost always of an unpleasant nature (Donnelly et al. 1989, Smith et al. 1987). Although it has sometimes been suggested that post-viral anosmia may show spontaneous remission (Goodspeed et al. 1986a), there is no hard evidence to support such a claim. Similarly, there is no evidence to support the suggestion that these patients are unlikely to recover (Marshall and Attia 1987). Although anecdotal reports suggest that recovery may be possible, there have been no longitudinal studies nor any attempt to correlate the degree of olfactory loss with the likelihood of recovery. Thus, not only is the mechanism of viral-induced anosmia poorly understood, but the prognosis for this disorder is completely uncertain. Given the relative frequency of this etiology in the production of olfactory loss, further research on the mechanisms and time course of post-viral anosmia is sorely needed.

D. Head Trauma

About 5% of patients with head injury suffer a resulting olfactory loss (Leigh 1943, Sumner 1964, Zusho 1982, Costanzo and Becker 1986), and these patients make up about 10 - 20% of those presenting with a primary olfactory complaint (Goodspeed et al. 1987, Henkin 1981, Smith et al. 1987). The loss of smell sensitivity following head trauma is sometimes accompanied by dysosmia, involving unpleasant smell sensations (Costanzo and Becker 1986, Donnelly et al. 1989, Schechter and Henkin 1974, Leigh 1943). As in patients with nasal and/or sinus disease, olfactory impairment in head injury patients is more likely to be severe, with a much greater percentage (about 80%) showing anosmia than hyposmia (Goodspeed et al. 1986a, Smith et al. 1987). Post-traumatic anosmia is commonly thought to result from shearing of the olfactory nerve filaments or from contusions following a sudden blow to the head (Costanzo and Becker 1986). The olfactory epithelium in patients with post-traumatic smell loss exhibits marked degenerative changes suggestive of a severed olfactory nerve (Moran et al. 1985). Intracranial hemmorhage and cerebral ischemia have also been shown in laboratory animals to lead to degeneration of the olfactory epithelium, even though the olfactory nerve is intact (Nakashima et al. 1983, 1984). Thus, a number of mechanisms may be involved in olfactory dysfunction following head trauma.

Since the olfactory system is remarkable in its ability to regenerate (Graziadei 1973, Graziadei and Monti-Graziadei 1983), there is potential for recovery of olfactory function after head injury. Indeed, there is typically recovery of the neuroepithelium in laboratory animals following olfactory axotomy (Monti-Graziadei and Graziadei 1979, Monti-Graziadei et al. 1980, Simmons and Getchell 1981). Recordings from cells in the hamster olfactory bulb demonstrate functional recovery within nine months following olfactory nerve transection (Costanzo 1985). However, the prognosis for the recovery of smell in humans after traumatic injury is generally poor, with various estimates of 15% (Zusho 1982), 33% (Costanzo and Becker 1986) and 39% (Sumner 1964). The time course for recovery is also somewhat questionable, with about 80% of recovery occurring within six months in one study (Sumner 1964) but only after about 20 months in another (Costanzo and Becker 1986). These differences may reflect the way in which olfactory function was measured in these studies and the criteria for recovery. Nonetheless, recovery from post-traumatic anosmia in humans may proceed over a period as long as five years (Costanzo and Becker 1986, Sumner 1964). It has been suggested that early recovery may be due to mechanisms such as the disappearance of blood clots or edema, and that later recovery may reflect the regeneration of neural elements (Sumner 1975). Further research is needed to delineate these mechanisms. Some patients may experience parosmia or phantosmia during the recovery process (Goodspeed et al. 1986a, Leigh 1943), although the mechanisms for these phenomena are unknown. It has been suggested that the occurrence of parosmia may reflect a system that is minimally functional (Wright 1987). Consistent with this hypothesis is the fact that head injury patients reporting parosmia or phantosmia score significantly higher on the University of Pennsylvania Smell Identification Test (UPSIT) than those without these symptoms (Smith et al. 1987).

E. Idiopathic Smell Dysfunction

A large group of patients with a primary olfactory complaint do not fall easily into any diagnostic category and the cause of their smell loss remains unknown. These patients have been termed idiopathic and comprise 17.9% (Smith et al. 1987), 19% (Henkin 1981) and 25.9% (Goodspeed et al. 1987) of patients with olfactory dysfunction in different studies. Although a specific syndrome of idiopathic hypogeusia and hyposmia, with or without dysgeusia and dysosmia has been described (Henkin et al. 1971), this categorization appears to reflect patients for which no specific etiology can be established (Goodspeed et al. 1987). Many of these patients may have olfactory symptoms as a result of one of a vast number of systemic diseases or neurologic or endocrine disorders (Doty 1979, Feldman et al. 1986, Jafek 1982, Kimmelman 1986, Schiffman 1983a, 1983b) that are not being detected during the olfactory evaluation. As our ability to provide a differential diagnosis for the cause of olfactory dysfunction improves, the number of patients with idiopathic olfactory disturbances will probably decrease (Feldman et al. 1986, Gent et al. 1987, Liston 1984).

F. Other Etiologies

Most patients (about 80%) presenting with a primary olfactory complaint will fall into one of the four etiologic categories described above. The remaining patients are diagnosed with one of several other causes. For example, in the study reported by Goodspeed et al. (1987), an additional 4.5% of their patients were considered to have multiple causes, such as a patient with a temporally related head trauma who also experienced a viral-like upper respiratory infection at about the same time. For these patients, the establishment of one definitive cause for their olfactory loss was impossible. In several studies, small numbers of patients (5 - 14%) have been grouped into a miscellaneous category, which reflects such diverse etiologies as seizure disorders, cerebrovascular accidents, brain surgery, endocrine disorders, and depression (Goodspeed et al. 1987, Henkin 1981, Smith et al. 1987).

A few patients (1 - 2%) present with a primary olfactory complaint stemming from exposure to toxic chemicals (Deems et al. in press, Goodspeed et al. 1987, Henkin 1981, Smith et al. 1987). A number of environmental and industrial agents have been implicated in anosmia or hyposmia, including: benzene, benzol, butyl acetate, carbon disulfide, ethyl acetate, formaldehyde, hydrazine, menthol, methyl bromide, paint solvents, oil of peppermint, radiation, trichloroethylene, and a variety of industrial dusts such as cadmium and nickel (Amoore 1986, Doty 1979, Halpern 1985). Many of the studies implicating these various chemicals in olfactory dysfunction suffer from methodological problems, so it is hard to draw any general conclusions from the data (Doty 1979).

However, in one well-controlled study, 27.3% of workers in an alkaline battery plant who were exposed to high levels of cadmium were found to be hyposmic or anosmic, compared to 4.8% of control subjects from a neighboring non-chemical factory (Adams and Crabtree 1961). Recent investigation of chemical workers exposed to acrylate and methacrylate vapors shows a dose-response relationship between olfactory dysfunction and cumulative exposure (Schwartz et al. 1989). Experimental studies demonstrate that a 1% solution of zinc sulfate will destroy the receptor cells and supporting cells of the olfactory epithelium of mice (Cancalon 1982). Recovery of the receptor cells and olfactory cilia is seen four to six weeks later. Little is known, however, about recovery following industrial exposure.

Cigarette smoke has been shown to alter the ultrastructure of the olfactory epithelium in certain strains of mice, resulting in a number of changes, including reduced or absent olfactory cilia (Matulionis 1974). Although there has been relatively little work in this area, there is some suggestion that smoking results in decreased olfactory sensitivity (Hubert et al. 1980, Thumfart et al. 1980, Venstrom and Amoore 1968). However, other investigators have found no effect of smoking on olfactory function, but have shown a clear reduction in the perception of nasal pungency, mediated through the trigeminal system (Cain 1981, Cometto-Muniz and Cain 1982, Dunn et al. 1982). More recent data show that when smoking history is taken into account there is a significant long-term, but reversible, effect of smoking on the ability to smell (Frye et al. 1990). Earlier studies, which simply dichotomized subjects according to current smoking behavior, may have missed seeing the real effects of smoking on olfactory function.

Exposure of the olfactory mucosa to ionizing radiation produces marked deficits in olfactory detection thresholds, which recover slowly after termination of radiation therapy (Ophir et al. 1988). More studies of the effects of environmental pollutants on olfactory sensitivity are needed to clearly establish those that might be implicated in patients presenting with a history of toxic exposure.

G. Smell Dysfunction Secondary to Other Diseases

Disorders of smell have been associated with a number of conditions, including endocrine, neurological, psychiatric, and nutritional disorders (Doty 1979, Doty and

Kimmelman 1986, Doty and Snow 1987, Estrum and Renner 1987, Feldman et al. 1986, Henkin 1981, Kimmelman 1986, Leopold 1986, Schiffman 1983a, 1983b). Although the occurrence of these disorders in patients presenting with a primary smell complaint is rare, these conditions must always be considered in the evaluation of such a patient. Disorders that have been reported to affect the sense of smell are listed in Table 1, which categorizes them into endocrine disorders, neurological disorders, nutritional disorders, local diseases or mechanical obstruction, psychiatric disorders, tumors, and infectious processes. Listed with each of these conditions is a reference, which reflects either a research study demonstrating the association of that particular factor with an olfactory disturbance, or a review article covering a wide range of studies. The article by Amoore (1986) on the effects of chemical exposure on olfaction is an example of the latter. Some of these conditions were discussed above as common causes of olfactory impairment. Others represent relationships described in the literature but uncommon in patients presenting primarily with smell symptoms (Goodspeed et al. 1987). Because many of these studies have measured only thresholds, which alone do not provide a complete picture of a patient's sensory impairment (Bartoshuk 1978, Bartoshuk and Marks 1986), these relationships must be carefully evaluated. Some of the earlier studies involved only a patient's subjective report of smell symptoms and are even less reliable than those reporting the results of threshold testing. In some cases, such as in cystic fibrosis, for example, earlier reports (Henkin and Powell 1962) have been contradicted by more recent studies (Hertz et al. 1975, Weiffenbach and McCarthy 1984).

A number of endocrine disorders have been reported to influence smell function. For example, olfactory impairment is seen in several kinds of gonadal dysfunction. Patients with hypogonadotropic hypogonadism (Kallmann's syndrome) often exhibit a congenital anosmia (Kallmann et al. 1944), which has been shown to follow an autosomal dominant mode of inheritance with incomplete expressivity (Santen and Paulsen 1973). Although Kallmann's syndrome usually occurs in men, some women with hypogonadotropic hypogonadism have also been described as anosmic (Tagatz et al. 1970), as have those suffering from primary amenorrhea (Marshall and Henkin 1971). Patients with chromatin negative gonadal dysgenesis (Turner's syndrome) have been reported to have smell impairment (Henkin 1967b). Changes in smell sensitivity also occur during the human menstrual cycle, with peaks in sensitivity at mid-cycle, during the mid-luteal phase, and sometimes in the latter part of the menses (Doty et al. 1981). Since these changes are also seen in women on oral contraceptives, it is not clear whether these fluctuations are modulated by gonadal hormones or are solely under the control of central neural factors (Doty 1979, Schneider et al. 1958).

TABLE 1 : Disorders associated with olfactory dysfunction

Disorder	Reference
Endocrine	
Adrenal cortical insufficiency	Henkin (1975)
Cushing's syndrome	Kallmann et al. (1944)
Cystic fibrosis	Weiffenbach and McCarthy (1984)
Diabetes mellitus	Jorgensen and Buch (1961)
Kallmann's syndrome	Kallmann et al. (1944)
Primary amenorrhea	Marshall and Henkin (1971)
Pseudohypoparathyroidism	Henkin (1968)
Turner's syndrome	Henkin (1967b)
Neurologic	
Alzheimer's disease	Doty et al. (1987)
Epilepsy	Eskenazi et al. (1986)
Familial dysautonomia	Henkin and Kopin (1964)
Head trauma	Sumner (1964)
Huntington's chorea	Moberg et al. (1984)
Multiple sclerosis	Wender and Szemza (1971)
Parkinson's disease	Ansari and Johnson (1975)
Temporal lobectomy	Eskenazi et al. (1983)
Nutritional	
Chronic renal failure	Schiffman et al. (1978)
Cirrhosis of the liver	Burch et al. (1978)
Cyanocobalamin (B_{12}) deficiency	Rundles (1946)
Korsakoff's psychosis	Jones et al. (1975)
Local diseases/Mechanical obstruction	
Adenoid hypertrophy	Ghorbanian et al. (1983)
Allergic rhinitis	Fein et al. (1966)
Atrophic rhinitis (ozena)	Strandbygard (1954)
Deformity secondary to trauma	Doty and Kimmelman (1986)
Exposure to toxic chemicals	Amoore (1986)
Laryngectomy	Mozell et al. (1986b)
Leprosy	Barton (1974)
Malignancy of paranasal sinuses with extension	Doty and Kimmelman (1986)
Nasal polyposis	Fein et al. (1966)
Nasal surgery	Champion (1966)

TABLE 1 : Disorders associated with olfactory dysfunction (continued)

Disorder	Reference
Local diseases/Mechanical obstruction (continued)	
Sinusitis	Ryan and Ryan (1974)
Sjögren's syndrome	Henkin et al. (1972)
Tumors of nasopharynx with extension	Doty and Kimmelman (1986)
Vasomotor rhinitis	Griffith (1976)
Psychiatric	
Depression	Pryse-Phillips (1971)
Olfactory reference syndrome	Pryse-Phillips (1971)
Schizophrenia	Kerekovic (1972)
Intracranial tumors	
Aneurysms of the anterior communicating bifurcation	Jefferson (1961)
Frontal lobe glioma	Elsberg (1935)
Olfactory groove meningioma	Bakay and Cares (1972)
Suprasellar meningioma	Elsberg (1935)
Temporal lobe tumors	Furstenberg et al. (1943)
Intranasal tumors	
Adenocarcinoma	Skolnik et al. (1966)
Esthesioneuroepithelioma	Takahashi et al. (1987)
Inverted papilloma	Skolnik et al. (1966)
Melanoma	Skolnik et al. (1966)
Neuroblastoma	Joachims et al. (1975)
Squamous-cell carcinoma	Skolnik et al. (1966)
Viral and infectious	
Acute viral hepatitis	Henkin and Smith (1971)
Herpes simplex	Twomey et al. (1979)
Influenza-like infections	Henkin et al. (1975)

Disorders of thyroid or parathyroid function also have been associated with smell impairment. Patients with untreated primary hypothyroidism have been shown to have elevated smell thresholds, which return to normal after treatment with thyroxine (McConnell et al. 1975), although more recent data show no such deficit (Lewitt et al. 1989). Pseudohypoparathyroidism, which is characterized by a lack of response to parathyroid hormone, has been associated with increased smell thresholds (Henkin 1968). Chemosensory impairment has also been related to adrenal cortical function. The hyperadrenocorticism characteristic of Cushing's syndrome has been associated with increased olfactory thresholds, whereas decreased detection thresholds have been shown in patients with adrenal cortical insufficiency (Henkin 1975). Finally, smell sensitivity appears to be reduced in some patients with diabetis mellitus (Settle 1986). Thus, olfactory impairment has been suggested in a number of endocrine disorders, any of which could be of etiologic significance in patients with smell complaints.

Olfactory dysfunctions are sometimes observed in patients with neurologic impairment. Head trauma was discussed above as one of the major etiologic factors in patients with primary olfactory complaints. In addition, when neurological disease involves peripheral or central olfactory pathways, smell function can be compromised. Smell dysfunction has been reported for patients with familial dysautonomia (Henkin and Kopin 1964), although the mechanism for the reduced smell sensitivity to thiophene and pyridine seen in some of these patients is not clear. The demyelinating effects of multiple sclerosis have been shown to produce deficits in taste and perhaps in smell (Catalanotto et al. 1986, Cohen 1965, Doty et al. 1984a, Pinching 1977, Wender and Szemza 1971), although there are questions regarding olfactory deficits in multiple sclerosis. Early work in this area is conflicting, with some studies (measuring odor identification) suggesting olfactory impairment (Wender and Szemza 1971) and others (measuring detection thresholds) no deficit at all (Ansari 1976). These differences may reflect the effects of multiple sclerosis on the higher cognitive functions involved in odor identification (Pinching 1977), although recent work with the UPSIT (Doty et al. 1984a) shows only a small number of multiple sclerosis patients with smell impairment. Thus, smell may be compromised in multiple sclerosis only in some patients, probably depending upon the particular distribution of the disease.

Recent work has shown that a number of central neural disorders seem to involve olfactory deficits, including epilepsy (Eskenazi et al. 1986), Parkinson's disease (Ansari and Johnson 1975, Doty et al. 1988, Quinn et al. 1987, Ward et al. 1983), and Alzheimer's disease (Doty et al. 1987, Serby 1986). Patients with temporal lobe epilepsy are slightly deficient in odor identification and odor recognition memory and these deficiencies are exacerbated by temporal lobectomy, with greater impairment for the nostril ipsilateral to the lesion (Eskenazi et al. 1986). There does not appear to have been any difference between normal controls and epileptic patients in butanol threshold measures (Eskenazi et al. 1986). Since these patients were also deficient in recognition memory for amorphous shapes, these impairments may have more to do with odor memory processing than with a sensory impairment.

Several studies suggest that patients with Parkinson's disease show deficits in olfactory ability. Early work by Ansari and Johnson (1975) demonstrated that 10 out of 22 Parkinsonian patients had increased thresholds for amyl acetate. This deficit was confirmed by Ward et al. (1983) using an odor identification task, in which 49% of 72 patients were unable to identify the odor of coffee and 35% the odor of cinnamon. Parkinsonian patients who were normal on a picture identification test similar to the UPSIT in format were tested for their detection and recognition of odors (Doty et al. 1988). Of these 81 patients, 48 had scores on the UPSIT below the 10th percentile for their age and sex, and an additional 15 had scores between the 10th and 25th percentile. These patients also showed marked impairment on a phenethyl alcohol detection threshold task. Deficits in odor identification in these patients were independent of age, which produces a steady decline in performance on the UPSIT beyond age 60 (Doty et al. 1984b). Olfactory thresholds have been shown to be unrelated to pharmacological manipulation of dopaminergic or cholinergic status in Parkinsonian patients (Quinn et al. 1987), suggesting little or no role for these neurotransmitters in the olfactory impairment.

Finally, it has recently been established that Alzheimer's disease is associated with olfactory deficits in both odor identification and threshold (Doty et al. 1987, Serby 1986). Relative to age-, gender-, and race-matched controls, patients with well-defined Alzheimer's disease were deficient on the UPSIT and in a phenethyl alcohol threshold test (Doty et al. 1987). All of these patients scored in the normal range on the picture identification test, which demonstrates that they had the cognitive ability to take the UPSIT. Only two of the 34 Alzheimer's patients were aware of an olfactory impairment before testing. The olfactory deficits seen in Alzheimer's disease are probably related to the existence of neuritic plaques and neurofibrillary tangles throughout olfactory-related brain structures (Esiri and Wilcock 1984, Pearson et al. 1985). In fact, several dementia-related disorders, which result in lesions in the olfactory pathways, are accompanied by olfactory deficits, including Huntington's chorea (Moberg et al. 1984), Parkinson's disease (Ansari and Johnson 1975, Doty et al. 1988, Quinn et al. 1987), and Korsakoff's psychosis (Jones et al. 1975, Mair et al. 1986). It has even been proposed that the olfactory system could be the route of invasion of etiologic factors responsible for some of these diseases (Shipley 1985).

Several kinds of nutritional deficiencies have been implicated in smell dysfunction. Disturbances of taste or smell in malnutrition or in other systemic conditions, such as pellagra or pernicious anemia, have been attributed to vitamin or trace metal imbalances (Green 1971, Rundles 1946). Early work by Henkin and his colleagues (Henkin et al. 1967, Henkin and Bradley 1969, Schecter et al. 1972) suggested that reduced serum copper levels resulted in depressed taste sensitivity, which could be reversed by $CuSO_4$ or $ZnSO_4$ administration. As a result, it has become relatively standard practice to prescribe $ZnSO_4$ for both taste and smell deficiences, regardless of their etiologies (Estrum and Renner 1987). However, there is little evidence that zinc therapy is an effective treatment for these conditions (Price 1986). Double-blind clinical trials using a cross-over design, in which $ZnSO_4$ was compared to placebo, showed no significant effects of zinc in the treatment of hypogeusia (Henkin et al. 1976). In cases of zinc deficiency due to renal disease (Atkin-Thor et al. 1978, Mahajan et al. 1980) and hepatic cirrhosis (Weisman et al. 1979), double-blind studies have shown that decreased taste acuity can be restored by $ZnSO_4$ treatment. Thus, when zinc-deficient states are accompanied by hypogeusia, correction of the zinc deficiency often restores taste sensitivity, but there is no reason to assume that any particular case of taste or smell dysfunction is the result of zinc deficiency or that it would be helped by $ZnSO_4$ therapy (Price 1986).

Many of the remaining disorders listed in Table 1 represent factors discussed above as common causes of smell dysfunction, such as allergic rhinitis, influenza-like infections, sinusitis, and exposure to toxic chemicals. A few additional disorders, such as various intranasal or intracranial tumors, are well known as potential causes of olfactory loss or distortion (Doty 1979, Feldman et al. 1986, Kimmelman 1986, Leopold 1986), although their occurrence in patients presenting only with olfactory symptoms is relatively rare (Goodspeed et al. 1986a, 1987). Similarly, although olfactory symptoms may be manifested in psychiatric disturbances (Amsterdam et al. 1987, Kerekovic 1972, Pryse-

Phillips 1971), these are not typically the kind of patients presenting with primary smell complaints.

V. CONCLUSION

It is clear that disturbances of smell may result from a wide variety of underlying etiologies (Table 1), making the diagnosis and treatment of these disorders a difficult and challenging problem. As more specific information becomes available from the various taste and smell research centers, these patients should become easier to manage and whenever possible to cure.

ACKNOWLEDGMENTS

This chapter was prepared while the senior author was supported by Jacob Javits Neuroscience Investigator Award NS-23524 from the National Institute of Neurological and Communicative Disorders and Stroke.

REFERENCES

Adams RG, Crabtree N (1961) Anosmia in alkaline battery workers. Brit J Indust Med 18: 216-222

Amoore JE (1986) Effects of chemical exposure on olfaction in humans. In: Barrow, CS (ed) Toxicology of the nasal passages. Hemisphere Publishing Co., Washington DC, p 155

Amsterdam JD, Settle RG, Doty RL, Abelman E, Winokur A (1987) Taste and smell perception in depression. Biol Psychiat 22: 1481-1485

Ansari KA (1976) Olfaction in multiple sclerosis. Eur Neurol 4: 138-145

Ansari, KA, Johnson A (1975) Olfactory function in patients with Parkinson's disease. J Chronic Diseases 28: 493-497

Atkin-Thor E, Goddard BW, O'Nion J, Stephen RL, Kolff WJ (1978) Hypogeusia and zinc depletion in chronic dialysis patients. Am J Clin Nutr 31: 1948-1951

Bakay L, Cares HL (1972) Olfactory meningiomas. Report on a series of twenty-five cases. Acta Neurochir 26: 1-12

Barton RPE (1974) Olfaction in leprosy. J Laryngol Otol 88: 355-361

Bartoshuk LM (1978) The psychophysics of taste. Am J Clin Nutr 31: 1068-1077

Bartoshuk LM, Marks LE (1986) Ratio scaling. In: Meiselman HL, Rivlin RS (eds) Clinical measurement of taste and smell. MacMillan, New York, p 50

Burch RE, Sackin DA, Ursick JA, Jetton MM, Sullivan JF (1978) Decreased taste and smell acuity in cirrhosis. Arch Intern Med 138: 743-746

Cain WS (1981) Olfaction and the common chemical sense: Similarities, differences, and interaction. In: Moskowitz HR, Warren CB (eds) Odor quality and chemical structure. American Chemical Society, Washington DC, p 109

Cain WS, Gent JF (1986) Use of odor identification in clinical testing of olfaction. In: Meiselman HL, Rivlin RS (eds) Clinical measurement of taste and smell. Macmillan, New York, p 170

Cain WS, Gent J, Catalanotto FA, Goodspeed RB (1983) Clinical evaluation of olfaction. Am J Otolaryngol 4: 252-256

Cain WS, Gent JF, Goodspeed RB, Leonard G (1988) Evaluation of olfactory dysfunction in the Connecticut Chemosensory Clinical Research Center. Laryngoscope 98: 83-88

Cancalon P (1982) Degeneration and regeneration of olfactory cell induced by $ZnSO_4$ and other chemicals. Tissue Cell 14: 717-733

Catalanotto FA, Dore-Duffy P, Donaldson JO (1986) Taste and smell in multiple sclerosis. In: Meiselman HL, Rivlin RS (eds) Clinical measurement of taste and smell. Macmillan, New York, p 519

Champion R (1966) Anosmia associated with corrective rhinoplasty. Brit J Plast Surg 19: 182-185

Cohen L (1965) Disturbance of taste as a symptom of multiple sclerosis. Brit J Oral Surg 2: 184-185

Cometto-Muniz JE, Cain WS (1982) Perception of nasal pungency in smokers and nonsmokers. Physiol Behav 29: 727-731

Costanzo RM (1985) Neural regeneration and functional reconnection following olfactory nerve transection in hamster. Brain Res 361: 258-266

Costanzo RM, Becker DP (1986) Smell and taste disorders in head injury and neurosurgery patients. In: Meiselman HL, Rivlin RS (eds) Clinical measurement of taste and smell. MacMillan, New York, p 565

Deems DA, Doty RL, Settle RG, Moore-Gillon V, Shaman P, Mester AF, Kimmelman CP, Brightman VJ, Snow JB Jr (in press) Smell and taste disorders: A study of 750 patients from the University of Pennsylvania Smell and Taste Center. Arch Otolaryngol Head Neck Surg

Donnelly JW, Cain WS, Scott A (1989) Parosmia among patients with olfactory complaints. Chem Sens 14: 695

Doty RL (1979) A review of olfactory dysfunctions in man. Am J Otolaryngol 1: 57-79

Doty RL, Deems DA, Stellar S (1988) Olfactory dysfunction in parkinsonism: a general deficit unrelated to neurologic signs, disease stage, or disease duration. Neurology 38: 1237-1244

Doty RL, Kimmelman CP (1986) Smell and taste and their disorders. In: Asbury AK, McKhann GM, McDonald WI (eds) Diseases of the nervous system: clinical neurobiology, vol I. W B Saunders, Philadelphia, p 466

Doty RL, Reyes PF, Gregor T (1987) Presence of both odor identification and detection deficits in Alzheimer's disease. Brain Res Bull 18: 597-600

Doty RL, Shaman P, Dann M (1984a) Development of the University of Pennsylvania Smell Identification Test: a standardized microencapsulated test of olfactory function. Physiol Behav 32: 489-502

Doty RL, Shaman P, Applebaum SL, Giberson R, Siksorski L, Rosenberg L (1984b) Smell identification ability: changes with age. Science 226: 1441-1443

Doty RL, Snow JB Jr (1987) Olfaction. In: Goldman J (ed) The principles and practice of rhinology. John Wiley, New York, p 761

Doty RL, Snyder PJ, Huggins GR, Lowry LD (1981) Endocrine, cardiovascular, and psychological correlates of olfactory sensitivity changes during the human menstrual cycle. J Comp Physiol Psychol 95: 45-60

Douek E, Bannister LH, Dodson HC (1975) Recent advances in the pathology of olfaction. Proc Roy Soc Med 68: 467-470

Dunn JD, Cometto-Muniz JE, Cain WS (1982) Nasal reflexes: reduced sensitivity to CO_2 irritation in cigarette smokers. J Appl Toxicol 2: 176-178

Elsberg CA (1935) The sense of smell. XII. The localization of tumors of the frontal lobe of the brain by quantitative olfactory tests. Bull Neurol Inst NY 4: 535-543

Esiri MM, Wilcock GK (1984) The olfactory bulbs in Alzheimer's disease. J Neurol Neurosurg Psychiat 47: 56-60

Eskenazi B, Cain WS, Novelly RA, Friend KB (1983) Olfactory functioning in temporal lobectomy patients. Neuropsychologia 21: 365-374

Eskenazi B, Cain WS, Novelly RA, Mattson R (1986) Odor perception in temporal lobe epilepsy patients with and without temporal lobectomy. Neuropsychologia 24: 553-562

Estrum SA, Renner G (1987) Disorders of taste and smell. In: The otolaryngology clinics of North America, vol 20. no 1. W B Saunders, Philadelphia, p 133

Fein BT, Kamin PB, Fein NN (1966) The loss of sense of smell in nasal allergy. Ann Allergy 24: 278-283

Feldman JI, Wright HN, Leopold DA (1986) The initial evaluation of dysosmia. Am J Otolaryngol 4: 431-444

Frye RE, Schwartz BS, Doty RL (1990) Dose-related effects of cigarette smoking on olfactory function. J Am Med Assoc 263: 1233-1236

Furstenberg AC, Crosby E, Farrior B (1943) Neurologic lesions which influence the sense of smell. Arch Otol 48: 529-530

Gent JF, Cain WS, Bartoshuk LM (1986) Taste and smell measurement in a clinical setting. In: Meiselman HL, Rivlin RS (eds) Clinical measurement of taste and smell. Macmillan, New York, p 107

Gent JF, Goodspeed RB, Zagraniski RT, Catalanotto FA (1987) Taste and smell problems: validation of questions for the clinical history. Yale J Biol Med 60: 27-35

Ghorbanian SN, Paradise JL, Doty RL (1983) Odor perception in children in relation to nasal obstruction. Pediatrics 72: 510-516

Goodspeed RB, Catalanotto FA, Gent JF, Cain WS, Bartoshuk LM, Leonard G, Donaldson JO (1986a) Clinical characteristics of patients with taste and smell disorders. In: Meiselman HL, Rivlin RS (eds) Clinical measurement of taste and smell. MacMillan, New York, p 451

Goodspeed RB, Gent JF, Catalanotto FA (1987) Chemosensory dysfunction: clinical evaluation results from a taste and smell clinic. Postgrad Med 81: 251-260

Goodspeeed RB, Gent JF, Catalanotto FA, Cain WS, Zagraniski RT (1986b) Corticosteroids in olfactory dysfunction. In: Meiselman HL, Rivlin RS (eds) Clinical measurement of taste and smell. MacMillan, New York, p 514

Graziadei PPC (1973) Cell dynamics in the olfactory mucosa. Tiss Cell 5: 113-131

Graziadei PPC, Monti-Graziadei GA (1983) Regeneration in the olfactory system of vertebrates. Am J Otolaryngol 4: 228-233

Green RF (1971) Subclinical pellagra and idiopathic hypogeusia. J Am Med Assoc 218: 1303

Griffith IP (1976) Abnormalities of smell and taste. Practitioner 217: 907-913

Halpern BP (1985) Environmental factors affecting chemoreceptors: an overview. In: Hayes AW (ed) Toxicology of the eye, ear, and other special senses. Raven Press, New York, p 195

Henkin RI (1967a) The definition of primary and accessory areas of olfaction as the basis for a classification of decreased olfactory acuity. In: Hayashi T (ed) Olfaction and taste, vol II. Pergamon Press, Oxford, p 235

Henkin RI (1967b) Abnormalities of taste and olfaction in patients with chromatin negative gonadal dysgenesis. J Clin Endocrinol Metab 27: 1436-1440

Henkin RI (1968) Impairment of olfaction and the losses of sour and bitter in pseudohypoparathyroidism. J Clin Endocrinol Metab 28: 624-628

Henkin RI (1975) The role of adrenal corticosteroids in sensory processes. In: Blaschko H, Smith AD, Sayers G (eds) Handbook of physiology, vol 6. American Physiological Society, Washington DC, p 209

Henkin RI (1981) Olfaction in human disease. In: English GM (ed) Otolaryngology, vol 2. J B Lippincott, Philadelphia, p 1

Henkin RI, Bradley DF (1969) Regulation of taste acuity by thiols and metal ions. Proc Natl Acad Sci USA 62: 30-37

Henkin RI, Keiser HR, Jaffee IA, Sternlieb I, Scheinberg IH (1967) Decreased taste sensitivity after D-penicillamine reversed by copper administration. Lancet 2: 1268-1271

Henkin RI, Kopin IJ (1964) Abnormalities of taste and smell thresholds in familial dysautonomia: improvement with methacholine. Life Sci 3: 1319-1325

Henkin RI, Larson AL, Powell RD (1975) Hypogeusia, dysgeusia, hyposmia, and dysosmia following influenza-like infection. Ann Otol 84: 672-682

Henkin RI, Powell GF (1962) Increased sensitivity of taste and smell in cystic fibrosis. Science 138: 1107-1108

Henkin RI, Schechter PJ, Friedewald WT, Demets DL, Raff M (1976) A double-blind study of the effects of zinc sulfate on taste and smell dysfunction. Am J Med Sci 272: 285-299

Henkin RI, Schechter PJ, Hoye R, Mattern CFT (1971) Idiopathic hypogeusia with dysgeusia, hyposmia, and dysosmia: a new syndrome. J Am Med Assoc 217: 434-440

Henkin RI, Smith FR (1971) Hyposmia in acute viral hepatitis. Lancet 1: 823-826

Henkin RI, Talal N, Larson AL, Mattern CFT (1972) Abnormalities of taste and smell in Sjögren's syndrome. Ann Intern Med 76: 375-383

Hertz J, Cain WS, Bartoshuk LM, Dolan TF (1975) Olfactory and taste sensitivity in children with cystic fibrosis. Physiol Behav 14: 89-94

Hotchkiss WT (1956) Influence of prednisone on nasal polyposis with anosmia. Arch Otolaryngol 64: 478-479

Hubert HB, Fabsitz RR, Feinleib M, Brown KS (1980) Olfactory sensitivity in humans: genetic versus environmental control. Science 208: 607-609

Jafek BW (1982) Anosmia and ageusia. In: Gates GA (ed) Current therapy in otolaryngology - head and neck surgery 1982-1983. C V Mosby., St Louis, p 279

Jafek BW, Eller PM (1989) Post-viral olfactory dysfunction. Chem Sens 14: 713

Jefferson M (1961) Anosmia and parosmia. Practitioner 187: 715-718

Joachims HZ, Altman MM, Mayer SW (1975) Olfactory neuroblastoma. J Laryngol Otol 89: 335-343

Jones BA, Moskowitz HR, Butters N (1975) Olfactory discrimination in alcoholic Korsakoff patients. Neuropsychologica 13: 173-179

Jorgensen MB, Buch NH (1961) Studies on the sense of smell and taste in diabetics. Acta Otolaryngol Stockh 53: 539-545

Kallmann FJ, Schoenfeld WA, Barrera SE (1944) The genetic aspects of primary eunuchoidism. Am J Ment Defic 48: 203-236

Kerekovic M (1972) The relationship between objective disorders of smell and olfactory hallucinations. Acta Otorhinolaryngol Belg 26: 518-523

Kimmelman CP (1986) Disorders of taste and smell. American Academy of Otolaryngology - Head and Neck Surgery Foundation, Washington DC

Leigh AD (1943) Defects of smell after head injury. Lancet 1: 38-40

Leopold DA (1986) Physiology of olfaction. In: Cummings CW, Fredrickson JM, Harker LA, Krause CJ, Schuller DE (eds) Otolaryngology - head and neck surgery, vol 1. General, face, nose, paranasal sinuses. C V Mosby, St Louis, p 527

Liston SL (1984) Olfactory disorders. In: Holt GR, Mattox DE, Gates GA (eds) Decision making in otolaryngology. C. V. Mosby, St Louis, p 28

Lewitt MS, Laing DG, Panhuber H, Corbett A, Carter JN (1989) Sensory perception and hypothyroidism. Chem Sens 14: 537-546

Mahajan SK, Prasad AS, Lambujon J, Abbasi AA, Briggs WA, McDonald FD (1980) Improvement of uremic hypogeusia by zinc: a double-blind study. Am J Clin Nutr 33: 1517-1521

Mair RG, Doty RL, Kelly KM, Wilson CS, Langlais PJ, McEntee WJ, Vollmecke TA (1986) Multimodal sensory discrimination deficits in Korsakoff's psychosis. Neuropsychologia 24: 831-839

Marshall JR, Henkin RI (1971) Olfactory acuity, menstrual abnormalities and oocyte status. Ann Intern Med 75: 207-211

Marshall KG, Attia EL (1987) Anosmia, hyposmia and dysosmia. In: Marshall KG, Attia EL (eds) Disorders of the nose and paranasal sinuses. PSG Publishing Co., Littleton, Massachusetts, p 235

Matulionis DH (1974) Ultrastructure of olfactory epithelia in mice after smoke exposure. Ann Otol 83: 192-201

McConnell RJ, Menendez CE, Smith FR, Henkin RI, Rivlin RS (1975) Defects of taste and smell in patients with hypothyroidism. Am J Med 59: 354-364

Moberg PJ, Pearlson GD, Speedie LJ, Lipsey JR, Folstein SE (1984) Deficits in olfactory, but not visual or verbal recognition, in early affected Huntington's disease patients. Soc Neurosci Abstr 10: 318

Monath TP, Cropp CB, Harrison AK (1983) Mode of entry of a neurotropic arbovirus into the central nervous system: reinvestigation of an old controversy. Lab Invest 48: 399-410

Monti-Graziadei GA, Graziadei PPC (1979) Neurogenesis and neuron regeneration in the olfactory system of mammals. II. Degeneration and reconstitution of the olfactory sensory neurons after axotomy. J Neurocytol 8: 197-213

Monti-Graziadei GA, Karlan MS, Bernstein JJ, Graziadei PPC (1980) Reinnervation of the olfactory bulb after section of the olfactory nerve in monkey (Samiri sciureus). Brain Res 189: 343-354

Moran DT, Jafek BW, Rowley JC, Eller PM (1985) Electron microscopy of olfactory epithelia in two patients with anosmia. Arch Otolaryngol III: 122-126

Mozell MM, Hornung DE, Scheehe PR, Kurtz DB (1986a) What should be controlled in studies of smell? In: Meiselman HL, Rivlin RS (eds) Clinical measurement of taste and smell. Macmillan, New York, p 154

Mozell MM, Schwartz DN, Youngentob SL, Leopold DA, Hornung DE, Sheehe PR (1986b) Reversal of hyposmia in laryngectomized patients. Chem Sens 11: 397-410

Nakashima T, Kimmelman CP, Snow JB Jr (1983) Progressive olfactory degeneration due to ischemia. Surg Forum XXXIV: 566-568

Nakashima T, Kimmelman CP, Snow JB Jr (1984) Effect of olfactory nerve section and hemorrhage on the olfactory neuroepithelium. Surg Forum XXXV: 562-564

Ophir D, Gross-Isseroff R, Lancet D, Marshak G (1986) Changes in olfactory acuity induced by total inferior turbinectomy. Arch Otolaryngol Head Neck Surg 112: 195-197

Ophir D, Guterman A, Gross-Isseroff R (1988) Changes in smell acuity induced by radiation exposure of the olfactory mucosa. Arch Otolaryngol Head Neck Surg 114: 853-855

Pearson RCA, Esiri MM, Hiorns RW, Wilcock GK, Powell TPS (1985) Anatomic correlates of the distribution of the pathological changes in the neocortex in Alzheimer's disease. Proc Natl Acad Sci USA 82: 4531-4534

Pinching AJ (1977) Clinical testing of olfaction reassessed. Brain 100: 377-388

Price S (1986) The role of zinc in taste and smell. In: Meiselman HL, Rivlin RS (eds) Clinical measurement of taste and smell. Macmillan, New York, p 443

Pryse-Phillips W (1971) An olfactory reference syndrome. Acta Psychiat Scand 47: 484-509

Quinn NP, Rossor MN, Marsden CD (1987) Olfactory thresholds in Parkinson's disease. J Neurol Neurosurg Psychiat 50: 88-89

Rundles RW (1946) Prognosis in the neurologic manifestations of pernicious anemia. Blood 1: 209-219

Ryan RE Sr, Ryan RE Jr (1974) Acute nasal sinusitis. Postgrad Med 56: 159-162

Santen RJ, Paulsen CA (1973) Hypogonadotropic eunuchoidism. I. Clinical study of the mode of inheritance. J Clin Endocrinol Metab 36: 47-54

Schechter PJ, Friedewald WT, Bronzert DA, Raff MS, Henkin RI (1972) Idiopathic hypogeusia: A description of the syndrome and a single blind study with zinc sulfate. Int Rev Neurobiol Suppl 1: 125-140

Schechter PJ, Henkin RI (1974) Abnormalities of taste and smell after head trauma. J Neurol Neurosurg Psychiat 37: 802-810

Schiffman SS (1983a) Taste and smell in disease. I. New Engl J Med 308: 1275-1279

Schiffman SS (1983b) Taste and smell in disease. II. New Engl J Med 308: 1337-1343

Schiffman SS, Nash ML, Dackis C (1978) Reduced olfactory discriminations in patients on chronic renal hemodialysis. Physiol Behav 21: 239-242

Schneider RA, Castiloe JP, Howard RP, Wolf S (1958) Olfactory perception thresholds in hypogonadal women: changes accompanying administration of androgen and estrogen. J Clin Endocrinol Metab 18: 379-390

Schwartz BS, Doty RL, Monroe C, Frye R, Barker S (1989) Olfactory function in chemical workers exposed to acrylate and methacrylate vapors. Am J Pub Health 79: 613-618

Scott AE, Cain WS, Clavet G (1988) Topical corticosteroids can alleviate olfactory dysfunction. Chem Sens 13: 735

Scott AE, Cain WS, Leonard G (1989) Nasal/sinus disease and olfactory loss at the Connecticut Chemosensory Clinical Research Center (CCCRC). Chem Sens 14: 745

Seiden AM, Litwin A, Smith DV (1989) Olfactory deficits in allergic rhinitis. Chem Sens 14: 746-747

Seiden AM, Smith DV (1988) Endoscopic intranasal surgery as an approach to restoring olfactory function. Chem Sens 13: 736

Serby M (1986) Olfaction and Alzheimer's disease. Prog Neuro-Psychopharmacol Biol Psychiat 10: 579-586

Settle RG (1986) Diabetes mellitus and the chemical senses. In: Meiselman HL, Rivlin RS (eds) Clinical measurement of taste and smell. Macmillan, New York, p 487

Shipley MT (1985) Transport of molecules from nose to brain: Transneuronal anterograde and retrograde labeling in the rat olfactory system by wheat germ agglutinin-horseradish peroxidase applied to the nasal epithelium. Brain Res Bull 15: 129-142

Simmons PA, Getchell TV (1981) Physiological activity of newly differentiated olfactory receptor neurons following olfactory nerve section. J Comp Neurol 197: 237-257

Skolnik EM, Massari FS, Tenta LT (1966) Olfactory neuro-epithelioma: review of the world literature and presentation of two cases. Arch Otolaryngol 84: 649-653

Smith DV (1988) Assessment of patients with taste and smell disorders. Acta Otolaryngol Stockh Suppl 458: 129-133

Smith DV, Frank RA, Pensak ML, Seiden AM (1987) Characteristics of chemosensory patients and a comparison of olfactory assessment procedures. Chem Sens 12: 698

Snow JB Jr (1969) The classification of respiratory viruses and their clinical manifestations. Laryngoscope 79: 1485-1493

Strandbygard E (1954) Treatment of ozena and rhinopharyngitis chronica sicca with vitamin A. Arch Otolaryngol 59: 485-490

Sumner D (1964) Post-traumatic anosmia. Brain 87: 107-120

Sumner D (1975) Disturbances of the sense of smell and taste after head injuries. Part II. Injuries of the brain and skull. In: Vinken PJ, Bruyn GW (eds) Handbook of clinical neurology, vol 24. American Elsevier, New York, p 1

Tagatz GT, Fialkow PJ, Smith D, Spadoni L (1970) Hypogonadotropic hypogonadism associated with anosmia in the female. New Engl J Med 283: 1326-1329

Takahashi H, Ohara S, Yamada M, Ikuta F, Tanimura K, Honda Y (1987) Esthesioneuroepithelioma: A tumor of true olfactory epithelium origin. Acta Neuropathol Berl 75: 147-155

Thumfart W, Plattig KH, Schlicht N (1980) Taste and smell sensitivity in the elderly human. Z Gerontol 13: 158-188

Twomey JA, Barker CM, Robinson G, Howell DA (1979) Olfactory mucosa in herpes simplex encephalitis. J Neurol Neurosurg Psychiat 42: 983-987

Venstrom D, Amoore JE (1968) Olfactory threshold in age, sex or smoking. J Food Sci 33: 264-265

Ward CD, Hess, WA, Calne DB (1983) Olfactory impairment in Parkinson's disease. Neurol 33: 943-946

Weiffenbach JM, McCarthy VP (1984) Olfactory deficits in cystic fibrosis: distribution and severity. Chem Sens 9: 193-199

Weisman K, Christensen E, Dreyer V (1979) Zinc supplementation in alcoholic cirrhosis: a double-blind clinical trial. Acta Med Scand 205: 361-366

Wender M, Szmeja Z (1971) Examination of hearing, vestibular system function, taste and olfactory systems in patients with disseminated sclerosis. Neurol Neurochir Pol. 64: 767-772

Wright HN (1987) Characterization of olfactory dysfunction. Arch Otolaryngol Head Neck Surg 113: 163-168

Zusho H (1982) Post-traumatic anosmia. Arch Otolaryngol 108: 90-92

15

EPIDEMIOLOGY AND ITS APPLICATION TO OLFACTORY DYSFUNCTION

BRIAN S SCHWARTZ

I. INTRODUCTION

Although accurate estimates of the incidence, prevalence, and risk factors associated with disorders of vision and hearing have been determined using modern epidemiologic techniques (Eagles 1973, Hinchcliffe 1973, Surjan et al. 1973, Fraser 1974, Angle and Wissmann 1980, Taylor 1981, The National Eye Institute Symposium on the Epidemiology of Eye Diseases and Visual Disorders 1983), this is not the case for disorders of the sense of smell. For example, there are few published estimates of the prevalence of olfactory dysfunction in the general population and no published estimates of the yearly incidence of this condition. Even the few available estimates are derived from studies troubled by subject selection and sampling biases, as well as a lack of consensus as to what, in fact, constitutes olfactory dysfunction. These problems, along with the dearth of information on this topic, likely reflects, at least in part, a general lack of awareness of many chemosensory researchers as to the rigorous epidemiologic tools that are available for addressing such issues.

The primary goals of this Chapter are; 1) to provide the reader with a basic introduction to principles of epidemiology applicable to the study of the human capacity to smell, both in health and disease, 2) apply such principles to a review of published studies in the field, and 3) discuss the risk factors for olfactory dysfunction that have emerged from epidemiologic investigations. A summary is provided which addresses current research needs and issues in this important emergent field. The reader is referred elsewhere for more detailed general discussions of many of the principles outlined in this Chapter (Mausner and Kramer 1985, Lilienfeld and Lilienfeld 1980, Rothman 1986, Kleinbaum et al. 1982).

II. EPIDEMIOLOGIC PRINCIPLES

Epidemiology, a field whose name is derived from the Greek words for "upon the people", is the study of the distribution and determinants of diseases and injuries in human populations (Mausner and Kramer 1985). In accord with standard practice, the term "disease" is used generically throughout this Chapter as the primary outcome variable under consideration e.g., olfactory sensitivity, anosmia, dysosmia, hyposmia, etc. Diseases are not randomly distributed throughout populations, since they may differ in frequency within population subgroups. Associations are thus observed between factors and diseases in the subgroups and inferences are made about cause and effect. For example, the epidemiologist may examine the various characteristics of those with disease, compare these observations to those in persons without disease, and ask the question, "Do these characteristics distinguish the diseased from the nondiseased?" The characteristics can be demographic, biological, social, economic, educational, personal or genetic, among others (Lilienfeld and Lilienfeld 1980).

Epidemiology is a relatively young science; the main body of epidemiologic principles and research has been generated in the last 20 years (Rothman 1986). The application of these methods to the study of olfactory function and dysfunction is even earlier in its infancy, and as will be seen later in this Chapter, less than half a dozen published studies have used rigorous epidemiologic principles in the evaluation of olfactory dysfunction.

Epidemiologic research has generally been divided into two types, experimental and observational (analytic). In the former, exposures, which is a generic term that can include characteristics distributed in the population by such factors as birth e.g., race or genetic variables, choice e.g., personal habits such as cigarette smoking, or opportunity e.g., education or socioeconomic status, are assigned to the subjects being studied. Since perhaps the majority of exposures of interest to medical science cannot be assigned because of ethical considerations e.g., toxic chemicals and radiation, most epidemiologic studies aside from medical therapeutics are observational. In observational studies, the researcher observes how exposures are distributed in the population.

Although there are many possible epidemiologic designs, only three observational designs will be briefly discussed; cohort, case-control, and cross-sectional (Kleinbaum et al. 1982). In the cohort study, persons with and without exposure to factors of interest are followed for the development of disease. The follow-up period can be from some past date to the present, termed an historical or retrospective cohort study, or from the present date to some time in the future, termed a prospective cohort study. Such studies provide incidence data (see below) because of the passage of time. An example of a cohort study would be the follow-up of persons with normal olfactory ability who smoke cigarettes (the exposure) and do not smoke cigarettes to determine which ones develop olfactory dysfunction.

In the case-control study, persons with the disease of interest (cases) are compared to persons without the disease of interest (controls) on several well-defined risk factors. A measure of the strength of the association between the exposure and disease can be obtained by calculating the odds ratio, which is defined below. In general, case-control studies are best performed with incident cases of disease i.e., the number of new cases occurring during a specified time period in a specified population at risk, not prevalent cases, which are the number of cases in existence at a certain time in a designated population (Kleinbaum et al. 1982, Rothman 1986). Case-control studies have several important advantages, including investigation of multiple exposures, relatively low cost, and relatively short time for completion. It should be noted that, as the passage of time is not involved in case-control studies, incidence rates can not be determined. In the past, several case-control studies (not of olfactory dysfunction) were poorly designed and subsequently discredited. However, it is now widely recognized that carefully designed studies of this design can provide excellent information for causal inference (Rothman 1986). An example of a case-control study would be the comparison of persons with and without olfactory dysfunction and the calculation of odds ratios associated with exposure to cigarettes, the number of upper respiratory infections in the past year, medications, and other variables. The odds ratio is a measure of what has been termed the "relative risk" and will be discussed below.

The final type of study presented here, the cross-sectional study, is a study of prevalence where exposure and disease status are measured at one point in time. This is probably the most common type of observational epidemiologic study of morbidity, as opposed to mortality, because of its ease of completion. Odds ratios can be calculated for the strength of the association between multiple factors and the disease of interest. It is important to note that these studies are subject to several potential biases such as survival effects and exposure status changes because of disease development occurring before the study, and because it is sometimes difficult to know what came first, exposure or disease. Thus, although this type of study can provide data useful for causal inference, the potential for sources of bias must be recognized.

After collecting observations and identifying associations between factors and diseases in studies of various designs, the epidemiologist must make inferences about cause and effect. An in-depth review of causal inference is beyond the scope of this Chapter; the reader is referred to some excellent discussions of the topic (Hill 1965, Rothman 1988). Briefly, the researcher considers: the strength of the observed association (see below); the consistency of the observation from study to study; the specificity of the association; the temporality (does the cause precede the effect in time); the presence of a dose-response relation (more exposure to the presumed cause leads to an increased risk of disease); the biological plausibility of the association; experimental data in animals or humans (the latter are usually not available), and the coherence of the association i.e., a cause and effect interpretation of the association does not conflict with what was previously known of the history and biology of the disease (Rothman 1986).

In considering the strength of the association the epidemiologist utilizes one or several epidemiologic measures of effect that compare disease occurrence in exposed and unexposed groups. The most common epidemiologic measures are incidence rate, cumulative incidence, and prevalence (Kleinbaum et al. 1982). Incidence rate is the most important and useful measure in epidemiology, but is often more difficult and costly to ascertain. Typically, the incidence rate is calculated by counting up the number of previously non-diseased individuals who become diseased in a specified period of time and dividing this number by the person-time (one person followed one year provides one person-year of observation) observation in the entire non-diseased group. This is referred to as the incidence density method of calculating incidence rates. Cumulative incidence, usually calculated for diseases with short follow-ups, is the proportion of individuals in a group that develops disease in a specified period of time. For example, if ten persons out of a group of 100 non-diseased individuals develops disease over a two-year period, the two-year cumulative incidence is 0.1. Finally, prevalence is distinctly different from the previous two measures in that the passage of time is not involved in its calculation. Prevalence is the proportion of individuals with disease in a population at one point in time.

Epidemiologic effect measures can be of either 1) absolute effects, which are differences in incidence rate, cumulative incidence, or prevalence between the exposed and non-exposed groups, or 2) relative effects which involve ratios of these measures (Rothman 1986). In the past, these ratio measures have been confused and imprecisely referred to as a group as the "relative risk." However, several epidemiologists have made great efforts to more carefully name and define the different effect measures (Greenland and Thomas 1982, Rothman 1986). Although a detailed discussion is beyond the scope of this Chapter, suffice it to say that there are ratios of *risks, rates,* and *prevalences*. Risk is a proportion, a unitless number from 0 to 1, and is most commonly measured with cumulative incidence; the ratio of the cumulative incidence of disease in exposed and non-exposed groups is called the cumulative incidence ratio. Rates have units of $time^{-1}$; the most common rate ratio is the incidence density ratio. Prevalence is also a proportion and is obtained when disease and exposure status are measured at one point in time; prevalence ratios are obtained by dividing the prevalence of disease in exposed subjects by the prevalence in non-exposed subjects. Finally, the distinction between probability (p) and odds (p/1-p) must be recognized; if the odds of disease in the exposed is divided by the odds of disease in the non-exposed, then an odds ratio is obtained. This is the effect measure which is calculated in epidemiologic studies using the case-control design, and under certain circumstances (Greenland and Thomas 1982) approximates the cumulative incidence ratio or the incidence density ratio.

III. APPLICATION OF EPIDEMIOLOGIC PRINCIPLES TO STUDIES OF OLFACTORY DYSFUNCTION

A. Definition of Olfactory Dysfunction

An important obstacle to the rigorous epidemiologic study of olfactory dysfunction has been the lack of an agreed upon definition of the condition. There has been little consensus as to the types of tests appropriate for defining decreased olfactory ability and the cutoffs that should be used on any one test. Furthermore, although there are many available diagnostic tests, few of these are standardized, well-validated, or reliable.

The testing of olfactory function can include assessment of olfactory identification ability, odor detection thresholds, suprathreshold odor intensity perception, odor discrimination (the ability to distinguish qualitatively among odorants), odor pleasantness, and other abilities (see Chapter 5 this book). Although an extensive literature exists on many of these areas of olfactory testing, many questions remain unresolved, including; 1) what single test or combination of tests best operationally defines olfactory dysfunction? 2) What are the ranges of normal and abnormal scores for each of the tests? 3) How can test results from one type of assessment e.g., olfactory identification, be compared to test results from another type e.g., odor detection thresholds, across studies? 4) What is the definition of anosmia or hyposmia? The resolution of these questions and others would lay the foundation for the commencement of large scale epidemiologic studies of olfactory dysfunction.

Characteristics of some of the tests make them less suitable for epidemiologic use. Odor detection thresholds, for example, are subject to marked interindividual and intraindividual variation, have not been adequately standardized, and are typically time-consuming (Perry et al. 1980, Doty et al. 1986b, Deems and Doty 1987). To illustrate the first point, a study of olfactory thresholds by Perry et al. (1980) revealed large standard deviations around the mean minimally perceptible odor concentration in subjects grouped in age strata, suggesting marked interindividual variation. A similar large interindividual variation was observed by Deems and Doty (1987) who reported a 100-500 fold interquartile range of threshold concentrations among the subjects studied. These and other studies have failed to report any reliability testing within individuals, so intraindividual variation has not been discernible in several studies. Olfactory threshold testing is sensitive to a number of procedural factors (Doty et al. 1986b) and has been reported by some investigators to have problems with reliability (Cain and Rabin 1989).

An example of a threshold test that has been used epidemiologically is the OlfactoLabs phenylethyl-methyl-ethyl carbinol threshold test (PM-Carbinol Olfactory Threshold Test)

(Amoore 1986a, Gullickson et al. 1988). For example, the test was used epidemiologically in the evaluation of the olfactory function of painters and plumbers (Gullickson et al. 1988). This test has been standardized and some normative data are available; however, these data are not available by age or gender. Evaluation of the reliability of this test appears to be incomplete at present. As this is a threshold test, the other concerns raised about threshold tests are also relevant. It would thus seem, based on current information, that olfactory thresholds should not form the basis of tests of olfactory function to be used epidemiologically.

Several odor identification tests have been used epidemiologically, including the University of Pennsylvania Smell Identification Test (UPSIT) (Doty et al. 1984b) and a component of the test developed by Cain et al. (1983). The former test has been standardized, is rapid, quantitative and reliable, and is relatively inexpensive (Doty et al. 1985, Doty et al. 1986a). Normative data from thousands of subjects also allow for the development of population-based definitions of "normal" and "abnormal" (Doty 1989).

Cain et al. (1983) have combined data from an odor threshold test and an odor identification test to provide a composite test that they have used primarily in the clinical setting. The test is designed for monorhinic administration by an examiner which probably makes its use in large scale epidemiologic studies impractical. Studies of this test are difficult to assess, as the definition and selection of patients and controls have not been reported (Cain et al. 1983, Cain 1989) and age- and gender-specific normative data are not available. Furthermore, to date the reliability of this test has not been fully ascertained. Studies of the test, however, suggested that the two components of the test provided largely similar information (Cain 1989) and that perhaps, in view of studies that reported poor test-retest reliability for the olfactory threshold portion (Heywood and Costanzo 1986), this component of the test could be dropped. This would save approximately 20 min of the time required for the measurement of olfactory detection thresholds (Cain 1989) and make the test more amenable to epidemiologic use.

In the ensuing discussion, and because there are no widely agreed upon definitions of olfactory dysfunction, specific tests and definitions are outlined as necessary.

B. Clinical Epidemiology

Clinical epidemiology is the application of the principles and methods of "classical" epidemiology, used to study populations and to the clinical care of individual patients (see Fletcher et al. 1988, Sackett et al. 1985). Particularly relevant to the study of olfactory dysfunction are the clinical epidemiologic principles related to the diagnosis, prognosis (or natural history), and treatment of olfactory disorders.

1. Diagnosis

Considering the diagnosis of olfactory dysfunction, it is important to assess the quality of observations and measurements and how the prevalence of disease in a population affects the probability that a given test result is, in fact, a true positive or true negative result. Considering the quality of the data leads us to the issues of reliability and validity of our measurements.

Reliability, also often referred to as consistency, precision, or reproducibility, can be agreement between different observers or tests (interobserver), or the same observer or test at different points in time (intraobserver). In assessing reliability, it is important to employ a statistic that accounts for chance agreement (Fleiss 1981). For categorical data, the kappa statistic is most often used, and for continuous data the intraclass correlation can be employed (Snedecor and Cochran 1980). These measures are probably preferred to the Pearson chi square test for categorical data and the Pearson correlation coefficient for continuous data in that they correct for chance agreement and thus do not inflate the actual reliability of the two measures.

Validity is the "truth" of the measurement, and is closely related to accuracy and bias (Fletcher et al. 1988). Although there are several types of validity (Aday 1989), such as content, criterion, and construct, the most common form employed in medical diagnosis is criterion validity. The performance of the test under consideration is compared with that of a criterion measure, or "gold standard", that is the currently accepted definition of the disease under study. It is essential to note that a test may be both valid and reliable, neither valid nor reliable, or one (valid or reliable) without the other. Three important issues in the validity of a test are sensitivity, specificity, and positive predictive value.

Briefly, sensitivity is the proportion of persons with disease who have a positive test result, specificity is the proportion of persons without disease who have a negative test result, and positive predictive value is the proportion of all positive tests that are true positives (Fletcher et al. 1988, Sackett et al. 1985).

Positive predictive value takes us to our final point concerning diagnosis: Changes in the prevalence of the disease under study do not affect sensitivity or specificity (if the gold standard is valid) but do change the positive predictive value. For increasingly rare diseases (decreasing prevalences) the positive predictive value of a given test, with a given sensitivity and specificity, decreases; that is, the proportion of all positive tests that are true positives becomes smaller. It is important to realize this in interpreting the results of tests applied to large populations.

Suffice it to say that clinical epidemiologic issues surrounding diagnosis have rarely been applied to olfactory disorders. While intraobserver reliability has been partially addressed for some tests of olfactory dysfunction (Doty et al. 1984a, 1985), interobserver (between tests) reliability has not been studied in several important ways e.g., how well do several different tests of olfactory function agree in their ability to classify a general population into normal and abnormal groups. Furthermore, as a widely accepted gold standard of olfactory dysfunction has not been achieved, with a few exceptions, none of the issues of test validity have been addressed. Gent et al. (1987) reported on the sensitivity, specificity, and positive predictive value of a number of questions for the clinical history of taste and smell disorders. Using a clinical definition of olfactory dysfunction, the question "Do you have trouble smelling?" had a sensitivity of 0.95 (95% of persons diagnosed with olfactory dysfunction answered "yes" to this question) but a specificity of only 0.64 (only 64% of persons diagnosed as normal answered "no" to the question). Not surprisingly, the specificity of this question was much lower for nasal/sinus disease or post-upper respiratory infection as the cause of the olfactory dysfunction (0.18 and 0.14, respectively). The authors' conclusion, "It seems that a good way to screen for olfactory deficit is simply to ask patients to describe their ability to smell", may be overly optimistic. It is known that clinic-based referral populations are a poor setting in which to determine sensitivity and specificity (Sackett et al. 1985, Fletcher et al. 1988), thus these estimates are probably significantly inflated. It would be expected that patients referred to specialized smell and taste clinics would state that they have a problem with their sense of smell and then be found to have abnormal tests of olfactory ability. This is unlikely to be the case for the general population, which is the setting in which the sensitivity and specificity of this question should be established. In fact, data from paint-manufacturing workers suggested that self-reported olfactory ability (normal or abnormal) was neither a sensitive nor specific indicator of olfactory function (Schwartz unpublished data). Nevertheless, this approach to diagnostic issues i.e., establishment of sensitivity, specificity, and positive predictive value, in olfactory dysfunction, should be applied to all diagnostic tests in the field.

Finally, Goodspeed et al. (1987) reported that the total circulating eosinophil count may be a useful screening test for nasal/sinus disease in patients complaining of chemosensory dysfunction. The sensitivity of this test in their clinic-based population was 0.66 and the specificity was 0.67. Even with these optimistic estimates of the test's characteristics, such values for sensitivity and specificity would make a poor screening test.

2. *Prognosis*

Only a few points will be made about prognostic issues in olfactory disorders. Prognosis consists of the probabilities of a set of outcomes following a symptom, sign, or disease (Longstreth et al. 1987). What we want to know about, the natural history of the condition from its start to end, is often substituted for by the clinical course of the disease from diagnosis, and influenced by intervention. In the interpretation of disease prognosis, it is important to consider issues of bias in the selection of diseased and non-diseased subjects. For example, it is well known that clinic-based studies often suggest much worse disease outcome, higher disease rates, and skewed causal factors than general population based studies (Sackett et al. 1985). This is probably due to selection bias in the subjects studied.

To date, studies of olfactory dysfunction have not been studied in general populations but rather in subjects referred to specialized clinics (Cain 1989, Deems et al. submitted), or in working populations (Schwartz et al. 1989, Schwartz et al. 1990). With few exceptions e.g., Deems et al. (1990), there are very few data on the natural history or clinical course of olfactory disorders.

3. Treatment

The most relevant issue of the treatment of olfactory dysfunction is the efficacy of such treatment. The benchmark type of study for the evaluation of therapeutic efficacy is the randomized controlled trial. Subjects are randomly assigned to treatment or no treatment groups in a "double-blind" design where neither the patient nor the investigators are aware of the group assignment. There are many variations in such experimental designs, such as crossover or self-controlled designs, that are better suited to the study of some treatments (Louis et al. 1984).

There are no therapies that have been subjected to such epidemiologically rigorous evaluation in the treatment of olfactory dysfunction. Scott (1989b) has reviewed the therapeutic options in a variety of smell and taste disorders, as well as the deficiencies in several studies on the topic. This review showed that, few studies quantitatively measured olfactory function, many studies included very few subjects or lacked controls, and there were no published randomized controlled trials that quantitatively measured olfactory function. Likewise, other therapeutic issues such as compliance with various treatment regimens, side effects, and safety have not been similarly studied. This lack of carefully designed studies of treatment is not surprising in view of the fact that most studies of medical therapeutics in all specialties have not been thorough, controlled, randomized trials. As such, the treatment of olfactory disorders is not different from the treatment of other disorders in its study to date.

C. Descriptive Epidemiology

Despite claims that an estimated 1.5 to 2 million Americans experience some type of olfactory dysfunction (Report of the Panel on Communicative Disorders to the National Advisory Neurological and Communicative Disorders and Stroke Council 1979) and that 1% of Americans are anosmic (Doty et al. 1986a), there are few epidemiologic data which substantiate these claims. An extensive review of the literature failed to reveal any studies which have evaluated olfactory function in normal subjects over time, and thus incidence rate data are not available for olfactory dysfunction.

Some limited prevalence data have been reported. Heywood and Costanzo (1986) used the

Connecticut Chemosensory Clinical Research Center (CCCRC) test to assess the olfactory function of 65 normal subjects who had no history of smell or taste dysfunction, cold or congestion at the time of testing, or previous olfactory testing. The study revealed that 23% of these "normals" were identified as having abnormal olfactory ability; the majority of these were categorized as having mild hyposmia and none were classified as anosmic. A similar study performed by Cain et al. (1983), with the same definitions of olfactory dysfunction identified 10% of 65 "normal" subjects as having abnormal olfactory ability. These proportions of normal individuals with abnormal olfaction, who were presumed to approximate a "general" population, may not be useful epidemiologically. As Heywood and Costanzo (1986) pointed out, the poor test-retest reliability on one component of the CCCRC test, and problems with the normative data for the test, may explain the unexpected high prevalence of abnormal olfactory ability in these "normal" populations and the discrepancy in the estimates from the two studies.

Doty and coworkers (Doty et al. 1984a, Doty et al. 1984b, Doty 1989, Frye et al. 1990, Schwartz et al. 1989) have generated many data that are useful epidemiologically. Defining anosmia as a score of 19 or less, and major dysfunction as a score below the tenth percentile for age on the UPSIT, this group reported in one study that the majority of persons older than 65 years of age evidenced marked olfactory dysfunction. Almost 50% of persons over the age of 80 years were anosmic or nearly anosmic, another 30% evidenced major olfactory impairment, approximately 25% of persons between 65 and 80 years were anosmic or nearly anosmic, and another 35% of this group had major dysfunction (Doty et al. 1984a). These prevalence data also suggested that very few persons between the ages of 20 and 50 years were anosmic (had scores below 19). These normative data from 1955 subjects were generated from the study of hospital employees, residents of homes for the elderly, persons attending regional health fairs and other public events, primary and secondary grade school students, and children enrolled at summer camps. Although these are the best data approximating a general population to date, important potential selection factors may have produced normative data biased in one direction or the other. For example, without an estimation of the percent of subjects who participated at each site relative to the total eligible to participate at each testing site, it is difficult to know whether a self-selection bias may have been present.

In a recent cross-sectional study of 731 chemical workers whose mean age was 42.9 years (range 17-69), the mean overall UPSIT score (SD) was 36.4 (4.0) (Schwartz et al. 1989). Nine workers (1.2%) scored 19 or below and thus could be classified as anosmic or nearly anosmic, which is in good agreement with the earlier prevalence data. The tenth percentile on the UPSIT for the plant was 32, which is in good agreement with what would be expected for the ages of these workers from the normative data. The data also suggested that, if major dysfunction is defined as an UPSIT score less than the tenth percentile by age on the normative data, another 10% of these workers evidenced major dysfunction at the time of the testing.

Another cross-sectional study of the olfactory ability of 185 male paint-manufacturing workers with a mean age of 41.6 years (range 24 - 63), revealed a mean UPSIT score and (SD) of 34.2 (5.9) (Schwartz et al. 1990). Nine workers (4.9%) scored 19 or less on the UPSIT and thus would be classified as anosmic or nearly ansomic. The tenth percentile for the two plants was 28, which is significantly lower than the normative data suggests would be expected in this population of workers (the tenth percentile for males in this age range should be approximately 32-34). Compared to the normative data, the UPSIT scores for males between the ages of 20 and 55 years, indicated that 27.6% of these workers would be classified as having a major dysfunction. As will be discussed below, these UPSIT scores suggesting poor olfactory ability in the workers at the two plants were probably due to chronic inhalational exposure to mixed hydrocarbon solvents. The data revealed a high prevalence of olfactory dysfunction in the two paint manufacturing plants studied and also proved the UPSIT normative data to be useful in this context.

The foregoing discussion of the descriptive epidemiology of olfactory dysfunction would suggest the following conclusions. First, there is no information on the yearly incidence of olfactory dysfunction in any population. Second, the prevalence of anosmia, mainly generated by assessment of olfactory identification ability with the UPSIT, is approximately 1%, and this prevalence increases with age, resulting in the majority of persons older than 65 years evidencing some olfactory impairment. Third, approximately 10% of persons evidence major olfactory impairment in cross-sectional studies, but this proportion can be much higher in selected working populations. This estimate, however, is somewhat circular. Until an agreed upon definition of major impairment is developed, the tenth percentile on available normative data is being used to define hyposmia, and thus 10% of persons in cross-sectional studies should fall at or below this level. Fourth, little is known about the natural history of these olfactory deficits as there are very few prospective data.

IV. RISK FACTORS FOR OLFACTORY DYSFUNCTION

Several reviews have been written on the factors thought to cause anosmia or decreased olfactory ability (Schneider 1967, Doty 1979, Schiffman 1983, Feldman et al. 1986, Estrem and Renner 1987. Important causes of olfactory dysfunction in these reviews included drugs, inhaled chemicals, nervous system disorders, psychiatric disorders, nutritional deficiencies, trauma, therapeutic irradiation of the head and neck, congenital anomalies, endocrine disorders, neoplasms, local obstructive processes, the aging process, and viruses. The evidence on which these claims are based is highly variable. First, many are based on case reports, which from an epidemiologic viewpoint, cannot support causal inferences and thus constitute weak evidence that would need to be further evaluated in rigorously designed studies (Kleinbaum et al. 1982). The best data on which these claims rely are cross-sectional studies. However, there are no studies of the cohort or case-control design, which are thought to provide stronger evidence for causal inference (Kleinbaum et al. 1982) than those that have been performed to date. Second, many studies fail to quantitatively measure olfactory function. Third, a large source of the research reports on these factors are from smell and taste clinics with highly selected referral populations. The studies are often the descriptive epidemiology of case series (without controls), a design that is not considered to provide much evidence for causal inference (Kleinbaum et al. 1982, Rothman 1986). Such patients likely represent the most severe, irreversible, and recognizable of patients with olfactory dysfunction. Furthermore, there are likely to be selection factors by medical insurance coverage. For example, the average socioeconomic status and educational level of these clinic patients has been shown to be high (Deems et al. submitted). Thus, several of these "causes" of olfactory dysfunction would have to be considered preliminary and as such would need corroborating evidence.

A. Non-occupational Factors

Several studies provide some evidence on a number of important risk factors for olfactory dysfunction. The best studied non-occupational risk factors are age, age-related neurologic diseases including Alzheimer's disease and parkinsonism, gender, cigarette smoking, head trauma, nasal/sinus disease and upper respiratory infections. These will be discussed in order below.

The effect of age on olfactory function has been the subject of several reviews and research reports (Doty et al. 1984b, Cain and Stevens 1988, Doty 1988, Doty and Snow 1988, Wysocki and Gilbert 1988) and the topic is covered in this book (Chapter 9). Briefly, although the effects of age are somewhat unresolved (Cain and Stevens 1988, Wysocki and Gilbert 1988), it is thought that age effects performance on a number of tests of olfactory function. Furthermore, a number of age-related diseases, such as Alzheimer's disease and parkinsonism, are known to be associated with olfactory deficits (Ansari and Johnson 1975, Doty et al. 1987, Doty et al. 1988b, Koss et al. 1988, Talamo et al. 1989). It is not known whether the effects associated with age are due to the aging process itself or are the cumulative effect of a lifetime of exposure to viruses, airborne chemicals, or cigarette smoke. It seems clear, however, that the olfactory dysfunction associated with Alzheimer's and Parkinson's diseases are not an effect of age itself. Many epidemiologic issues surrounding the effects of age are unresolved, mainly concerning the age-specific incidence of olfactory dysfunction. Doty et al. (1984b) reported that the <u>prevalence</u> of olfactory dysfunction is quite high in persons over the age of 65 years and approaches 80% in persons over the age of 80 years. Until large-scale prospective studies are performed, many of the effects of age will remain unresolved. However, age is an important risk factor for olfactory dysfunction and thus is an important potential confounding variable that should be controlled for in epidemiologic studies.

In almost all studies the effects of gender have been remarkably consistent: women perform better than men on several different tests of olfactory ability at all ages (Doty 1986, Wysocki and Gilbert 1988). This makes gender an important potential confounding variable to control for in epidemiologic studies. The lifetime cumulative incidence in males and females and the relative risk of olfactory dysfunction associated with being male at different ages, among other epidemiologic issues, are not known.

Cigarette smoking and its effects on olfactory function has been the subject of several published reports (Moncrieff 1957, Joyner 1964, Venstrom and Amoore 1967, Ahlstrom et al. 1987, Frye et al. 1990). Although some controversy remains, the weight of the evidence indicates that cigarette smoking adversely affects the sense of smell. For example, a recent cross-sectional study of 638 subjects reported that current smokers were almost twice as likely to exhibit an olfactory deficit as persons who never smoked; the odds ratio for cigarette smoking adjusted for confounding variables was 1.9 ($p = 0.05$) (Frye et al. 1990). Furthermore, a dose-response relation was revealed in the data for both previous and current smokers: As lifetime smoking dose in pack-years increased UPSIT scores significantly decreased. In previous smokers the data suggested that olfactory function improved in relation to time elapsed since cessation of smoking. The effects of cigarette smoking, therefore, may be reversible, at least to some extent, but cross-sectional studies are not well-suited to evaluate this question.

Head trauma has been found to be an important risk factor for olfactory dysfunction. It has been estimated that 5-10% of persons suffering head trauma experience loss of smell after their injury (Leigh 1943, Sumner 1964, Costanzo and Becker 1986, Scott 1989a). The risk seems to increase with the severity and location of the head injury, with injury to the occipital region providing the highest risk. Olfactory loss with head trauma is generally severe and up to 80% of persons with olfactory dysfunction after head injury are anosmic (Costanzo and Becker 1986). In Sumner's classic study of 1167 patients with head injury, 39% of persons with post-traumatic anosmia recovered, but if the duration of amnesia after head injury exceeded 24 hr, 90% of persons had permanent anosmia (Sumner 1964).

The influence of nasal/sinus disease on olfactory function has generally been studied in the clinical setting, and thus few data of epidemiologic use are available. In one study from the University of Connecticut over 600 patients were evaluated for olfactory dysfunction. This group, however, may not have been a consecutive series and may have included a disproportionate number of persons amenable to therapeutic intervention (Scott 1989a). Nasal/sinus disease was clinically thought to be the cause of olfactory dysfunction in 30% of these patients, and in 74% of the patients the olfactory loss was complete. Most of the patients with nasal/sinus disease and olfactory dysfunction were male (65%)

and 74% of persons were between the ages of 35 and 65 years. In a consecutive case series of 750 patients at the University of Pennsylvania, Deems et al. (submitted) have reported that 16.4% of men and 13.0% of women had nasal/sinus disease as the apparent cause of their chemosensory deficit. The nasal/sinus disease patients were intermediate in their olfactory loss between head trauma patients and upper respiratory infection patients and were less likely than these latter two groups to complain of parosmias. When retested months to years later, the nasal/sinus disease patients had a slight but significant improvement in their olfactory function.

The situation with olfactory loss after upper respiratory infection is similar to that with nasal/sinus disease in that data are available from the clinical setting only. Henkin et al. (1971, 1975) reported that of 143 consecutive patients evaluated at the National Institutes of Health for smell and taste disorders, 59% were thought to have these problems because of a preceding viral illness. This syndrome was termed post-influenza-like hypogeusia and hyposmia (PIHH) and the viral illness was described by those affected as "the worst cold I ever had" (Henkin et al. 1975). The authors, without the benefit of data, estimated "an incidence (of PIHH) of 1 in 400 in the United States, approximately 500,000 patients" (Henkin et al. 1975), but may have meant a prevalence of 1 in 400. Scott (1989a) reported that 16% of the patients studied at the University of Connecticut had olfactory dysfunction thought secondary to an upper respiratory infection; these patients were generally older, predominantly female (73%), and usually hyposmic in their olfactory loss. Deems et al. (submitted) reported that 29.5% of women and 20.8% of men studied could have their olfactory dysfunction attributed to a preceding upper respiratory infection. This upper respiratory infection group was older than the nasal/sinus disease or head trauma patients and did not evidence any improvement in deficits over time upon retesting.

It should be noted that the incidence and risk of olfactory dysfunction after many of the aforementioned "exposures" is not known. For example, the risk of olfactory dysfunction after a single upper respiratory infection, the rate of development of olfactory dysfunction in patients newly diagnosed with nasal/sinus disease, the "relative risk" of olfactory dysfunction associated with each of these conditions (with the exception of cigarette smoking), and the percent of persons in the general population with olfactory dysfunction attributable to each of these potential causes, remains unstudied.

B. Occupational Factors

Occupational and environmental exposures are an important cause of olfactory dysfunction and the subject has been extensively reviewed by Amoore (1986b). Although there is a large literature, many of the published studies are reports of cases or case series and cannot provide information for causal inference. Few of these studies measured exposure quantitatively, few employed rigorous epidemiologic methods, and the quantitative measurement of olfactory function has generally not been performed (Amoore 1986b, Schwartz et al. 1989). There have been no studies of the case-control design which use incident cases of olfactory dysfunction and no studies of the cohort design. Two recent cross-sectional studies do provide evidence that airborne exposure to chemicals in the occupational setting can cause olfactory loss.

Exposure to acrylate and methacrylate vapors, two chemicals (monomers) used to make acrylic plastics and thus in widespread use, has been implicated in toxicologic studies in rodents to cause dose-related and specific histologic damage to the olfactory neuroepithelium (Schwartz et al. 1989). A recent study of 731 workers at a chemical facility which manufactures these chemicals revealed that workers who ever worked in areas with exposure to these chemicals were nearly three times as likely to evidence olfactory dysfunction as workers without such exposure (exposure odds ratio adjusted for potential confounding variables was 2.8, $p < 0.05$). The relative risk was much higher in persons who never smoked (odds ratio adjusted for confounding variables was 13.5, $p < 0.05$), but only slightly elevated for current smokers and not elevated for previous smokers. A relative risk of 13.5 is substantial; to put this in perspective, the relative risk of lung cancer associated with cigarette smoking is between 7 and 10. This differential effect of chemical exposure in never smokers, current smokers, and previous smokers was of great interest, and has been found in other studies (see below). The data also revealed a dose-response relationship between the risk of olfactory dysfunction, measured using the UPSIT, and increasing lifetime cumulative exposure to these chemicals (Table 1).

TABLE 1 : Crude and adjusted odds ratios for the association of increasing acrylate/methacrylate exposure and olfactory dysfunction in 154 chemical workers (from Schwartz et al. 1989)

Cumulative Exposure Score*	Crude Odds Ratio	Adjusted** Odds Ratio
0	1.0	1.0
0.1 - 4.9	1.7	1.4
5 - 13	2.6	4.2
> 13.1	1.9	2.4

Test for linear trend of odds ratios across exposure categories (from logistic regression, controlling for confounding variables): $p < 0.05$.

* The cumulative exposure score was obtained as follows: 1) Jobs at the plant were assigned one of three exposure scores (0, 1, or 2), corresponding to no, low, or high exposure, respectively. 2) The job exposure score was next multiplied by the duration (years) the job was held. 3) These products were then summed for all jobs ever held at the plant.

** Adjusted for cigarette smoking history, ethnic/racial group, medications, age, educational level, medical history, gender, and work shift tested.

The data were less clear on the issue of the reversibility of these deficits (the cross-sectional design of the study was not suited to answering this question), as the crude (not adjusted for confounding variables) exposure odds ratios decreased with increasing duration since last exposure to the chemicals but the adjusted odds ratios (controlling for confounding variables) showed no decrease.

Hydrocarbon solvents, chemicals with many uses e.g., paint thinners, degreasing agents, and cleaning agents, have been implicated to cause neurotoxic syndromes in the central and peripheral nervous system (Baker and Fine 1986, Spencer and Schaumberg 1985). Several reports have implicated solvents to cause olfactory dysfunction in humans (Emmett 1976, Ryan et al. 1988). A recent cross-sectional study of 187 paint-manufacturing workers for whom detailed occupational exposure data were available, revealed dose-related declines in olfactory function (on the UPSIT) associated with increasing lifetime exposure to mixed organic solvents (Schwartz et al. 1990). The exposure variable equally weighted intensity of exposure (concentration in parts per million [ppm] for each job the worker ever held) and duration of exposure (years each job was held). The data revealed a significant and large dose-related decline in UPSIT scores with increasing exposure to solvents in never smokers only ($p = 0.01$) after adjustment for confounding variables (Table 2). Previous and current smokers did not evidence this dose-related decline but had UPSIT scores lower than would be expected for their age. The differential effect of chemical exposure in never smokers and ever smokers was statistically significant (p-value for interaction term was 0.03). The effect of solvent

TABLE 2 : Adjusted* mean UPSIT scores in Smoking by exposure groups in 187 paint manufacturing workers (from Schwartz et al. 1990)

Lifetime Solvent Exposure, ppm-years	Smoking Group	
	Never** (N)	Ever (N)
< 68	35.3 (24)	34.5 (37)
68 - 170	33.8 (26)	34.6 (36)
> 170	31.1 (20)	34.8 (42)

* Adjusted for Wechsler Adult Intelligence Scale - Revised Vocabulary score and age
** Two never smokers were missing data.

Note: Smoking - exposure interaction, $p = 0.03$; test for linear trend across exposure groups, $p = 0.01$.

exposure on olfactory test scores in never smokers was large; for exposures > 170 ppm-years (N = 20 never smokers), the mean UPSIT score was 31.1, which is in the abnormal range for this age group. Analysis of the olfactory deficits by UPSIT suggested that the olfactory loss may have been confined to a subset of odorants when compared to workers examined by Schwartz et al. (1989). However, this finding will have to be confirmed in future studies before it can be concluded that the effects of inhaled solvents are not general.

The differential effect of chemical exposure on different categories of smokers has thus been suggested by two studies. The authors hypothesized that cigarette smoke may induce olfactory mucosal cytochrome P-450 enzymes and thus degrade inhaled solvents to inactive metabolites before they exert their toxic effect on the olfactory mucosa (Schwartz et al. 1990). It should be noted that induced cytochrome enzyme systems could metabolize inhaled chemicals to more toxic compounds, and thus the effects of some chemicals in cigarette smokers may be exaggerated. Further study will be necessary to resolve this issue as there are many potential sites of action for the toxic effects of inhaled chemicals on the olfactory system.

Ryan et al. (1988), in a study of the effects of solvents on the central nervous system which revealed an association between parosmia (defined as an unpleasant sensation of odors; the authors termed this as cacosmia) and disorders of learning and memory, postulated that solvents affect rhinencephalic structures in the brain. As limbic structures are thought to be very sensitive to toxic insults and because learning, memory, and olfaction are served by centers in close proximity in the central nervous system, this hypothesis was interesting and plausible. A recent study, however, revealed no association between solvent-associated olfactory dysfunction and deficits in learning and memory in subjects in whom solvent exposure, learning and memory, and olfactory function were all quantitatively measured (Schwartz et al. in press). In the latter instance, memory was measured with the Benton Visual Retention test, Wechsler Memory Scale (WMS), Delayed Logical Memory test, WMS Visual Reproductions test, and the WMS Logical Memory test. Learning was measured with the Rey Auditory Verbal Learning test, Symbol-Digit Learning test, Serial Digit Learning test (interviewer-administered with verbal stimuli), and a computer-administered Serial Digit Learning test (visual stimuli).

Other studies have evaluated olfactory function in workers exposed to solvents. Ahlstrom et al. (1986) measured the detection thresholds and perceived intensity responses to four odorants of 20 tank cleaners exposed to a mixture of industrial solvents, and of 40 controls (20 office workers and 20 watchmen). Although this study controlled for smoking by matching on current smoking status, this may have been an inadequate control for the effects of cigarette smoking (Frye et al. 1990, Schwartz et al. 1990). The study assessed current solvent exposure only and found an association between such exposure and elevated odor detection thresholds for n-butanol and oil vapor. The effects of current exposure could not be distinguished from the effects of cumulative exposure to these agents.

In a recent study, Sandmark et al. (1989) failed to find an association between solvent exposure and olfactory dysfunction (on the UPSIT) in 54 painters and 42 unexposed controls. However, the study had several important limitations. In particular, there were no quantitative measurements of solvent exposure; workers were classified as exposed or unexposed on the basis of current estimated exposure only. In addition, smoking was controlled for as current or not current, which would fail to detect the possible differential effect of solvent exposure in never smokers.

Harada et al. (1983) examined the olfactory function of workers exposed to sulfur dioxide and ammonia in a chemical plant that manufactured a variety of acids and fertilizers. One thousand workers were employed at the plant; 76 of 94 workers organized by a trade union returned a health questionnaire. Twenty of these 76 workers complained of olfactory disorders and had higher olfactory detection thresholds to five substances when measured using the standardised "T & T Olfactometer" (Tagaki 1989) than a comparison group. Interpretation of the data was hampered by a lack of exposure data and selection biases in exposed and control workers.

Olfactory dysfunction has also been associated with exposure to nickel and cadmium dusts, consisting of nickel hydroxide and cadmium oxide, in the manufacture of alkaline batteries (Adams and Crabtree 1961). Results from this study of 106 battery workers and 84 age-matched controls was hampered by the lack of a standardized quantitative test of olfactory function, the use of an odorant (phenol) that has irritant effects, and only partial control of important confounding variables such as control of the intensity of cigarette exposure, non-cumulative cigarette exposure, and no adjustment for differences in age distribution, etc. The data revealed a prominent reduction in olfactory sensitivity in the battery workers. The authors also assessed the workers' self-report of olfactory ability (good, diminished, or none) and found remarkable agreement between this self-report and the threshold for the detection of phenol. Although the latter study had several methodologic problems, a recent study of cadmium-exposed workers also reported decreased olfactory ability after such exposure (Rose et al. 1986). This more recent study assessed cadmium exposure with urinary levels of the metal as well as a marker of cadmium-induced kidney damage (beta-2-microglobulinuria) and found lower olfactory function scores in workers with high urinary cadmium levels and microglobulinuria. Previous research in cadmium workers has indicated a correlation between external exposure to cadmium and the two biological measures of internal dose used by Rose et al. (Lauwerys 1983).

Multiple chemical sensitivities (MCS), a syndrome characterized by complaints of a heightened sense of smell and somatic hypersensitivity to environmental chemicals, is a disorder of increasing importance in environmental and occupational medicine (Cullen 1987). Doty et al. (1988a) reported that 18 patients operationally defined to have MCS had equivalent olfactory thresholds, higher nasal resistance, and higher Beck Depression Inventory scores than 18 controls matched on age, ethnic background, smoking habits, and gender. Future studies of MCS and olfaction are needed to more fully define the olfactory ability of these patients.

V. RESEARCH ISSUES AND DIRECTION

The application of epidemiologic methods and principles to the study of olfactory disorders has been very incomplete to date. As such, there are many important unresolved areas of epidemiologic research and many areas of controversy that need to be resolved before large scale epidemiologic studies can be commenced. What follows are some general suggestions for proceeding in epidemiologic studies of olfactory disorders.

1) Obtain consensus on epidemiologically useful definitions of olfactory dysfunction, including such terms as anosmia and hyposmia or major and minor dysfunction. These could include clinical or test-based definitions of the olfactory disorders.
2) Evaluate the reliability (within test and between tests) and validity (including sensitivity and specificity) of available diagnostic tests against the gold standard of olfactory dysfunction defined above. Calculate the positive predictive value of each test in populations with varying prevalences of olfactory dysfunction.
3) Assess the prevalence of olfactory dysfunction in general populations, not clinic-based, self-selected, or referral populations.

4) Follow the normal individuals from such prevalence studies for the development of olfactory disorders to determine the incidence of these disorders.
5) Perform well designed case-control studies of olfactory dysfunction to identify and quantitate the risk associated with several possible risk factors for these disorders. A number of risk factors are of particular interest, including cigarettes, medications, degenerative neurological disorders, medications, and viral illnesses, since the magnitude of the risk associated with these "exposures" has not been sufficiently measured to date.
6) Follow individuals exposed and not exposed to the risk factors identified in case-control studies to determine the incidence of olfactory dysfunction associated with exposure to the risk factors ie., in the cohort study design.
7) Subject potential therapies for olfactory disorders to rigorous scientific evaluation in randomized controlled clinical trials.

VI. SUMMARY AND CONCLUSIONS

The epidemiologic study of olfactory dysfunction is in its infancy. The available data are generally from cross-sectional epidemiologic studies or from clinical settings, each with its own set of biases and problems in interpretation. Despite these limitations in the data, it can be reasonably concluded that the prevalence of olfactory dysfunction increases with age; that head injury, cigarette smoking and exposure to many airborne chemicals are important risk factors for the development of olfactory loss; that upper respiratory infection and nasal/sinus disease are likely causes of olfactory dysfunction, although the magnitude of the risk and the yearly incidence are unknown, and that olfactory loss is associated with many disease states, especially Alzheimer's disease and parkinsonism. The magnitude of the relative risk associated with cigarette smoking appears to be moderate while the relative risk associated with exposure to some airborne chemicals may be very high. There is a great need for case-control studies to estimate the relative risk of multiple potential factors, and for prospective studies to evaluate the incidence of olfactory dysfunction in different populations and subjects with different exposures.

REFERENCES

Adams RG, Crabtree N (1961) Anosmia in alkaline battery workers. Brit J Ind Med 18: 216-221

Aday LA (1989) Designing and conducting health surveys. Jossey-Bass, San Francisco

Ahlstrom R, Berglund B, Berglund U (1987) A comparison of odor perception in smokers, nonsmokers, and passive smokers. Am J Otolaryngol 8: 1-6

Ahlstrom R, Berglund B, Berglund U, Lindvall T, Wennberg A (1986) Impaired odor perception in tank cleaners. Scand J Work Environ Health 12: 574-581

Amoore JE (1986a) Clinical olfactometry: improved convenience in squeeze-bottle kits; and a portable olfactometer. Chem Sens 11: 576-577

Amoore JE (1986b) Effects of chemical exposure on olfaction in humans. In: Barrow CS (ed) Toxicology of the nasal passages. Hemisphere Publishing, Washington DC, p 155

Ansari KA, Johnson A (1975) Olfactory function in patients with Parkinson's disease. J Chronic Dis 28: 493-497

Angle J, Wissmann DA (1980) The epidemiology of myopia. Am J Epidemiol 111: 220-228

Baker EL, Fine LJ (1986) Solvent neurotoxicity: the current evidence. J Occup Med 28: 126-129

Cain WS (1989) Testing of olfaction in a clinical setting. Ear Nose Throat J 68: 316-328

Cain WS, Gent J, Catalanotto FA, Goodspeed RB (1983) Clinical evaluation of olfaction. Am J Otolaryngol 4: 252-6

Cain WS, Rabin MD (1989) Comparability of two tests of olfactory functioning. Chem Sens 11: 479-485

Cain WS, Stevens JC (1988) Uniformity of olfactory loss in aging. Ann NY Acad Sci 510: 29-38

Costanzo RM, Becker DP (1986) Smell and taste disorders in head injury and neurosurgery patients. In: Meiselman HL, Rivlin RS (eds) Clinical measurement of taste and smell. MacMillan Publishing Co., New York

Cullen MR (ed) (1987) Workers with multiple chemical hypersensitivities. State Art Rev Occup Med 2: 655-806

Deems DA, Doty RL (1987) Age-related changes in the phenyl ethyl alcohol odor detection threshold. Trans Penn Acad Ophthalmol Otolaryngol 39: 646-650

Deems DA, Doty RL, Settle RG, Moore-Gillon V, Shaman P, Mester AF, Kimmelman CP, Brightman VJ, Snow JB (submitted) Smell and taste disorders: a study of 750 patients from the University of Pennsylvania Smell and Taste Center.

Doty RL (1979) A review of olfactory dysfunctions in man. Am J Otolaryngol 1: 57-79

Doty RL (1986) Gender and endocrine-related influences on human olfactory perception. In: Meiselman HL, Rivlin RS (eds) Clinical measurement of taste and smell. MacMillan Publishing Co., New York

Doty RL (1988) Influence of age and age-related diseases on olfactory function. Ann NY Acad Sci 510: 76-86

Doty RL (1989) The Smell Identification Test™ Administration Manual, 2nd edn. Sensonics Inc., Haddonfield, NJ

Doty RL, Deems DA, Frye RE, Pelberg R, Shapiro A (1988a) Olfactory sensitivity, nasal resistance, and autonomic function in patients with multiple chemical sensitivities. Arch Otolaryngol Head Neck Surg 114: 1422-1427

Doty RL, Deems DA, Stellar S (1988b) Olfactory dysfunction in parkinsonism: a general deficit unrelated to neurologic signs, disease stage, or disease duration. Neurology 38: 1237-1244

Doty RL, Gregor T, Monroe CB (1986a) Quantitative assessment of olfactory function in an industrial setting. J Occup Med 28: 457-460

Doty RL, Gregor TP, Settle RG (1986b) Influence of intertrial interval and sniff-bottle volume on phenyl ethyl alcohol odor detection thresholds. Chem Sens 11: 259-264

Doty RL, Newhouse MG, Azzalina JD (1985) Internal consistency and short-term test reliability of the University of Pennsylvania Smell Identification Test. Chem Sens 10: 297-300

Doty RL, Reyes PF, Gregor T (1987) Presence of both odor identification and detection deficits in Alzheimer's disease. Brain Res Bull 18: 597-600

Doty RL, Shaman P, Applebaum SL (1984a) Smell identification ability: changes with age. Science 226: 1441-1443

Doty RL, Shaman P, Dann M (1984b) Development of the University of Pennsylvania Smell Identification Test: a standardized microencapsulated test of olfactory function. Physiol Behav (monogr) 32: 489-502

Doty RL, Snow JB (1988) Age-related alterations in olfactory structure and function. In: Margolis FL, Getchell TV (eds) Molecular neurobiology of the olfactory system. Plenum Press, New York

Eagles EL (1973) A longitudinal study of ear disease and hearing sensitivity in children. Audiology 12: 438-445

Emmett EA (1976) Parosmia and hyposmia induced by solvent exposure. Brit J Ind Med 33: 196-198

Estrem SA, Renner G (1987) Disorders of smell and taste. Otolaryngol Clinics 20: 133-147

Feldman JE, Wright HN, Leopold DA (1986) The initial evaluation of dysosmia. Am J Otolaryngol 4: 431-444

Fleiss JL (1981) Statistical methods for rates and proportions, 2nd edn. John Wiley and Sons, New York

Fletcher RH, Fletcher SW, Wagner EH (1988) Clinical epidemiology: The essentials, 2nd edn. Williams and Wilkins, Baltimore

Fraser GR (1974) Epidemiology of profound childhood deafness. Audiology 13: 335-341

Frye R, Schwartz BS, Doty RL (1990) Chronic dose-related effects of cigarette smoking on olfactory function. J Am Med Assoc 263: 1233-1236

Gent JF, Goodspeed RB, Zagraniski RT, Catalanotto FA (1987) Taste and smell problems: validation of questions for the clinical history. Yale J Biol Med 60: 27-35

Goodspeed DB, Gent JF, Catalanotto FA (1987) Chemosensory dysfunction: clinical evaluation results from a taste and smell clinic. Postgrad Med 81: 251-260

Greenland S, Thomas DC (1982) On the need for the rare disease assumption in case-control studies. Am J Epidemiol 116: 547-553

Gullickson G, Cone J, Quinlan P, Guerriero J, Rosenberg J (1988) Loss of the sense of smell in painters and plumbers. Am Public Health Assoc 116th Ann Meet, Abstr

Harada N, Fujii M, Dodo H (1983) Olfactory disorder in chemical plant workers exposed to SO_2 and/or NH_3. J Sci Labor 59: 17-23

Henkin RI, Larson AL, Powell RD (1975) Hypogeusia, dysgeusia, hyposmia, and dysosmia following influenza-like infection. Ann Otolaryngol 84: 672-682

Henkin RI, Schechter PJ, Hoye R, Mattern CFT (1971) Idiopathic hypogeusia with dysgeusia, hyposmia, and dysosmia. J Am Med Assoc 217: 434-440

Heywood PG, Costanzo RM (1986) Identifying normosmics: a comparison of two populations. Am J Otolaryngol 7: 194-199
Hill AB (1965) The environment and disease: association or causation? Proc R Soc Med 58: 295-300
Hinchcliffe R (1973) Epidemiology of sensorineural hearing loss. Audiology 12: 446-452
Joyner RE (1964) Effects of cigarette smoking on olfactory acuity. Arch Otolaryngol 80: 576-579
Kleinbaum DG, Kupper LL, Morganstern H (1982) Epidemiologic research: principles and quantitative methods. Van Nostrand Reinhold Company, New York
Koss E, Weiffenbach JM, Haxby JV, Friedland RP (1988) Olfactory detection and identification performance are dissociated in early Alzheimer's disease. Neurology 38: 1228-1232
Lauwerys RR (1983) Industrial chemical exposure: Guidelines for biological monitoring. Biomedical Publications, Davis, California, p 17
Leigh AD (1943) Defect of smell after head injury. Lancet i: 138-140
Lilienfeld AM, Lilienfeld DE (1980) Foundations of epidemiology. Oxford University Press, New York
Longstreth WT, Koepsell TD, van Belle G (1987) Clinical neuroepidemiology. II. Outcomes. Arch Neurol 44: 1196-1202
Louis TA, Lavori PW, Bailar JC III (1984) Crossover and self-controlled designs in clinical research. New Engl J Med 310: 24-31
Mausner JS, Kramer S (1985) Epidemiology - An introductory text. WB Saunders Company, Philadelphia
Moncrieff RW (1957) Smoking: its effects on the sense of smell. Am Perfumer 60: 40-43
The National Eye Institute Symposium on the Epidemiology of eye diseases and visual disorders (1983) Am J Epidemiol 117: 129-300
Perry JD, Frisch S, Jafek B, Jafek M (1980) Olfactory detection thresholds using pyridine, thiophene, and phenyl ethyl alcohol. Otolaryngol Head Neck Surg 88: 778-782
Report of the Panel on Communicative Disorders to the National Advisory Neurological and Communicative Disorders and Stroke Council. Washington DC, US Public Health Service (1979), p 319 (NIH publication no. 79-1914)
Rose CS, Heywood PG, Costanzo RM (1986) Impairment of olfactory function in workers chronically exposed to cadmium fumes. Assoc Chemoreception Sci Ann Meet, Abstr
Rothman KJ (1986) Modern epidemiology. Little Brown and Company, Boston
Rothman KJ (ed) (1988) Causal Inference. Epidemiology Resources Inc., Chestnut Hill, Massachusetts
Ryan CM, Morrow LA, Hodgson M (1988) Cacosmia and neurobehavioral dysfunction associated with occupational exposure to mixtures of organic solvents. Am J Psychol 145: 1442-1445
Sackett DL, Haynes RB, Tugwell P (1985) Clinical epidemiology: A basic science for clinical medicine. Little Brown and Co., Boston
Sandmark B, Broms I, Lofgren L, Ohlson CG (1989) Olfactory function in painters exposed to organic solvents. Scand J Work Environ Health 15: 60-63
Schiffman SS (1983) Taste and smell in disease. New Engl J Med 308: 1275-1279, 1337-1343
Schneider RA (1967) The sense of smell in man - its physiologic basis. New Engl J Med 277: 299-303

Schwartz BS, Doty RL, Monroe CB, Frye R, Barker S (1989) Olfactory function in chemical workers exposed to acrylate and methacrylate vapors. Am J Public Health 79: 613-618

Schwartz BS, Ford DP, Bolla KI, Agnew J, Bleecker ML (in press) Solvent-associated olfactory dysfunction: not a predictor of deficits in learning and memory. Am J Psychol

Schwartz BS, Ford DP, Bolla KI, Agnew J, Rothman N, Bleecker ML (1990) Solvent associated decrements in olfactory function in paint manufacturing workers. Am J Ind Med 18: 697-706

Scott AE (1989a) Clinical characteristics of smell and taste disorders. Ear Nose Throat J 68: 297-315

Scott AE (1989b) Medical management of smell and taste disorders. Ear Nose Throat J 68: 386-392

Snedecor GW, Cochran WC (1980) Statistical methods, 7th edn. Iowa State University Press, Ames

Spencer PS, Schaumberg HH (1985) Organic solvent neurotoxicity: facts and research needs. Scand J Work Environ Health 11: 53-60

Sumner D (1964) Post-traumatic anosmia. Brain 87: 107-120

Surjan L, Devald J, Palfalvi L (1973) Epidemiology of hearing loss. Audiology 12: 396-410

Tagaki SF (1989) Human olfaction. University of Tokyo Press, Tokyo.

Talamo BR, Rudel R, Kosik KS, Lee VMY, Neff S, Adelman L, Kauer JS (1989) Pathological changes in olfactory neurons in patients with Alzheimer's disease. Nature 337: 736-739

Taylor H (1981) Racial variation in vision. Am J Epidemiol 113: 62-80

Venstrom D, Amoore JE (1967) Olfactory thresholds in relation to age, sex or smoking. J Food Sci 33: 264-265

Wysocki CJ, Gilbert AN (1988) National Geographic Smell Survey: effects of age are heterogenous. Ann NY Acad Sci 510: 12-28

16

INFLUENCE OF DRUGS ON SMELL FUNCTION

ROBERT G MAIR

LOREDANA M HARRISON

1. INTRODUCTION

Until recently, a book on olfaction would have been unlikely to dedicate a Chapter exclusively to neurochemistry or pharmacology. Over the past decade, however, several seemingly disparate trends have conspired to end this neglect. First, neurochemists have localized a rich variety of neuroactive substances in specific types of olfactory neurons (Halász and Shepherd 1983, Macrides and Davis 1983). The apparent chemical specificity of different cell types, particularly within the olfactory bulb and primary olfactory cortex, suggest that the olfactory system is functionally organized at a molecular or neurochemical level. Second, there has been significant progress towards understanding molecular aspects of olfactory transduction. Recent evidence suggests that there may be common molecular events involving G-proteins and cyclic nucleotide-gated ion channels that lead to neuronal depolarization following odorant activation of membrane-bound protein receptors (Lancet 1986, Snyder et al. 1988). Third, there has been a large increase in clinical olfactory research in humans, demonstrating a variety of drug effects on olfactory perception and describing olfactory deficits in a number of disease states associated with altered neurochemical activity. And fourth, there is increasing interest among physiological psychologists in the neurobiology and psychopharmacology of olfactory learning and memory.

In spite of the progress that has been made describing basic neurochemical mechanisms, there have been relatively few reports on the behavioral pharmacology of olfaction. There are a number of possible reasons for this. Beyond the technical difficulties of studying odor-guided behaviors, the problem of establishing behavioral correlates of neurochemical activity is exacerbated by the lack of a clear conceptual framework for analyzing drug effects within the olfactory system. Drugs can affect olfactory behavior through a variety of mechanisms ranging from the regulation of nasal airflow, to the alteration of receptive or transductive mechanisms, the modulation of activity in afferent olfactory pathways, and the control of more general psychological processes such as attention, motivation, or learning.

This Chapter is concerned with the influence of drugs on behaviors related to olfactory perception. Although much of the literature concerns rodents, an attempt is made to relate results from animal studies to what is known about human olfaction. Agents acting at a number of levels can influence performance on odor-guided tasks. Some of these mechanisms involve processes that are not exclusive to olfaction. For instance, the capacity to perceive odorants is dependent on the state of the narrow nasal airways by which odorants reach the olfactory epithelium. The importance of nasal airflow to olfactory function is shown by the high incidence of conditions affecting nasal patency among patients complaining of anosmia or hyposmia (Deems et al. 1988, Leopold et al. 1988, Goodspeed et al. 1986). There are also several clinical reports in which people with olfactory deficits, presumably related to restricted nasal airflow, have been treated successfully with corticosteroids. Goodspeed et al. (1986), for example, reported that treatment with prednisone (50 mg P.O.) improved olfactory function in six of ten patients with histories of nasal or paranasal disease. The physiological control of nasal airflow and its relationship to olfactory function is complex and beyond the scope of this Chapter (see DeLong and Getchell 1987, Doty and Frye 1989, for reviews).

The present review will focus on neural mechanisms specifically related to olfaction. It will ignore processes such as the control of nasal airflow, or attention and learning that can influence odor perception but involve physiological functions that are not specifically confined to olfaction. In the first part of this Chapter the diversity of mechanisms by which drugs can influence olfactory perception are considered. For the sake of brevity, this section will be limited in its discussion of central mechanisms to the projections of the olfactory bulb to primary olfactory cortex. Without question odor-guided behaviors depend on the activity of higher neural centers, such as the mediodorsal thalamic nucleus and its orbitofrontal projections, or the lateral hypothalamus. However, these systems are involved in many functions that are not purely olfactory in nature and any adequate review of their psychopharmacology would be beyond the scope of this review. The second part of the Chapter focuses on the functional significance of one mechanism, the noradrenergic innervation of the olfactory bulb, to illustrate some of the problems involved in establishing the significance of a specific neurotransmitter system in odor-guided behaviors.

II. THE PHYSIOLOGICAL BASIS OF DRUG ACTIONS ON OLFACTION

A. Receptor Processes

Drugs can influence olfactory receptor function by affecting the normal turnover of receptor neurons, or by interfering with molecular processes involved in the transduction of chemical activity that occurs during odor reception into neural activity. The mature olfactory epithelium contains receptor cells in various stages of maturation (Graziadei 1973, Moulton et al. 1970). Convergent lines of evidence suggest that these receptor cells have a limited lifespan and must be replaced continuously to maintain normal olfactory function in adults (Gesteland 1986). There are several reports that drugs that block mitosis or interfere with cell development produce anosmia, presumably by disrupting the normal turnover of olfactory receptor neurons. One interesting example of a drug that affects receptor plasticity is the commonly used antidepressant amitriptyline. This drug is reported to interfere with the development of chick or rat olfactory receptor cell neurites in vitro (Farbman et al. 1988) and the normal laminar differentiation of rat olfactory bulb in vivo (Chuah and Hui 1986). Currently there appears to be no published evidence that treatment with amitryptyline affects olfactory function. Other studies have reported that human olfactory perception is impaired by treatment with antiproliferative medications that presumably block formation of receptor neurons by mitosis (Schiffman 1983).

There is now compelling evidence that olfactory transduction involves a sequence of events beginning with the interaction of odorant molecules with membrane-bound receptor molecules, activation of adenylate cyclase or other second messenger systems through processes involving G proteins, depolarization of the receptor cell membrane through cyclic nucleotide-gated ion channels, and finally generation of action potentials in the olfactory nerve (Firestein and Werblin 1989, Lancet 1986, Snyder et al. 1988). The apparent role of G proteins and cyclic nucleotide second messengers in the generation of the receptor potential, suggests that there are important similarities in the molecular aspects of transduction between olfaction and other sensory modalities. Although there is no direct evidence that cyclic nucleotide systems mediate olfactory functions in humans, the report that the potency of odorants in humans correlates positively with measures of their tendency to stimulate adenylate cyclase in olfactory receptors cells of animals, provides indirect support for this view (Doty et al. in press). At present there is no evidence that the second messenger systems involved in olfaction can be manipulated in any unique way by drug treatments. Presumably drugs affecting cyclic nucleotide activity may affect olfactory perception, however, there is scant evidence to support this possibility. The pharmacology of cyclic nucleotide systems is reviewed in many textbooks and is beyond the scope of this Chapter.

There are several reports that drugs can affect the activity of olfactory receptor cells in various animal species. The olfactory epithelium contains receptors for diazepam, adrenergic and cholinergic ligands, as well as L-carnosine (Getchell 1986). It has been argued that L-carnosine is the neurotransmitter of the receptor cell, however, there is controversy over its capacity to affect postsynaptic neurons (Margolis 1981). Physiological studies have also shown that receptor cell activity can be increased by exogenous application of acetylcholine or Substance P (Bouvet et al. 1988, Edwards et al. 1987). It is not clear whether these neuroactive substances affect receptor cells directly, or whether they have an indirect effect mediated by perireceptor processes such as glandular secretion of olfactory mucus (Getchell 1986). Shirley et al. (1987) reported that treatment with concanavalin A decreased EOG's evoked by a little more than half of 112 odorants tested. Because this effect was blocked by methyl mannoside, they argue that this lectin disrupts the activity of at least several odorant receptors by binding to a sugar-specific site on one or more cell surface proteins.

B. Olfactory Bulb

Axons from olfactory receptor cells travel along the first nerve through the cribriform plate and into the olfactory bulb where they terminate in areas of dense synaptic activity called glomeruli. The olfactory bulb is a simple cortical structure that receives input from both the receptor epithelium and from higher order areas within the central nervous system. Mitral cells form a prominent layer several hundred microns beneath the surface of the bulb, and with internal tufted cells, they provide the axons of the lateral olfactory tract that terminate in the primary olfactory cortex. These output neurons have primary dendrites that cross the external plexiform layer (EPL) and terminate in the glomeruli, and secondary dendrites that can extend for hundreds of microns laterally within the EPL. For every output neuron, the olfactory bulb contains over one hundred interneurons that form the basis for synaptic interactions at this level. Granule cells are the most prevalent type of interneuron and lack axons. Their cell bodies occupy the deep layers of the bulb and send processes through the mitral cell layer into the EPL. The olfactory bulb also contains several types of short axon interneurons that have cell bodies located in the glomerular layer, external plexiform layer and deep to the mitral cells in the internal plexiform and granule cell layers. The anatomical organization of the olfactory bulb has been the subject of a number of excellent reviews (Macrides and Davis 1983, Scott 1986, Shepherd 1972, 1979, see also Chapter 2 this book).

The olfactory bulb contains most known neurotransmitter candidates including amino acids, neuropeptides, biogenic amines and acetylcholine. Different types of neural elements within the bulb can be distinguished by the neuroactive substances they contain, including centrifugal axons that arise in specific parts of the brain and terminate in the bulb, morphologically defined classes of interneurons, and the output neurons having axons that terminate in the olfactory cortex. The close correlation between morphology and chemistry suggests that different neurotransmitter systems may be involved in distinct functions.

Amino acid neurotransmitters have been identified in several prominent classes of bulb neurons. Gamma amino butyric acid (GABA) has been identified as a neurotransmitter in two types of interneurons, the granule and periglomerular cells, where it is thought to have an inhibitory function (Halász and Shepherd 1983, Macrides and Davis 1983). Physiological (McLennan 1971, Nicoll 1971) and ultrastructural (Ribak et al. 1977) studies have provided evidence that GABA is released from reciprocal dendrodendritic synapses from granule and periglomerular cells onto mitral cells. Granule and periglomerular cells receive synaptic inputs from mitral cells as well as intrinsic processes from bulb interneurons and extrinsic processes arising in the central nervous system

(Scott 1986, Shepherd 1972). Pharmacological studies show that GABA depresses mitral cell activity, and that administration of GABA antagonists, such as picrotoxin or bicuculline, can block the actions produced by GABA, as well as a variety of other substances that are known to depress mitral cell activity (McLennan 1971, Nicoll 1971, Nowycky et al. 1981a, 1981b, Olpe et al. 1987). Thus it appears that GABAergic synapses may serve as a final common pathway by which mitral cell activity is influenced by a number of neural systems.

Behavioral studies have shown that mouse-killing behavior by rats, a widely utilized model of aggression, is sensitive to local manipulations of GABA activity in the olfactory bulb. Drugs that interfere with GABA synthesis (allylglycine) or block GABA receptors (bicuculline, picrotoxin) within the bulb can induce muricide (Mack, cited in Molina et al. 1986). In contrast, mouse-killing behavior is immediately inhibited by treatments that facilitate local GABA activity within the olfactory bulb by stimulating receptors, inhibiting reuptake, or blocking degradation by GABA-transaminase. Taken together, these data suggest that mouse-killing behavior is inversely related to the degree of GABAergic inhibition of mitral cell activity. It is not clear whether mouse-killing behavior is dependent on the general activity of the olfactory system, as controlled by GABAergic mechanisms, or whether it involves a more specific pathway that is mediated by either granule or periglomerular cells.

Although there is strong evidence that mitral cells contain glutamate and aspartate, there is some controversy over whether these amino acids constitute the primary neurotransmitter released by axons of the lateral olfactory tract (LOT) in primary olfactory cortex. Biochemical studies (Halász and Shepherd 1983, Macrides and Davis 1983) have shown that there are high concentrations of these amino acids in the piriform cortex that are reduced following bulbectomy or transection of the lateral olfactory tract. Further, both glutamate and aspartate are released by a calcium dependent mechanism on stimulation of the lateral olfactory tract, and both have been shown to excite neurons in the primary olfactory cortex when applied exogenously. Braitman (1986) argued against either of these amino acids being the neurotransmitter in this pathway after demonstrating that desensitizing piriform pyramidal cells to glutamate and aspartate did not affect the responses of these cells to LOT stimulation. Hori et al. (1981) made a similar argument based on their observation that 2-amino-4-phosphonobutyrate (APB), a presumed glutamate antagonist, blocks the response of primary olfactory cortical neurons to LOT stimulation but not to exogenously applied aspartate or glutamate. More recently, Ffrench-Mullen et al. (1985) raised the possibility that N-acetylaspartylglutamate is the endogenous transmitter of mitral cells based on evidence that this peptide is diminished in concentration in the primary olfactory cortex by bulbectomy, excites pyramidal cells in the piriform cortex when applied exogenously, and this excitation is blocked along with the response to stimulation of the LOT by APB.

There has been speculation recently based on indirect evidence that glutaminergic processes may be involved in the formation of odor memories. Lynch and Baudry (1984) have argued that memory formation may be related to long term potentiation produced by alterations in NMDA-type glutamate receptors by calcium activated proteases. Staubli et al. (1985) showed that leupeptin, a potent protease inhibitor known to block long term potentiation, interferes with olfactory discrimination learning in the same manner that it disrupts memory processes measured by spatial maze performance in rats. More recently Lincoln et al. (1988) provided evidence that NMDA-type receptors in the olfactory bulb may be critical for odor preference learning. When rat pups are exposed to an odorant accompanied by tactile stimulation, they develop a behavioral attraction to that odor that is correlated with an increased uptake of 2-deoxyglucose within glomeruli of the olfactory bulb. Lincoln et al. (1988) report that pups treated with 2-amino-5-phosphonovaleric acid, a specific blocker of NMDA-type receptors, fail to exhibit both this early form of olfactory learning and its associated change in 2-deoxyglucose uptake. Although these data are not conclusive, they raise the intriguing possibility that there may be a chemically specific system within the olfactory bulb that can mediate the formation of simple odor memories.

The mammalian olfactory bulb contains measurable quantities of a number of peptides that may function as neurotransmitters, including carnosine, substance P (SP), met-enkephalin (M-Enk), β-endorphin, β-lipotropin, vasopressin, oxytocin, gonadotropin releasing hormone (GnRH), thyrotropin releasing hormone (TRH), somatostatin, neurotensin, vasointestinal peptide, gastrin, cholecystokinin (CCK), and insulin (Halász and Shepherd 1983). Many of these peptides may be present in centrifugal fibers from the brain that terminate within the olfactory bulb. Immunohistochemical studies have identified specific types of interneurons within the bulb that contain somatostatin, SP and M-Enk (Bogan et al. 1982, Burd et al. 1982, Davis et al. 1982). In addition, the existence of intrinsic TRH containing neurons has been supported by wet assays showing that TRH is reduced by kainic acid lesions that destroy cell bodies in the bulb (Sharif 1988), but not by surgical destruction of centrifugal inputs to the bulb (Kreider et al. 1982). The functional significance of olfactory bulb peptides is poorly understood and, at present, is more a matter of conjecture than of scientific deduction. Substance P is thought to be localized within external tufted cells and possibly within centrifugal fibers. Olpe et al. (1987) showed that SP decreases the activity of neurons within the olfactory bulb. Since, the depression was blocked by bicuculline and picrotoxin and did not occur in a Ca^{++} free/high Mg^{++} medium that interferes with the physiological release of neurotransmitters, it was argued that this depression is mediated by the physiological release of GABA, presumably within periglomerular cells.

Cholecystokinin is considered to be a potential endogenous satiety signal. Since it is released in the gut during feeding, the concentration of CCK in circulation should be greatest at the end of a meal. Furthermore, peripheral administration of CCK has been shown to inhibit feeding in the rat (Schneider et al. 1983). Given the important role of olfaction in the control of feeding behavior, it may be argued that CCK neurons in the olfactory bulb provide a mechanism by which sensory aspects of satiety can be regulated. However, there are several reasons to question this possibility. First, direct administration of CCK within the brain does not have consistent effects on feeding (Schneider et al. 1983). Second, the cessation of eating produced by peripheral CCK is blocked by vagotomy and thus appears to involve some target outside the brain (Smith et al. 1981). Third, the concentration of CCK is higher in the cortex, striatum and hippocampus than in the olfactory bulb or hypothalamus (Dockray 1983), and thus CCK does not appear to be distributed with any selective preference to areas assumed to be important for olfactory regulation of feeding. Although these data are inconsistent with olfactory bulb CCK being critically important in the induction of satiety, the possibility remains that it may serve some other function related to the sensory regulation of appetite.

Gonadotrophin releasing hormone has been localized by immunohistochemistry in the main and accessory olfactory bulbs of the hamster (Macrides and Davis 1983, Phillips et al. 1982), but has been reported to be missing in the bulbs of rats and guinea pigs (Krey and Silverman 1983). Behavioral studies have implicated GnRH in the production of sexual receptivity in rodents (Shivers et al. 1983, Moss and Dudley 1984). Given the critical role of chemosensory cues in stimulating the reproductive processes of rodents, it is plausible that the GnRH containing fibers in the bulb are involved in modulation of sexual receptivity. As appealing as this hypothesis may be, however, it is difficult to reconcile with the absence of GnRH fibers in the bulbs of several rodent species, and with the observation that lordosis is stimulated in rodents by direct application of GnRH in the mesencephalic central gray and blocked by the antibody to GnRH in this same site (Shivers et al. 1983). Like CCK, GnRH is a neuropeptide associated with an olfactory-related behavior that apparently does not depend critically on the peptidergic innervation of the olfactory bulb. At present, it must be considered possible, but not proven, that the GnRH innervation has some role in the regulation of reproduction.

In addition to amino acids and peptides, the olfactory bulb contains dopamine (DA), norepinephrine (NE), serotonin (5-HT) and acetylcholine (ACh) (Macrides and Davis 1983, Halász and Shepherd 1983, and Fig 1). Histochemical studies have provided evidence that some periglomerular and superficial tufted cells are dopaminergic (Davis and Macrides 1983, Halász et al. 1985, Halász and Shepherd 1983; Macrides and Davis, 1983). Apart from these apparent DA-containing neurons, these putative neurotransmitters are thought to be contained in centrifugal fibers from NE containing neurons in the locus coeruleus, DA containing neurons in the substantia nigra and the ventral tegmental area (Lindvall and Björklund 1983), 5-HT containing neurons in the dorsal and median raphe (Steinbusch and Nieuwenhuys 1983), and cholinergic nerve cells in the medial septum-diagonal band complex in the basal forebrain (Macrides and Davis 1983). Norepinephrine, dopamine, serotonin and acetyl choline constitute the known sources of extrathalamic modulation of the cerebral cortex (Foote and Morrison 1987). That is, each arises from cell bodies in nuclei outside major sensory and motor pathways, each of the transmitters appear to alter the activity of postsynaptic neurons in a manner consistent with a modulatory role, and although their patterns of innervation differ, each innervate widespread areas of cortex in a manner inconsistent with the mediation of a single behavioral function. The innervation of the primitive cortex within the olfactory bulb by these same systems, raises the possibility that the bulb may be subject to control by extrinsic neurochemical systems thought to regulate the neocortex.

Exogenous application of ACh has long been known to affect the activity of cells in the olfactory bulb. In early studies (von Baumgarten et al. 1963, Bloom et al. 1964) ACh was found to affect the activity of about 25% of bulb neurons, increasing activity in about 7% and decreasing activity in the remainder. The horizonal limb of the diagonal band (HDB) provides the major cholinergic innervation of the main olfactory bulb (Macrides and Davis 1983). Recently, Nickell and Shipley (1988) reported that while single electric shocks of the HDB have little effect on mitral cell activity, several seconds of shocks delivered at a rate approximating theta (10 Hz) produce a potentiated state in which the bulbar responses to subsequent HDB shocks are greatly augmented. While the physiological significance of this phenomenon is not certain, it is consistent with the hypothesis that rhythmic activity of cholinergic neurons in the medial septal-diagonal band helps coordinate the theta rhythms of the olfactory bulb and hippocampus during bouts of exploratory sniffing associated with olfactory learning (Macrides 1976). Little is known about the effects of cholinergic compounds on olfactory function. It is of note that olfactory function is severely impaired in Alzheimer's disease (Doty et al. 1987), a condition associated with degeneration of forebrain cholinergic systems (Price 1986). However, there is other non-cholinergic pathology associated with this disease that might well account for its associated olfactory deficits (Mair et al. 1986, Reyes et al. 1985).

Early reports indicated that exogenous application of the monoamines NE and 5-HT decreases the activity of olfactory bulb neurons (von Baumgarten et al. 1963, Bloom et al. 1964). In isolated turtle olfactory bulb, application of DA or the DA agonist apomorphine has been reported to decrease the field potentials produced by orthodromic or antidromic stimulation of mitral cells (Halász et al. 1985, Nowycky et al. 1983). Opposite effects were noted for the DA antagonist fluphenozine. Dopamine and apomorphine were also observed to reduce the suppression produced by paired pulse stimulation. Since this suppression is believed to be mediated by dendrodendritic synapses between mitral cells and interneurons (Shepherd 1972), it was argued that DA may modulate mitral cell activity by some interneuronal mechanism. Further, the agonist properties of apomorphine and antagonist effects of fluphenozine are consistent with an effect mediated at a D-2 receptor. The apparent relationship between DA and inhibitory processes within the olfactory bulb, is supported by the recent evidence that DA is co-localized with GABA within superficial interneurons which are thought to inhibit the activity of mitral cells (Gall et al. 1987).

Behavioral studies have indicated that the odor detection performance of rats is improved by low doses but impaired by high doses of amphetamine (Doty and Ferguson-Segall 1987), a drug known to potentiate the effects of dopamine and other monoamines by enhancing the release and blocking reuptake mechanisms. Doty and Risser (1989) have clarified this somewhat ambiguous result by showing that the D-2 agonist quipirole impairs olfactory sensitivity in rats in a dose dependent fashion, an effect diminished by the D-2 receptor blocker spiperone. Thus there is convergence between electrophysiological and behavioral evidence that stimulation of D-2 receptors diminishes activity of mitral cells and decreases performance of rats in tasks measuring olfactory sensitivity.

Parallels between human and animal responses to these drugs are not particularly clear. The apparent hyposmic actions of D-2 agonists would predict that patients treated with neuroleptics should exhibit signs of altered olfactory sensitivity, however, there are no reports that this occurs. Patients with Parkinson's disease, a condition associated with the degeneration of dopaminergic neurons have been reported to be impaired on several measures of olfactory function (Doty et al. 1988a). This observation is in apparent conflict with the simplistic notion that increased dopaminergic activity impairs olfactory sensitivity. The resolution of this conflict is not certain. It may be that Parkinson's disease impairs olfaction through some other non-dopaminergic mechanism, or that the systemic dopaminergic deficits of this disorder disrupts olfactory functions at some other level of the nervous system (Moore and Bloom 1978).

III. NOREPINEPHRINE IN THE OLFACTORY BULB

A. A Case Study

Norepinephrine is probably the most thoroughly studied of the neuroactive substances within the olfactory bulb. There is an extensive literature describing the anatomical, physiological, and behavioral aspects of this system. In addition, there is considerable clinical interest in this system because of the number of conditions associated with diminished NE activity in which olfactory discrimination is also impaired. These conditions include Korsakoff's disease (Mair et al. 1980, 1986), normal aging (Doty et al. 1984), Parkinson's disease (Doty et al. 1988a, Doty et al. 1989), and Alzheimer's disease (Doty et al. 1987).

In the classic studies of the autonomic system, pharmacologists developed a number of standard criteria for establishing that a substance acts as a neurotransmitter (Cooper et al. 1982). First, the substance in question must be synthesized and stored within those neurons in which it has been localized. Enzymes, precursors and substances relevant to its synthesis must also be present. Second, physiological stimulation of these neurons must result in the release of the putative transmitter, and in an interaction with specific postsynaptic receptor sites. Third, mechanisms for inactivation or uptake of the substance should be present. Lastly, exogenous application of the putative transmitter to those areas in which it has been localized must mimic the activity produced by electrical or physiological stimulation. Simply conforming to these criteria does not establish the behavioral and pharmacologic actions of a neurotransmitter. Ideally, there are at least two other criteria that must be met. First, it is necessary to know something about how and when a neurotransmitter is released in an awake behaving animal. Since the effects of neurotransmitters are generally dose-dependent, it is important to understand both the conditions under which it is released and the effective amounts in which it is released. Second, pharmacological manipulations affecting the activity of that neurotransmitter must be shown to alter behavioral performance. Since drugs can have diverse effects, it is important to rule out possible side effects of any treatments.

There is convergent evidence that NE functions as a neurotransmitter in the olfactory bulb. Wet chemical assays in a number of laboratories with a variety of species have demonstrated that the olfactory bulb contains significant concentrations of NE and its metabolites (Halász and Shepherd 1983). Figure 1 shows typical results from our laboratory indicating concentrations of NE and other parent amines and their major metabolites in the olfactory bulb of young adult Long Evans rats.

Observation of tissue treated with histofluorescence methods has demonstrated the existence of catecholaminergic (CA) fibers densely innervating the granule cell layer of the olfactory bulb (Fallon and Moore 1978, Lindvall and Björklund 1974). Several other observations have provided evidence that these fluorescing fibers are the axons of NE containing neurons originating in the locus coeruleus (LC). First, LC lesions deplete the NE content in the ipsilateral olfactory bulb (Fallon and Moore 1978). Second, in rodents these deep fibers exhibit histochemical evidence of being noradrenergic, namely uptake of tritiated NE and labeling with antibodies to dopamine beta hydroxylase (Halász et al. 1985, Macrides and Davis 1983). Third, there is extensive labeling of LC neurons following local injection of horseradish peroxidase into the olfactory bulb or epithelium (Shipley and Adamek 1984, Shipley et al. 1985).

Electrophysiological studies have demonstrated that spontaneous mitral cell activity is decreased by exogenous application of NE (Bloom et al. 1964, McLennan 1971, Salmoiraghi et al. 1964, von Baumgarten et al. 1963), an effect blocked by dibenamine and phentolamine and thus considered to be mediated by an alpha adrenergic receptor (Bloom et al. 1964). There is somewhat conflicting evidence that suggests a complex interaction between the release of GABA and NE in the olfactory bulb. In cats, McLennan (1971) demonstrated that the suppression of mitral cell activity by exogenous NE is blocked by bicuculline, and he argued that NE excites granule cells that inhibit mitral cells through GABAergic dendrodendritic synapses. In isolated turtle olfactory bulb, Jahr and Nicoll (1982) reported that NE decreases paired pulse IPSP's, an effect they ascribe to NE blocking the release of GABA from granule cells. On the other hand, Gervais (1987) has argued that GABA affects the release of NE through a presynaptic mechanism based on the observation that application of GABA increases the release of NE in superfused rat olfactory bulb slices.

There are a number of lines of indirect evidence that NE affects the responses of olfactory bulb neurons to odorous stimulation. Early studies demonstrated that stimulation of midbrain areas associated with ascending NE fibers (Lindvall and Björklund 1974) suppresses olfactory bulb activity (Yamamoto and Iwama 1961) and modulates responses evoked by olfactory stimulation (Mancia et al. 1962). Similarly, peripheral stimulation known to activate the LC (Aston-Jones and Bloom 1981) was shown to have comparable effects (Yamamoto and Iwama 1961, Mancia 1962), whereas cortical activation by stimulation of the intralaminar thalamic nuclei did not. More recently, Gervais and Pager (1979) recorded multiunit activity from the mitral cell layer of chronically implanted unanesthetized rats. They reported that cells were nonresponsive to odorous stimulation during deep slow wave sleep and paradoxical sleep, both levels of arousal associated with the lowest amounts of LC activity (Aston-Jones and Bloom 1981). Gervais and Pager (1983) reported that unilateral depletion of olfactory bulb NE by local infusion of 6-hydroxydopamine impairs the modulation of multiunit bulb responses to food odors according to nutritional state, without disrupting the more general responsiveness of the olfactory bulb to odorous stimulation. In another intriguing line of research, Gray et al. (1986) reported that electroencephalogram (EEG) measures of olfactory bulb activity show evidence of slower habituation during subacute applications of NE by osmotic minipumps to the rabbit olfactory bulb. As opposite effects were observed during subacute application of propranolol, Gray et al. (1986) argued that NE modifies olfactory habituation by activating a beta receptor. However, since others have shown that the effects of exogenous NE on mitral cell activity in rabbits are antagonized by alpha blockers, but not by the beta blocker dichloroisoproterenol (Bloom et al. 1964), the mechanism by which NE affects this EEG habituation is not certain.

B. Turnover of Olfactory Bulb Norepinephrine in Awake Behaving Animals

To understand the role of olfactory bulb NE in odor-guided behavior, it is necessary to determine the conditions under which it is released in awake behaving animals. Chase and Kopin (1968) used a push-pull cannula to measure the efflux of tritiated NE in cerebrospinal fluid surrounding the olfactory bulb of the rat. Although they found evidence of increased NE release during odorous stimulation, they observed comparable results for metabolically inert substances and thus found no evidence for selective neurotransmitter release. Brennels (1974) used a cortical cup to show that tritiated NE efflux is increased selectively in rabbit olfactory bulb during electrical stimulation of the medial olfactory tract, the surface of the bulb near the medial olfactory tract, or the lateral olfactory tract. These data help to establish NE as a neurotransmitter in the bulb, however, they do not indicate whether the increased release of NE was evoked by antidromic excitation of mitral cells, direct activation of NE containing axons that terminate in the olfactory bulb, or as an indirect result of stimulating some other centrifugal input to the bulb.

Recently, Mair et al. (1987) measured turnover of monoamines in the olfactory bulb and cerebellum in awake behaving rats during three behavioral conditions: odor-guided foraging for sucrose pellets hidden in pine shavings, complex motor manipulation to remove a ball from a polyurethane tube, and resting in the home cage (control). On the sixth day of training rats were given an injection of alpha-methyl-paratyrosine (AMPT) (200 mg/ kg ip) 30 min before training to inhibit the synthesis of new catecholamines. The turnover of the catecholamines NE and DA were determined by measuring the reduction in these parent amines following AMPT treatment. The turnover of 5-HT was measured by the concentration of its primary metabolite 5-hydroxyindole acetic acid (5-HIAA). Indices of monoamine turnover were compared between the three AMPT treated groups and the saline treated controls that were allowed to rest in their home cage. The results for the olfactory bulb are shown in Figure 1.

There were several trends in the data. First, the concentration of olfactory bulb NE differed significantly among groups ($F(3,44)=4,44$, $p=0.008$) with the olfactory and motor training groups showing signs of similar turnover that were significantly greater (lower NE) than the saline treated animals (Dunnett's t-test). Thus there was evidence of increased NE turnover for both the odor and motor tasks. Second, the relative rates of NE turnover during the three tasks were similar in the cerebellum and olfactory bulb. Although differences between groups for cerebellar NE did not reach statistical significance, this finding is inconsistent with the existence of gross regional differences in the turnover of NE in the different tasks. Third, there was a significant reduction in 5-HIAA, consistent with diminished turnover of 5-HT, that was restricted to the olfactory bulb of the odor-guided foraging group. Thus 5-HT, but not NE, may exhibit a regionally specific change in turnover within the bulb during the odor-guided task. Taken together,

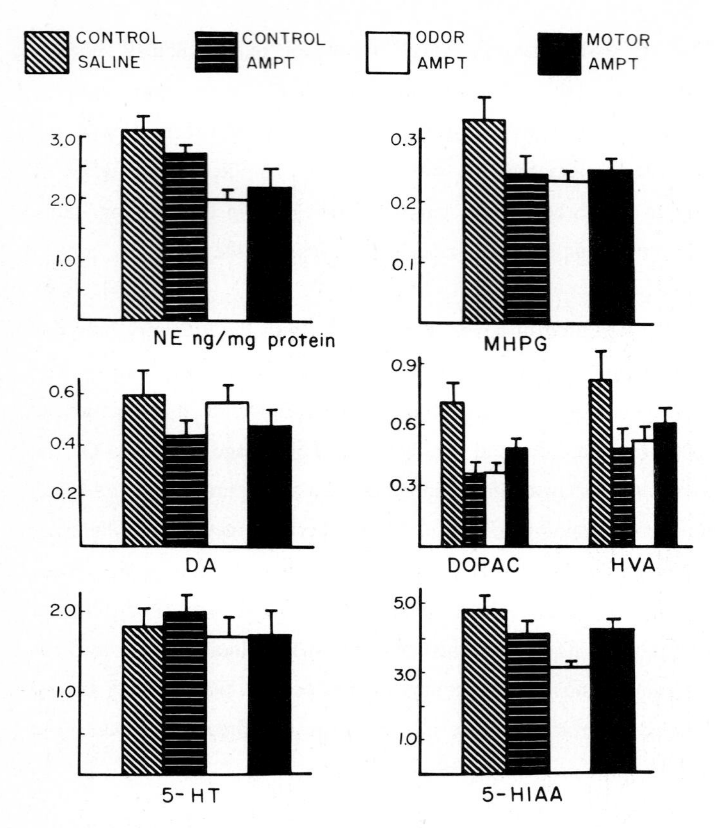

Fig 1. Mean concentrations in the olfactory bulb of norepinephrine (NE) and its metabolite 3-methoxy-4-hydroxy phenylglycol (MHPG); dopamine (DA) and its metabolites dihydroxyphenylacetic acid (DOPAC) and homovanillic acid (HVA), and serotonin (5-HT) and its metabolite 5-hydroxyindole acetic acid (5-HIAA). Concentrations are shown for control animals injected (IP) with saline , or with alpha-methylparatyrosine (AMPT); animals injected with AMPT 30 min prior to 30 min of motor training, and animals injected with AMPT 30 min prior to 30 min of odor-guided foraging. All animals were sacrificed 60 min after injection. See text for details.

these data suggest that the turnover of NE in the olfactory bulb is stimulated bybehavioral activity, whether or not the task involves the active sampling of odorants. Since the concentrations of all catecholamine metabolites were diminished by the AMPT treatment, it appears that this drug reduced catecholamine turnover at the time of sacrifice. Thus the data likely underestimates the actual physiological rate of NE turnover in the awake, behaving animal.

C. Behavioral Changes Following Depletion of Cortical and Bulbar Norepinephrine

Recent studies show that the depletion of olfactory bulb NE impairs the acquisition of conspecific odor discriminations. Rosser and Keverne (1985) demonstrated that depletion of olfactory bulb NE interferes with pregnancy block in female mice. Normally, if a pregnant female is exposed to the odor of any adult male but her mate, the pregnancy is spontaneously aborted. When Rosser and Keverne (1985) depleted NE prior to mating, by injection of 6-OHDA into the medial olfactory tract or the accessory olfactory bulb, exposure to the odor of the stud male also resulted in pregnancy loss. Since similar depletions made after mating neither prevented the formation of an olfactory memory for the stud male nor disrupted previously acquired responses to pheromonal cues, it was argued that there are critical temporal constraints during which NE is important to this type of learning. More recently, Kaba and Keverne (1988) observed comparable results by injecting the alpha blocker, phentolamine, into the accessory olfactory bulb to disrupt NE activity. Alpha-adrenergic blockade, during a four-hour period following mating, also disrupted the formation of an olfactory memory for the stud male.

There are a number of other reports that demonstrate that depletion of NE disrupts the acquisition of new conspecific odor-guided behaviors. In a study similar to that of Rosser and Keverne (1985), Pissonnier et al. (1985) reported specific periods of time i.e., 2-4 hr after giving birth, for the development of maternal recognition in sheep. Injections of 6-OHDA into noradrenergic pathways during this critical period prevented selective, odor-guided bonding between a ewe and her lamb, such that a depleted ewe adopted any lamb as her own. Depletion of NE following this critical period had no effect. Also, in an earlier study, Marasco et al. (1979) reported that systemic injections of 6-OHDA reduced rat pup responsiveness to conspecific odors, but had no effect on the development of preferences for botanical odors. Norepinephrine depleted neonates, housed in cages containing cedar lining, were tested for preferences to fresh cedar shavings versus soiled shavings from home cages. Depleted animals showed a preferential response to the cedar odor, demonstrating a normal attraction to familiar odors despite a reduction in responsiveness to conspecific odor cues.

The odor-guided behaviors shown to be sensitive to NE depletion have a number of common features that prevent their extension to olfaction in general. They are all biologically significant behaviors, they are learned in a short period of time, and they likely involve activity of the vomeronasal organ and its connections in the accessory olfactory bulb. Doty et al. (1988b) studied the effect of NE depletion on a previously learned measure of olfactory sensitivity. They trained rats in a go/no go discrimination of ethyl acetate vs clean air and then injected 6-OHDA into the olfactory bulb to produce a local depletion of NE. Following recovery from this treatment the rats exhibited no decrement in performance, and thus it was argued that olfactory bulb NE is not required for this task. These results demonstrate that NE is not needed for olfactory detection, however, they are not directly relevant to the types of olfactory deficits that others have described in NE depleted animals. First, Doty et al. (1988b) measured detection, not discrimination. Second, they measured performance on a previously learned problem and did not require animals to learn a new discrimination. Studies of conspecific behaviors (see above) have demonstrated that animals lacking bulb NE can perform behaviors based on previously learned olfactory information, but are impaired in their ability to establish new odor-guided behaviors.

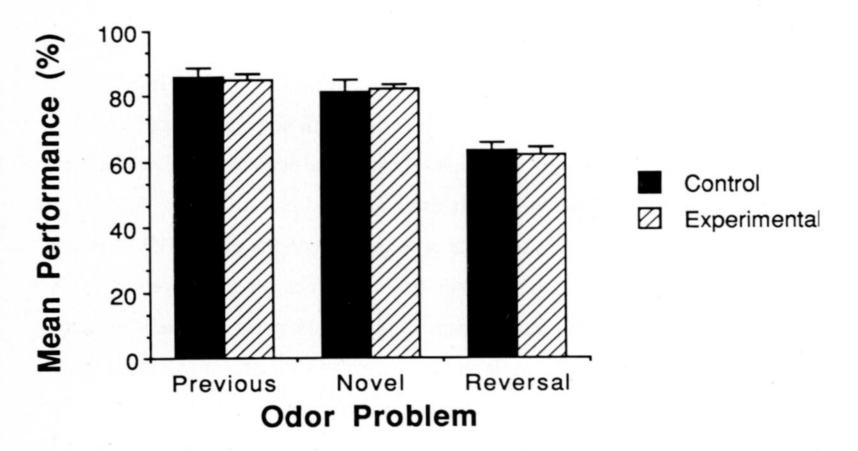

Fig 2. Performance of experimental versus control animals on a) a previously learned odor discrimination, b) a novel odor discrimination, and c) a reversal of the novel odor discrimination following depletion of norepinephrine in the rat cortex and olfactory bulbs. Histograms represent mean group performance as a function of percent correct responses made in each discrimination problem.

To learn whether depletion of NE impairs the acquisition of olfactory discriminations, Harrison and Mair (in preparation) trained rats to perform a two-choice olfactory discrimination task, depleted NE in cortex and olfactory bulb by injection of 6-OHDA into the dorsal noradrenergic bundle (DNB), and then measured the ability of animals to perform both previously learned and novel odor discriminations. To determine whether NE is necessary to perform a two-choice discrimination, performance was compared on the same problem before and after lesioning. The role of NE in olfactory discrimination learning was then analyzed by measuring the ability of NE depleted rats to acquire and reverse a novel discrimination problem. Since rats rapidly learn novel problems once they have mastered the procedures of this task, the speed with which this task was learned was comparable to that of the conspecific behaviors shown by others to be sensitive to the depletion of this neurotransmitter. The results of this study are shown in Figure 2.

As was expected no differences were found in experimental versus control performance during retraining on the odor discrimination task learned immediately prior to surgery. Treatment did not disrupt the ability to perform a discrimination learned prior to NE depletion. Contrary to expectation, however, no significant differences emerged between groups during acquisition of the novel odor discrimination. Injections of 6-OHDA into the DNB did not affect the ability of the animals to detect odorants, nor did they disrupt the acquisition of a new odor discrimination. Finally, no significant differences were observed in performance on the reversal of the novel discrimination task. These results are even more impressive given the widespread depletion of NE obtained by the dorsal bundle treatment. Neurochemical analyses showed significant reductions of NE in the cortex and olfactory bulb of 6-OHDA treated animals.

Taken together the data indicate that NE is not necessary for a variety of odor-guided tasks, including measures of sensitivity, pretrained discrimination, discrimination learning and reversal (Doty et al. 1988b, Harrison and Mair, in preparation). In this context, the impairment shown for tasks involving recognition of mates, offspring and territory must be regarded as being specific to these biologically salient behaviors. It has been argued that brain NE systems are important for selective control of attention under conditions of increased activation or arousal (Robbins and Everitt 1987). If this is the case, then the failure to observe deficits in the laboratory may reflect more on the limitations of the procedures used than on the importance of NE for the execution of natural odor-guided behaviors in the real world. At present, however, we must regard NE as a likely candidate neurotransmitter within the olfactory bulb of uncertain behavioral significance.

IV. SUMMARY

The olfactory system presents unique challenges to psychopharmacologists. Animal studies have provided evidence that drugs can affect olfactory function at a number of levels and through a variety of mechanisms, representing processes that are specific to olfaction and processes related to more general physiological or psychological functions. These include; 1) the transport of molecules through the airways of the nasal cavity to the olfactory epithelium; 2) the turnover and differentiation of receptor neurons; 3) transduction and coding mechanisms of receptor neurons; 4) the activity of interneurons and intrinsic circuits within the olfactory bulb; 5) the activity of centrifugal fibers that arise within the brain and terminate in the olfactory bulb; 6) the activity of higher order neurons within the primary olfactory cortex and its thalamo-neocortical and hypothalamic projections, and 7) processes related to the general regulation of attention, learning, and memory.

The diversity of neurochemical and physiological processes represented by these mechanisms suggests that human olfactory function may be altered by a surprising variety of agents. Until recently, pharmacologists largely ignored olfaction and related chemosensory functions in humans. In the last few years, clinical studies of olfaction have provided a number of examples that illustrate some of the variety of actions by which drugs can influence odor perception. For example, evidence has been presented that the turnover of receptor neurons can be disrupted by both anti-depressive as well as anti-proliferative medications. Other studies have shown olfactory impairments in a variety of clinical states associated with alterations in central catecholamine activity. At present, the catalogue of pharmacological actions is far from complete. It seems evident, however, that clinical studies of olfactory function must pay close attention to the possibility that apparent clinical impairments reflect drug-related and not disease-related deficits.

REFERENCES

Aston-Jones G, Bloom FE (1981) Activity of norepinephrine-containing locus coeruleus neurons in behaving rats anticipates fluctuations in the sleep-waking cycle. J Neurosci 1: 876-886

Bloom FE, Costa E, Salmoiraghi GC (1964) Analysis of individual rabbit olfactory bulb neuron responses to the microelectrophoresis of acetylcholine, norepinephrine and serotonin synergists and antagonists. J Pharmacol Exp Ther 146: 16-23

Bogan N, Brecha N, Gall C, Karten, HJ (1982) Distribution of enkephalin-like immunoreactivity in the rat main olfactory bulb. Neurosci 7: 895-906

Burd GD, Davis BJ, Macrides F (1982) Ultrastructural identification of substance P immunoreactive neurons in the main olfactory bulb of the hamster. Neurosci 7: 2697-2704

Bouvet JF, Delaleu JC, Holley A (1988) The activity of olfactory receptor cells is affected by acetylcholine and substance P. Neurosci Res 5: 214-223

Braitman DJ (1986) Desensitization to glutamate and aspartate in rat olfactory (prepyriform) cortex slice. Brain Res 364: 199-203

Brennels AB (1974) Spontaneous and neurally evoked release of labelled noreadrenaline from rabbit olfactory bulbs in vivo. J Physiol 240: 279-293

Chase TN, Kopin IJ (1968) Stimulus-induced release of substances from olfactory bulb using the push-pull cannula. Nature 217: 466-467

Chuah MI, Hui BS (1986) Effect of amitriptyline on laminar differentiation of neonatal rat olfactory bulb. Neurosci Lett 70: 28-33

Cooper JR, Bloom FE, Roth RH (1982) The biochemical basis of neuropharmacology, 4th edn. Oxford University Press, New York

Davis BJ, Burd GD, Macrides F (1982) Localization of methionine-enkephalin, substance P and somatostatin immunoreactivities in the main olfactory bulb of the hamster. J Comp Neurol 204: 377-383

Davis BJ, Macrides F (1983) The organization of centrifugal projections from the anterior olfactory nucleus, ventral hippocampal rudiment, and piriform cortex to the main olfactory bulb in the hamster: An autoradiographic study. J Comp Neurol 203: 475-493

Deems DA, Doty RL, Settle RG, Snow JB Jr (1988) Chemosensory dysfunction: Analysis of 750 patients from the University of Pennsylvania Smell and Taste Center. Chem Sens 13: 683

Delong RE, Getchell TV (1987) Nasal respiratory function - vasomotor and secretory regulation. Chem Sens 12: 3-36

Dockray GJ (1983) Cholecystokinin. In: Krieger DT, Brownstein MJ, Martin JB (eds) Brain peptides. John Wiley and Sons, New York, p 851

Doty RL, Deems D, Stellar S (1988a) Olfactory dysfunction in Parkinson's disease: A general deficit unrelated to neurologic signs, disease stage, or disease duration. Neurology 38: 1237-1234

Doty RL, Ferguson-Segall M (1987) Odor detection performance of rats to ethyl acetate following d-amphetamine treatment. Psychopharmacol 93: 87-93

Doty RL, Ferguson-Segall M, Lucki I, Kreider M (1988b) Effects of intrabulbar injection of 6-hydroxydopamine on ethyl acetate odor detection in castrate and non-castrate male rats. Brain Res 444: 95-103

Doty RL, Frye RE (1989) Nasal obstruction and chemosensation. Otolaryngol Clin North Am 22: 381-384

Doty RL, Kreiss D, Frye RE (in press) Odor intensity: Correlation with odorant-sensitive adenylate cyclase activity in cilia from frog olfactory receptor cells. Brain Res

Doty RL, Reyes P, Gregor T (1987) Presence of both odor identification and detection deficits in Alzheimer's disease. Brain Res Bull 18: 597-600

Doty RL, Riklan M, Deems DA, Reynolds C, Stellar S (1989) The olfactory and cognitive deficits of Parkinson's disease: Evidence for independence. Ann Neurol 25: 166-171

Doty RL, Risser J (1989) Influence of the D-2 dopamine receptor agonist quinpirole on rat odor detection performance before and after administration of spiperone. Psychopharmacol 98: 310-315

Doty RL, Shaman P, Applebaum SL, Giberson R, Sikorsky L, Rosenberg L (1984) Smell identification ability: Changes with age. Science 226: 1441-1443

Edwards DA, Mather RA, Shirley SG, Dodd GH (1987) Evidence for an olfactory receptor which responds to nicotine as an odorant. Experientia 43: 868-873

Fallon JH, Moore RY (1978) Catecholamine innervation of the basal forebrain. III. Olfactory bulb, anterior olfactory nuclei, olfactory tubercle and piriform cortex. J Comp Neurol 180: 553-544

Farbman AI, Gonzales F, Chuah MI (1988) The effect of amitriptyline on growth of olfactory and cerebral neurons in vitro. Brain Res 457: 281-286

Firestein S, Werblin F (1989) Odor-induced membrane currents in vertebrate olfactory receptor neurons. Science 244: 79-81

Ffrench-Mullen JMH, Koller K, Zaczek R, Coyle JT, Hori N, Carpenter DO (1985) N-Acetylaspartylglutmate: Possible role as the neurotransmitter of the lateral olfactory tract. Proc Natl Acad Sci USA 82: 3897-3900

Foote SL, Morrison JH (1987) Extrathalamic modulation of cortical function. Annu Rev Neurosci 10: 67-95

Gall CM, Hendry SHC, Seroogy KB, Jones EG, Haycock JW (1987) Evidence for coexistence of GABA and dopamine in neurons of the rat olfactory bulb. J Comp Neurol 266: 307-318

Gervais R (1987) Local GABAergic modulation of noradrenaline release in the rat olfactory bulb measured on superfused slices. Brain Res 400:151-154

Gervais R, Pager J (1979) Combined modulating effects of the general arousal and the specific hunger arousal on the olfactory bulb responses in the rat. EEG Clin Neurophysiol 16: 87-94

Gervais R, Pager J (1983) Olfactory bulb excitability selectively modified in behaving rats after local 6-hydroxydopamine treatment. Behav Brain Res 9: 165-179

Gesteland RC (1986) Speculations on receptor cells as analyzers and filters. Experientia 42: 287-291

Getchell TV (1986) Functional properties of vertebrate olfactory receptor neurons. Physiol Rev 68: 772-818

Goodspeed RB, Catalanotto FA, Gent JR, Cain WS, Bartoshuk LM, Leonard G, Donaldson JO (1986) Clinical characteristics of patients with taste and smell disorders. In: Meiselman HL, Rivlin RS (eds) Clinical measurement of taste and smell. MacMillan Publishing Company, New York, p 451

Gray CM, Freeman WJ, Skinner JE (1986) Chemical dependencies of learning in the rabbit olfactory bulb: Acquisition of the transient spatial pattern change depends on norepinephrine. Behav Neurosci 100: 585-596

Graziadei PPC (1973) Cell dynamics in the olfactory mucosa. Cell Tissue Res 5: 113-131
Halász N, Nowycky MC, Shepherd GM, Hokfelt T (1985) Catecholaminergic contributions to the neuronal machinery of the olfactory bulb: autoradiographic, immunohistochemical and evoked field potential studies. Chem Sens 10: 203-218
Halász N, Shepherd GM (1983) Neurochemistry of the vertebrate olfactory bulb. Neurosci 10: 579-619
Hori N, Auker CR, Braitman DJ, Carpenter DO (1981) Lateral olfactory tract transmitter: Glutamate, aspartate or neither? Cell Mol Neurobiol 1: 115-120
Jahr CE, Nicoll RA (1982) Noradrenergic modulation of dendrodendritic inhibition of the olfactory bulb. Nature 297: 227-228
Kaba H, Keverne EB (1988) The effect of microinfusions of drugs into the accessory olfactory bulb on the olfactory block to pregnancy. Neurosci 25: 1007-1111
Krey LC, Silverman AJ (1983) Luteinizng hormone releasing hormone. In: Kreiger DT, Brownstein MJ, Martin JB (eds) Brain peptides. John Wiley and Sons, New York, p 687
Kreider MS, Knight P, Winokur A, Kreiger N (1982) TRH concentration in rat olfactory bulb is undiminished by deafferentation. Brain Res 241: 351-354
Lancet D (1986) Vertebrate olfactory reception. Annu Rev Neurosci 9: 329-355
Leopold DA, Wright HN, Mozell MM, Youngentob SL, Hornung DE, Richman RA, Sheehe PR (1988) Clinical categorization of olfactory loss. Chem Sens 13: 708
Lindvall O, Björklund A (1974) The organization of the ascending catecholamine systems in the rat brains as revealed by the glyoxylic acid fluorescence method. Acta Physiol Scand Suppl 412, 1-48
Lindvall O, Björklund A (1983) Dopamine- and norepinephrine-containing neuron systems: Their anatomy in the rat brain. In: Emson PC (ed) Chemical neuroanatomy. Raven Press, New York, p 229
Lincoln J, Coopersmith R, Harris EW, Cotman CW, Leon M (1988) NMDA receptor activation and early olfactory learning. Brain Res. 467: 309-312
Lynch G, Baudry M (1984) The biochemistry of memory: a new and specific hypothesis. Science 224: 1056-1063
Macrides F (1976) Olfactory influences on neuroendocrine function in mammals. In: Doty RL (ed) Mammalian olfaction, reproductive processes and behavior. Academic Press, New York, p 29
Macrides F, Davis BJ (1983) The olfactory bulb. In: Emson, PC (ed) Chemical neuroanatomy. Raven Press, New York, p 391
Mair RG, Capra C, McEntee WJ, Engen T (1980) Odor discrimination and memory in Korsakoff's psychosis. J. Exp Psychol: Human Percept Perform 6: 445-458
Mair RG, Doty RL, Kelly KM, Wilson CS, Langlais PJ, McEntee WJ, Vollmecke TA (1986) Multimodal sensory discrimination deficits in Korsakoff's psychosis. Neuropsychologia 24: 831-839
Mair RG, Slade C, Langlais PJ (1987) Monoaminergic activity in olfactory bulb and cerebellum during olfactory and motor learning tasks. Chem Sens 12: 667
Mancia M, von Baumgarten R, Green JD (1962) Response patterns of olfactory bulb neurons. Arch Ital Biol 100: 449-462
Marasco E, Cornwell-Jones C, Sobrian SK (1979) 6-Hydroxydopamine reduces preference for conspecific but not other familiar odors in rat pups. Physiol Biochem Behav 10: 319-323

Margolis FL (1981) Neurotransmitter biochemistry of the mammalian olfactory bulb. In: Cagan RH, Hare MR (eds) Biochemistry of taste and olfaction. Academic Press, New York, p 369

McLennan H (1971) The pharmacology of inhibition of mitral cells in the olfactory bulb. Brain Res 29: 177-184

Molina V, Ciesielski L, Gobaille S, Mandel P (1986) Effects of the potentiation of the GABAergic neurotransmission in the olfactory bulbs on mouse-killing behavior. Pharmacol Biochem Behav 24: 657-664

Moore RY, Bloom FE (1978) Central catecholamine neuron systems: Anatomy and physiology of the dopamine systems. Annu Rev Neurosci 1: 129-170

Moss RL, Dudley C (1984) The challenge of studying the behavioral effects of neuropeptides. In: Iversen LL, Iversen SD, Snyder SH (eds) Handbook of psychopharmacology. Plenum Press, New York, p 397

Moulton DG, Celebi G, Fink RP (1970) Olfaction in mammals - two aspects: proliferation of cells in the olfactory epithelium and sensitivity to odours. In: Wolstenholme G, Knight J (eds) Taste and smell in vertebrates. JA Churchill, London, p 227

Nickell WT, Shipley MT (1988) Neurophysiology of magnocellular forebrain inputs to the olfactory bulb in the rat: Frequency potentiation of field potentials and inhibition of output neurons. J Neurosci 8: 4492-4502

Nicoll RA (1971) Pharmacological evidence for GABA as the transmitter in granule cell inhibition in the olfactory bulb. Brain Res 35: 137-149

Nowycky MC, Halász N, Shepherd GM (1983) Evoked field potential analysis of dopaminergic mechanisms in the isolated turtle olfactory bulb. Neurosci 8: 717-722

Nowycky MC, Mori K, Shepherd GM (1981a) GABAergic mechanisms of dendrodendritic synapses in isolated turtle olfactory bulb. J Neurophysiol 46: 639-648

Nowycky MC, Mori K, Shepherd GM (1981b) Blockade of synaptic inhibition reveals long-lasting synaptic excitation in isolated turtle olfactory bulb. J Neurophysiol 47: 649-658

Olpe HR, Heid J, Bittiger H, Steinmann MW (1987) Substance P depresses neuronal activity in the rat olfactory bulb in vitro and in vivo: Possible mediation via gamma-aminobutyric acid release. Brain Res 412: 269-274

Phillips HS, Ho BT, Linner JG (1982) Ultrastructural localization of LH-RH-immunoreactive synapses in the hamster accessory olfactory bulb. Brain Res 246: 193-204

Pissonnier D, Thiery JC, Fabre-Nys C, Poindron R, Keverne EB (1985) The importance of olfactory bulb noradrenaline for maternal recognition in sheep. Physiol Behav 35: 361-363

Price DL (1986) New perspectives on Alzheimer's disease. Annu Rev Neurosci 9: 489-512

Reyes PF, Golden GT, Fariello RG, Fagel L, Zalewska M (1985) Olfactory pathways in Alzheimers's disease: Neuropathological studies. Soc Neurosci Abstr 11: Abstr 54.10

Ribak CE, Vaughn JE, Saito K, Barber R, Roberts E (1977) Glutamate decarboxylase localization in neurons of the olfactory bulb. Brain Res 126: 1-18

Robbins TW, Everitt BJ (1987) Psychopharmacological studies of arousal and attention. In: Stahl SM, Iversen SD, Goodman EC (eds) Cognitive neurochemistry. Oxford University Press, Oxford, p 135

Rosser AE, Keverne EB (1985) The importance of central noradrenergic neurones in the formation of an olfactory memory in the prevention of pregnancy block. Neurosci 15: 1141-1147

Salmoiraghi GC, Bloom FE, Costa E (1964) Adrenergic mechanisms in rabbit olfactory bulb. Am J Physiol 207: 1417-1424

Schiffman SS (1983) Taste and smell in disease. New Engl J Med 308: 1275-1279

Schneider BS, Freidman JM, Hirsch J (1983) Feeding behavior. In: Kreiger DT, Brownstein MJ, Martin JB (eds) Brain peptides. John Wiley and Sons, New York, p 251

Scott JW (1986) The olfactory bulb and central pathways. Experientia 42: 223-232

Sharif NA (1988) Chemical and surgical lesion of rat olfactory bulb: Changes in thyrotropin-releasing hormone and other systems. J Neurochem 50: 388-394

Shepherd GM (1972) Synaptic organization of the mammalian olfactory bulb. Physiol Rev 52: 864-917

Shepherd GM (1979) The synaptic organization of the brain. Oxford University Press, New York

Shipley MT, Adamek GD (1984) The connections of the mouse olfactory bulb: A study using orthograde and retrograde transport of wheat germ agglutinin conjugated to horseradish peroxidase. Brain Res Bull 12: 669-688

Shipley MT, Halloran FJ, de la Torre J (1985) Surprisingly rich projection from locus coeruleus to the olfactory bulb in the rat. Brain Res 329: 294-299

Shirley SG, Polak EH, Mather RA, Dodd GH (1987) The effect of concanavalin A on the rat electro-olfactogram. Differential inhibition of odorant response. Biochem J 245: 175-184

Shivers BD, Harland RE, Pfaff DW (1983) Reproduction: The central nervous system role of luteinizing hormone releasing hormone. In: Krieger DT, Brownstein MJ, Martin JB (eds) Brain peptides. John Wiley and Sons, New York, p 389

Smith GP, Jerome C, Cushin BJ, Eterno R, Simansky KS (1981) Abdominal vagotomy blocks the satiety effect of cholecystokinin in the rat. Science 213: 1036-1037

Snyder SH, Sklar PB, Pevsner J (1988) Molecular mechanisms of olfaction. J Biol Chem 263: 13971-13974

Staubli U, Baudry M, Lynch G (1985) Olfactory discrimination learning is blocked by leupeptin, a thiol protease inhibitor. Brain Res 337: 333-336

Steinbusch HWM, Nieuwenhuys R (1983) The raphe nuclei of the rat brainstem: A cytoarchitectonic and immunohistochemical study. In: Emson P C (ed) Chemical neuroanatomy. Raven Press, New York, p 131

von Baumgarten R, Bloom FE, Oliver AP, Salmoiraghi GC (1963) Response of individual olfactory nerve cells to microelectrophorectically administered chemical substances. Pflugers Arch 227: 125-140

Yamamoto C, Iwama K (1961) Arousal reaction of the olfactory bulb. Jpn J Physiol 11: 335-345

17

EFFECTS OF ODORS ON MOOD AND BEHAVIOR: AROMATHERAPY AND RELATED EFFECTS

HARRY LAWLESS

I. INTRODUCTION

It goes without saying that any sense modality provides information to an organism so that environmental events may be perceived and acted upon. The sense of smell is no exception, and smells can influence behavior in many ways, some quite obvious. We smell burning food and change the setting on the oven. We smell mercaptans and call the gas company about a leak. A trained dairy judge smells methional, decides that a milk sample has been oxidized by light, concludes that the consumer will find this objectionable, and deducts an appropriate number of points from the grade. These are common examples of how humans use their noses as sources of useful information. Recently, the idea that odors can cause changes in emotional states or moods, and that such changes have physiological correlates, has been discussed in the literature on olfaction and perfumery.

In this Chapter the idea that the sense of smell affects human behavior, and more specifically, that smell may mediate such effects by changing moods or emotions, is examined. The Chapter is divided into five sections. The first section presents the general belief, widely held in the fragrance industry, that smells can affect mood. The second section discusses the history of the concept of "aromatherapy." The third section introduces potential mechanisms by which smells might affect moods. The fourth and largest section reviews the literature on physiological correlates of mood changes. The bulk of published findings concerns the effects of odors on encephalograms (EEG). The final section highlights unresolved issues and problems in measuring aromatherapy-related effects. In addition, some areas for further inquiry are suggested, including trigeminal effects, temporal patterns of stimulus presentation in conditioned associations to odors, and the application of aroma-supported relaxation training in systematic desensitization.

II. BELIEF IN THE EFFECTS OF ODORS ON MOOD

Can the experience of a familiar odor stop you "dead in your tracks?" Consider the following anecdote:

> "When ten to thirteen years of age I had much to do with horses and stables. Then nothing. At twenty years of age, I was one day walking along a country road and a cart, laden with stable manure was some 100 yards in front of me. The odor caused a sudden shock of memory of the years of my childhood, which thrilled me into *immobility*." (Laird 1935, p 129).

While a strong whiff of properly aged stable manure would probably get a reaction from most people, there is a certain poignant nostalgia to that recollection, an undeniable emotional tone.

The relationship of olfaction and emotion was recently discussed by van Toller (1988), who noted that in spite of the popular belief in the ability of fragrances to affect emotional state, and in spite of the apparent potency of pheromonal olfactory effects in animals, theorists have largely ignored any role that smells might play in human emotions. In fact, the most influential theory of human emotion has stressed the influence of social context i.e., that the interpretation of bodily activity in the context of what is socially learned and socially expected is a critical influence on the type and intensity of emotions we experience (Schacter and Singer 1962). Thus human-oriented theories of emotion have stressed cognitive influences and socially learned interpretations and tended to ignore the role of the "primitive" sense of smell.

The ability of odors to affect the behavior of animals within a given species is well documented in the literature on pheromones. Recent interest in the possible existence of human pheromones has led to much speculation and some research. The classical finding in this area indicates that women who live in close proximity tend to synchronize menstrual cycles, an effect which might be mediated by olfactory signals (McClintock 1971). Interest was further spurred by the finding that androstenone, an androgen metabolite suggested to be a pheromone in pigs, is found in human axillary sweat and urine. However, attempts to substantiate the existence of a human pheromonal sex attractant have been largely unsuccessful. For example, in studies by Filsinger and colleagues with androstenone, which smells unpleasant and urinous to about 50% of adult humans (the rest being mostly anosmic), subjects were asked to rate the sexual attractiveness of males and females as well as their own moods and feelings of sexiness in the presence of this odorant. The trends were consistent with the idea that smelling something unpleasant lowers ratings of attractiveness and subjects' reports of how sexy they themselves feel (Filsinger et al. 1984, Filsinger et al. 1985). While these studies show some effects of odors on behavior and perhaps mood, they run counter to the notion that androstenone has any effect as a sexual attractant for humans.

Recently, there have been efforts to evaluate whether smells can affect human moods and behaviors as well as attempts to develop commercial technologies based on such effects. These phenomena are loosely categorized under the heading "aromatherapy." While the historical roots of aromatherapy are a bit different from the current use of the term i.e. to signify an effect of fragrance on moods or emotions, the concept has been picked up with sufficient zeal by the popular press e.g., Alsop (1986), Freedman (1988), that the term is likely to persist with this more general interpretation. That fragrances affect moods and emotions is clearly part of the corporate zeitgeist of the fragrance business. In response to the question of whether a perfume can influence mood, the Haarmann and Reimer Book of Perfume states,

> "Definitely. When an occasion is boring or particularly stuffy or the mood in the office depressing, the stimulating fragrance of the right perfume can lighten the situation surprisingly quickly. A pleasant fragrance can both stimulate and calm, can bring a little happiness, relaxation, revitalization where none existed before." (Muller et al. 1984).

This goes beyond the simple use of a fragrance as a personal decoration and beyond the use of fragrances to counteract malodors. According to this philosophy, a fragrance serves multiple roles: as a personal signature/statement, a way to communicate ones own mood, and a way to influence both the mood of oneself and that of others. According to Muller et al., "Perfumes have a stimulating and correcting effect on moods" (p 129).

Furthermore, the creator of fine fragrances is viewed as a psychologist, an interpreter of human motivation and human culture." A successful perfumer must not only have a well-trained nose, he must also possess a high degree of sensitivity for the world of feelings, for moods and wishes, and for the secret desires of human beings." (Muller et al., p 55).

The potential economic impact of fragranced products that are designed to deliver some physiological or psychic benefits is attractive to a fragrance industry in an era of little growth (Alsop 1986). Given such commercial incentives, it seems prudent to substantiate aromatherapy claims with verification from independently-supported academic research. The current mood of the academic community regarding aromatherapy claims is more relaxed than aroused, in line with the Scottish verdict of "not proven" (Carsch 1987). In spite of this healthy skepticism, the first generation of consumer products with aromatherapy claims appeared in 1986 with the launch (by Avon) of a "relaxing" line of bath products. This was followed about a year later by a "stimulating" line (Freedman 1988). Both were based on an EEG technology developed with the support of Takasago Inc. (Carsch 1987), which measured event-related EEG potentials, called the contingent negative variation. These potentials were correlated with states of arousal and sedation.

III. HISTORICAL MEANING OF AROMATHERAPY

In addition to the bath products mentioned above, a variety of other products are currently available that are more attuned to the traditional practice of aromatherapy (Carsch 1987). Many of these products are designed to be used with massage e.g., as fragranced oils, since massage forms an important ingredient in the practice of traditional aromatherapy. The term "aromatherapy" has its roots in the early 20th-century observations of a cosmetic chemist, Rene-Maurice Gattefosse, who was aware of the antiseptic properties of some essential oils. When his hand was burned in a laboratory explosion, he immersed it in neat lavender oil, which apparently promoted healing (Tisserand 1977). This dermatological therapy was thus quite different from the present day use of "aromatherapy" to indicate a therapeutic influence via the sense of smell. However, practising aromatherapists who apply essential oils to the skin during massage may be taking advantage of both transdermal and inhaled routes of administration. A second historically important influence was that of Marguerite Maury, who developed techniques for the use of essential oils in massage. Dr Jean Valnet, a practising physician, also studied the therapeutic use of herbs, and continued the use of Gattefosse's term "aromatherapy." Tisserand (1977), a student of Valnet, outlined the principles of the field in the recently published "The Art of Aromatherapy," which summarizes the history of the art and its relation to herbalism, holistic medicine and oriental philosophy. Tisserand's book provides illuminating anecdotes, such as his own use of neat lavender oil to treat a second-degree burn of a neighbor's arm following an automobile radiator steam explosion.

The practice of aromatherapy can be viewed as an historical extension of herbal medicine, but with one major difference. Herbalists prefer to use the whole herb or at least the whole active part of the plant, while aromatherapists generally use the extracted essential oils. Both fields have a strong orientation toward holistic treatment, and borrow heavily from oriental philosophy. Tisserand, for example, categorizes different essential oils on the basis of their yin and yang alignments. Aromatherapy massage employs the principles of oriental "meridians" of energy flow. The emphasis on dermal application and massage places the sense of smell in a secondary role. In fact, the historical practice of aromatherapy may be thought of more as a "therapy using aromatics" than a "therapy using aromas." Thus the current thrust of the fragrance industry toward mood-altering smells (Alsop 1986, Freedman 1988) differs somewhat from the "art" as described by Tisserand.

In a more recent review concerning the use of essential oils as psychotherapeutic agents, effects classified as sedative vs stimulating were claimed for a variety of fragrance materials (Tisserand 1988). Much of this literature, however, was anecdotal. For example, Tisserand discussed the use of oils in clinical psychotherapy by Rovesti (1973), but admitted that Rovesti's claims are devoid of support from either controlled experimentation or controlled clinical trials. Animal research was also reviewed in this summary. For example, "sedative" effects were inferred from decrements in maze running performance after administration of various essential oils. However, the relevance of this literature to the assertions concerning essential oils as used in aromatherapy is also questionable. Tisserand himself points out that, pharmacological/behavioral investigations with animals often used doses about 100 times greater than those used in aromatherapy.

Furthermore, instead of being applied topically, the substances were usually administered orally or by intraperitoneal injection. In some cases, the doses were close to fatally toxic levels. The relevance of such experiments in substantiating "sedative" effects of fragrances is remote.

IV. POTENTIAL MECHANISMS FOR AN INFLUENCE OF SMELL ON MOOD

The connection between smell and emotion is expected on a number of grounds. One widely cited reason e.g., Lorig and Schwartz (1988), concerns the complex and numerous central olfactory projections to structures in the limbic system such as the amygdala and hippocampus (Price 1987), which are implicated in the modulation of emotion and memory. Such statements have the status of true generalizations. However, true generalizations are not the same as mechanistic hypotheses. For example, to say that "the limbic system is involved in emotion" is a lot like saying "the eyes are involved in vision." Details are lacking and further hypotheses and tests are not readily apparent from such statements. van Toller (1988) reminded olfactory theorists that the limbic system contains 53 regions and 35 associated tracts. In such a complex system, the multiplicity of connections and interactions do not lend themselves to simple theories concerning the relationships of smells to specific emotions.

A second common theme has to do with the ability of smells to evoke complex memories from the distant past of individuals, often with emotional, nostalgic recollection of another time and place. While it is true that not all odors evoke emotional recollections, the anecdotes are often compelling, as in the customarily cited example of Proust and his experiences with petites madeleines (Engen 1982). Entertaining examples were given by Laird (1935), for example:

> "I grew up in the Nevada desert in a small mining town. Since my seventeenth year my residence has been in California in the San Francisco bay area but I never have and never will learn to be happy in the fog and rain and dampness. I have spent several summers in the Tahoe district and each time have brought home a bunch of sage brush which I keep in a receptacle and not infrequently smell. When I do, visual and emotional sensations arise within me in considerable clarity

of the desert scene. A slight sniff doubles and redoubles that tranquil nostalgia." (p 127).

To the extent that an effect of odors on mood depends upon associative memory, a commercial fragrance product which is designed to affect moods must coincidentally tap a number of similar idiosyncratic memories, or some common experience that is culturally shared. To the extent that a large number of people have pleasant recollections associated with particular odors, for example, the smell of Christmas trees or of baby powder, similar nostalgic emotions may be evoked. It is not unreasonable to imagine that such an effect would be sufficiently similar among a group of individuals that a relatively uniform physiological reaction could be measured in a group of subjects under laboratory conditions.

A third potential means for odors to affect mood is the possibility of pharmacological action of some of the ingredients in essential oils. Entry into the bloodstream is clearly one of the bases for application of essential oils to the skin during aromatherapy massage (Tisserand 1977). While inhalation may not be the most efficient route to deliver pharmacological agents to the bloodstream, a recent European patent application cited this route as one means of delivering therapeutic doses of active ingredients from essential oils (Warren et al. 1986). Concentrations of active ingredients were approximated based on a model of bloodstream levels of cannabinols following inhalation. Whether the sense of smell is required or even involved at all in aromatherapy could be assessed with anosmic individuals or people restricted to mouth breathing. If mood changes can be observed in individuals who inhale odors of essential oils but do not smell them, then a more direct pharmacological mechanism is implied. The commercial implications of a direct pharmacological effect would be far-reaching. Products that are proven to act in such a manner would presumably fall under the regulation of agencies such as the Food and Drug Administration in the United States as drugs.

V. MEASUREMENT OF MOOD CHANGES TO ODORS

A currently popular theme in aromatherapy suggests that smells might reduce stress, and make people less anxious and more energized, etc. Testing such hypotheses requires that moods or emotions be measurable entities. There are two broad approaches to the measurement of mood states. First, subjective report of mood may be assessed using paper-and-pencil tests like the mood adjective checklists and rating scales used in clinical psychology. A second approach is to examine physiological correlates of emotional states. The cause-effect relationships of emotions and bodily responses has been hotly debated by psychologists since the James-Lange theory, which stated that emotions are a result of bodily activity (Grings and Dawson 1978). Regardless of whether the relationship between emotions and bodily activity is causative or merely correlational, almost all theories of emotions include the presence of bodily activity as a response to stimuli that are of emotional significance. Thus it seems reasonable to examine psychophysiological measures such as blood pressure, skin conductivity, heart and breathing rates and brain activity that relate to emotional reactions mediated through the autonomic and central nervous systems. Of course, any comprehensive theory concerning smells and emotions would assess both subjective reports and physiological responses, and establish the nature of the relationships between them. In order to provide information with an appearance of scientific objectivity, a recent surge of experimentation has examined physiological reactions to odors that may be related to states of altered mood, emotion, arousal and/or relaxation. This research is largely exploratory. Few reports have appeared in peer-reviewed journals and there is a surfeit of abstracts from professional society meetings. Most specific findings are unreplicated. Four major types of psychophysiological measures have been studied in order to assess the physiological correlates of the effects of aromas upon mood and emotion. These are, respiration, two types of encephalogram (EEG) measures, and blood pressure. One EEG measure, the contingent negative variation, assesses a specific waveform that occurs during speeded motor responses following a warning signal. The second EEG measure uses period analysis to examine different waveform frequency bands under different relaxing, stressful and odor-stimulated conditions. Each of these is considered below.

One psychophysiological measure of the smell-mood connection has been the analysis of respiratory cycles. In one study, subjects received lemongrass, lavender, eucalyptus and peppermint odorants. Strain-gauge measurements of respiration were evaluated using the averaged repetitive cycle method (Schwartz et al. 1986a). Subjects also rated the perceived intensity and pleasantness of the odors, as well as their own feelings of alertness, tenseness, relaxation and lightness. Smelling lavender and eucalyptus were associated with increases in rated alertness as well as increased respiration parameters. An interaction with odor pleasantness was also noted. Subjects who disliked lavender decreased their respiratory parameters, although it was still perceived as alerting. A related study using similar methodology found that stressful tasks such as serial arithmetic tended to decrease inspiration times, while increases in inspiration and expiration times were associated with a relaxing meditative condition (Schwartz et al. 1986b). These results suggest that breathing parameters may be useful indicators of tension, alertness and relaxation, and their potential interactions with more specific moods. However, odor hedonics may override respiration parameters in cases where the odor is strongly disliked and subjects actively inhibit sniffing or breathing.

Substantial research activity has been devoted to the evaluation of one particular EEG parameter, the contingent negative variation (CNV). Figure 1 shows the features of this event and the experimental arrangement. Subjects receive a warning signal such as a noise followed at a short interval by a second signal such as a light which they must terminate as rapidly as possible by pressing a switch. During these tasks, the CNV appears as a prolonged surface-negative component in the interval between the warning signal and the response signal. The signal is often discernible following an average of 10 or 20 presentations. The occurrence of the CNV is influenced by a number of factors related to the attitude and expectancies of the subject (Walter et al. 1964). In general, the CNV will only be seen in situations in which the subject is in a state of arousal and vigilant expectation. For example, diluting the correlation of the warning signal and response signal, telling the subjects there will be no more warnings, and making the response optional, all attenuate or eliminate the CNV, suggesting a strong relationship to attentiveness, operant responses and reinforcement contingencies. Since its discovery, the CNV has proven to be sensitive to a number of additional manipulations related to arousal and relaxation. Tecce et al., (1976) found that the CNV could be attenuated by distracting cognitive tasks imposed during the warning and response signal interval. They proposed that arousal related to distraction was responsible for this attenuation, and that there was an inverted U-shaped function relating arousal to CNV magnitude. That is, a little arousal would enhance selective attention, but a lot of arousal would interfere. The hypothesis that the CNV is related to arousal is consistent with observations that caffeine administration enhances, while nitrazapam administration attenuates the CNV (Ashton et al., 1974).

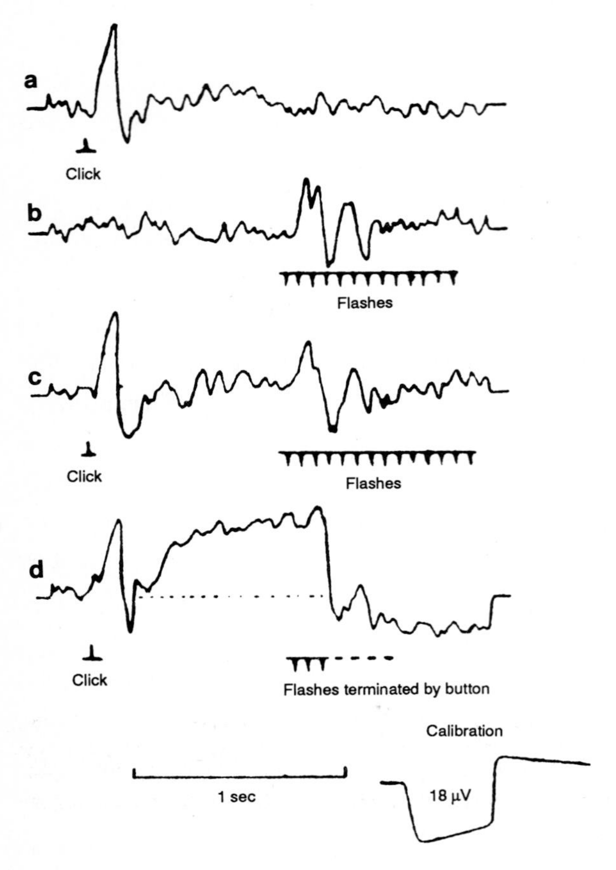

Fig 1. The contingent negative variation (CNV), from Walter et al. (1964). A. EEG response in the frontovertical region to clicks. B. Response to flickering light. C. Clicks followed by flicker. D. Clicks followed by flicker which is terminated as fast as possible by the subject pressing a button. The CNV follows the initial warning signal in the interval preceding the response signal. Average of 12 presentations. Reprinted by permission from NATURE Vol. 203, p 381, copyright 1964, MacMillan Magazines, Ltd.

To the extent that the CNV reflects changes in cortical arousal and/or depression, it seems reasonable that odors with potentially arousing or sedative effects on mood could be investigated using this phenomenon. A study reported by the Takasago Corporation (Torii and Fukuda 1985) showed that one component of the CNV was enhanced by presentation of an "excitatory" jasmine odor and attenuated for a "sedative" lavender odor. Experimental details of the procedure can be found in Torii et al. (1988). A typical record is shown in Figure 2.

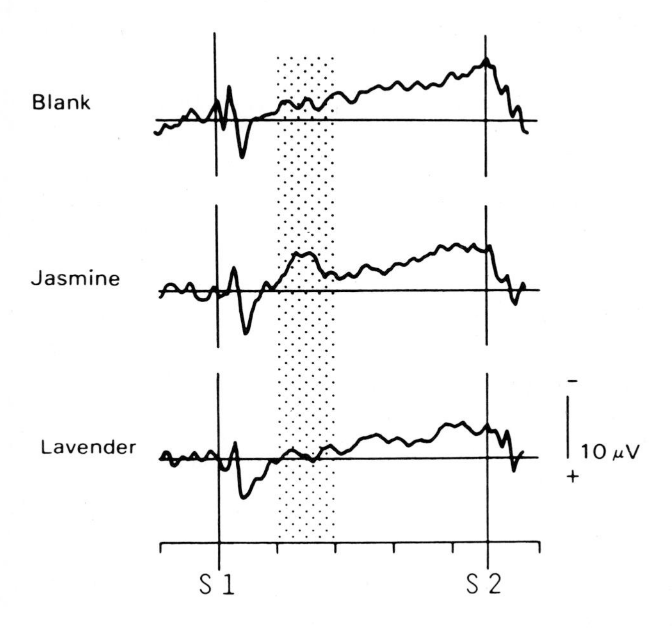

Fig 2. An example of CNV events as seen with the presentation of odors. S1 is a sound warning signal and S2 is a light that is terminated by a motor response. Average of 12 trials for a single subject. Some enhancement of an early component of the CNV is seen with the presentation of jasmine. From Torii et al. (1988); copyright 1988 by Routledge, Chapman and Hall, reproduced by permission.

An early component of the EEG, recorded from frontal electrodes and related to alertness, was affected by smell. Later components of the CNV were recorded from central and parietal electrodes, but it was theorized that these potentials were more closely tied to motor responses than the "expectancy-related" shifts seen from the frontal cortex (Torii et al. 1988). Torii et al., also found the stimulating effect of jasmine to be enhanced when subjects took a nap and then drank coffee before the assessments, as compared with a "sleepy" state measured before the nap. This is consistent with the idea that the CNV magnitude is positively related to arousal or alertness and negatively influenced by sedation. These authors also measured the skin potential level (SPL) as a control or covariate for the general level of arousal. Under conditions in which the SPL was stable and no changes were observed in SPL during odor stimulation, CNV changes were still observed, suggesting that the CNV might be more sensitive to the small changes in the level of alertness or arousal of the subject than the SPL. Another observation was that the enhancing effects of jasmine were not accompanied by increases in reaction time or heart rate, as occur with caffeine. Also, the sedating effects of lavender were not paralleled by changes in heart rate or reaction time, as are sometimes seen with drug effects on the CNV. Given this result, Torii et al., suggested that the effects of odors were mediated by different mechanisms than those of drugs such as caffeine or nitrazapam.

Experiments supporting the effects of odors on the CNV have been presented at several meetings of the Japanese Association for Smell and Taste Sciences. Kanamura et al. (1988) found jasmine odor to enhance and chamomile odor to depress the amplitude of the CNV relative to an odorless control. Furthermore, the average skin potential level was higher for jasmine and lower for chamomile than the control, which is consistent with the notion that jasmine odor is arousing. Sugano (1989) reported a number of olfactory-related modulations of brain activity, including EEGs, CNVs, and rheoencephalography (REG) measures. Lavender and alpha-pinene were reported to increase alpha waves of EEGs and amplitude of REGs while decreasing the CNV amplitude. Jasmine increased beta waves and CNV amplitude. The "sedative" effect of some terpene compounds was hypothesized to be related to relaxed feelings experienced when walking in a pine forest. Kanamura et al. (1989) reported CNV changes for a number of fragrances from the traditional perfume categories, and also investigated self-report of mood changes using six mood rating scales. For four of the 15 fragrances, a mood pattern emerged which was categorized as happy and excited. These fragrances produced CNV amplitudes which were higher than the controls. While the relationship between CNV changes and mood scales was less than perfect, the observed changes generally supported the theory that arousing odors enhance CNV magnitude. Figure 3 summarizes observations of CNV changes after presentation of different essential oils.

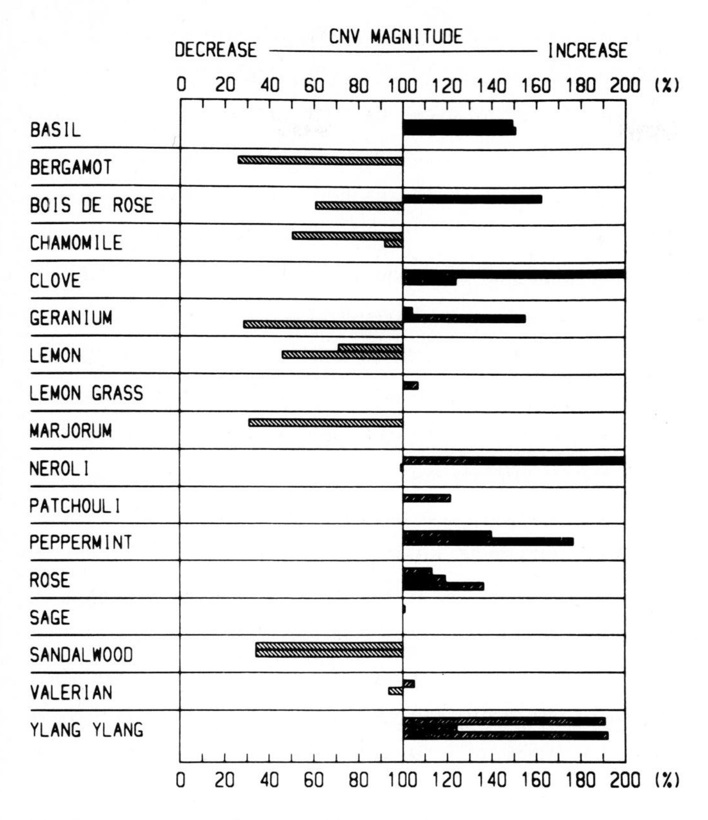

Fig 3. Summary of CNV effects as found for a variety of essential oils. Increases are indicative of heightened arousal and decreases of lowered arousal. From Torii et al. (1988); copyright 1988 by Routledge, Chapman and Hall, reproduced by permission.

Effects of aromas on other EEG measures have been reported elsewhere. Lorig and colleagues performed "period" analysis of EEG waveforms in response to a variety of stimuli including simple odorants, fine fragrances, various cognitively stressful or relaxing tasks, and to odor imagery. Period analysis differs from the more traditional spectral analysis of EEG data, in that it quantifies the number of waves that occur within a given epoch of measurement, rather than quantifying their amplitude. Period analysis has greater sensitivity to EEG changes during task performance (Lorig 1986) and was chosen as a potentially more informative measure of EEG activity associated with olfactory stimulation.

In one experiment, Lorig and Schwartz (1988) presented subjects with "spiced apple," eucalyptus and lavender odorants, and a distilled water control. Analysis of theta activity showed small decrements, particularly for the spiced apple odorant as compared with the theta activity in the baseline and lavender odor conditions. The eucalyptus condition was intermediate between the spiced apple and lavender conditions. No significant changes in alpha activity were observed, although responses to the odorless control tended to be lower than to odors. Subjects also filled out rating scales for odor pleasantness, intensity and 15 mood-related terms such as relaxed, tense, happy and bored. The spiced apple and eucalyptus conditions were associated with reports of significantly lower ratings for anxiety and tension than the ratings for the lavender condition. At first glance, these results seem to provide support for the idea that responses to a lavender odorant might be associated with theta activity, which in turn are associated with arousing or even anxiety-provoking stimuli. However, the mood scale data must be viewed in light of the large number of statistical comparisons. Observing differences in four out of 90 possible comparisons is close to the expected number of times a true null hypothesis would be rejected under alpha = .05. Such data provide weak evidence of mood changes which were available to the conscious perception of the subjects. A second experiment suggested that EEG changes need not parallel changes in self-report (Lorig and Schwartz 1988). In this experiment, five commercial fragrances consisting of floral-note perfumes were presented. Differences in alpha and theta activity in response to the different perfumes were noted as a function of time, anterior-posterior electrode placement and right/left hemisphere electrode placement. Odorants differed in perceived intensity, but none of the mood-related measures differed among odorants. In other words, despite significant EEG changes in response to different fragrances, mood self-reports showed no statistically significant changes.

Interpretation of these effects is further complicated by the finding that EEG changes of both alpha and beta waves occur as a function of nose vs mouth breathing in the absence of purposely presented odors (Lorig et al. 1988). Maintaining the idea that these EEG changes were linked to olfactory stimulation, Lorig and colleagues suggested that undetected odors in room air could be responsible. A follow-up study with varying concentrations of odors provided support for this idea (Lorig, Herman, Schwartz, Cain, unpublished data). Subjects who reported no detection of low concentrations of a lavender odorant, nonetheless showed increased beta activity in posterior relative to frontal regions, similar to the effect of nose breathing in the previous study. Only one self-report measure showed any effect; specifically, happiness ratings were lower when the lowest, undetected concentrations were presented. Once again, a significant finding in one out of 17 measures is about what one would expect by chance when applying 17 statistical tests to samples in which there was really no difference (when the null hypothesis is true and the alpha risk is set at 5%, a significant difference in one out of 20 tests is expected). A second study using undetected concentrations of Galaxolide, a musk compound, found decrements in alpha activity and lowered performance in a visual search task (Lorig et al. 1989). The connection between lowered alpha activity and impaired performance is somewhat confusing since lowered alpha activity may be associated with heightened arousal, which should improve sensorimotor performance. The authors drew the conservative conclusion that "these results support the hypothesis that undetected odors can alter central nervous system activity and extend this hypothesis to include the alteration of behavior."

However, proving that a stimulus is undetected or subliminal has both methodological and theoretical difficulties. In the framework of signal detection theory, one may question whether reports of "no detection" are truly indicative of zero sensation, or whether they are merely indicative of sensation intensities below the subject's criterion for responding. Two tenets of signal detection theory are germane. First, perceived sensation intensity grows continuously as the physical stimulus intensity is increased above the background noise. In this view, there is no real threshold in the sense of a discontinuity in perception. Second, subjective report is a function of both sensation intensity and the subject's criterion for response. In other words, although detection is inferred from behavior, the latter is both a function of perception and a decision process about appropriate responses. This approach renders both the idea of a subliminal stimulus and discussions of "perception without awareness" somewhat moot. van Toller (1988) also noted that subjects sometimes failed to report the presence of an odor in an experimental session while showing signs of physiological activity. At first glance, this may be interpreted as a case of perception without awareness. However, when the odor (in this case androstenone) was later re-presented during de-briefings, subjects often gave responses indicative of recognition, implying previous awareness. Failure to report can be caused by either a bias to withhold response, or by a failure to detect the stimulus. Only methods compatible with signal detection theory (such as forced-choice procedures) can unscramble the two, and mere reliance on uncued subjective report is inconclusive.

A final twist in the relationship between EEGs and behavior, concerns the potential mediation of aroma influences by the subjects' imagery of foods. In this regard, Lorig and Schwartz (1988-89) recorded EEG activity during a variety of cognitive tasks, some stressful and some relaxing, including sets of instructions to "remember your favorite dessert", and "remember your favorite main course." Conditions in which the favorite dessert was imagined, produced EEG changes associated with relaxation, decreased tension, and less boredom. Insofar as a spiced apple aroma may evoke pleasant food imagery, previously observed physiological and mood changes may be the result of the act of remembering pleasant food. This is noteworthy for several reasons. First, apple aroma and spice aromas (nutmeg, etc.) are major components in a common American dessert, apple pie. "Common" may be an understatement. "Apple pie" is used as a colloquial image for an item so familiar and characteristic to American culture that has become a virtual symbol, as in the phrase, "home, Mom and apple pie." It is possible that the choice of a spiced apple fragrance taps into culturally shared odor memories associated with home, motherhood, baking, holidays, pleasant food, etc. Second, as Lorig and Schwartz (1988-1989) pointed out, food imagery is used in systematic desensitization therapy, as competition for anxiety-producing stimuli in situations in which relaxation is desired. The realities of clinical intervention support the potency of food-related imagery in influencing moods and emotions.

Some odors may be able to influence cardiovascular function. For example, under what is assumed to be a moderately stressful situation of having one's olfactory function assessed, Doty et al. (1988) noted decreases in blood pressure and heart rate to presentations of phenethyl alcohol (a rose constituent). An effect of aromas on reduction in blood pressure was also mentioned without citation for a "spiced apple" fragrance in recent reports concerning the effects of odors on EEGs (Lorig and Schwartz 1987, Lorig and Schwartz 1988-89). In a recent European patent application, a variety of odorants and essential oils were claimed to reduce blood pressure under stressful circumstances (Warren et al. 1986). The "active" materials included a number of essential oils from the aromatherapists' battery such as nutmeg oil, mace extract, neroli oil, and valerian oil, as well as some single fragrance compounds, myristicin, elimicin and isoelimicin. Results were presented from four experiments which compared various combinations of these components to conditions which fragrances or diluents were presented without the "active" principles. A very sensitive experimental design was employed which involved within-subject difference scores comparing low and moderate stress conditions. Stress was manipulated by asking a variety of neutral or emotion-inducing personal questions, or by serial arithmetic tasks. Blood pressure was measured at baseline levels, during low and moderate stress conditions, and difference scores were computed from blood pressure changes during low to moderate stress conditions. Differences were also compared between conditions in which fragrances with and without "actives" were presented. Changes in mood, registered by self-report on a variety of scales were also assessed. The general pattern of results can be summarized as follows: When blood pressure is increased during a moderately stressful task or situation, changes (relative to low stress conditions) were generally smaller after the active ingredients had been smelled. Although different mood rating scales showed significant effects in different sub-experiments, the overall pattern was consistent with lowered anxiety and increased relaxation when odors with active ingredients were smelled. However, some caution is warranted regarding the interpretation of patent applications. These applications are legal documents meant to establish ownership of an invention and to indicate that the invention may work with some degree of plausibility. In some cases, pharmaceutical patents may include ingredients with various degrees of efficacy, and may be intentionally vague about the relative activity of cited ingredients, to preserve some developmental advantage over competitors.

VI. UNRESOLVED ISSUES AND FUTURE DIRECTIONS

While the amount of evidence for effects of odors on moods and their physiological correlates is mounting, a coherent theory is still lacking. Observations have been mainly empirical, some conflict (see Table 1), and it is difficult to find results tied to specific theoretical predictions. For example, why is lavender sedating in the Japanese reports of CNV changes but arousing (or at least less relaxing than spiced apple) in the period analysis of EEGs, yet capable of creating increased alertness as measured by respiratory parameters? Torii et al. (1988) indicated that the physiological results do not always fit comfortably with the properties ascribed to them in the aromatherapy literature. For example, rose is listed as sedating, but produced enhanced CNVs in at least some of their subjects. Apparent contradictions abound within the popular literature itself. Both jasmine and lavender, for example, are recommended for anxiety and depression (Tisserand 1977). The view of the practising aromatherapist is that "this demonstrates the versatility of essences as therapeutic agents, and their ability to respond [sic] to the needs of the individual" (p 100). Indeed, each essential oil is listed as a treatment for 10 to 20 ailments, with much duplication. Furthermore, the aromatherapist is warned to consider

TABLE 1 : Effects of lavender odor.

Measure	Result	Interpretation	Citation
Respiration	increased	arousal	Schwartz et al. 1986a
Mood ratings	increased alertness	arousal	Schwartz et al. 1986a
CNV	lower amplitude	sedation	Torii et al. 1988
SPL, heart rate reaction time	no effect	none	Torii et al. 1988
EEG period analysis	increased theta waves relative to apple[a] condition	unclear	Lorig and Schwartz 1988
Mood ratings	increased anxiety and tension relative to apple condition	arousal?	Lorig and Schwartz 1988
EEG, period analysis	increased beta waves to undetected lavender	arousal	Lorig et al. 1988

a See text for explanation of "apple" relevance.

the entire constellation of the client's physical and mental state in choosing from such a variety of options. The outside observer here is struck with a parallel to the "flexibility" seen in astrological predictions, in which the horoscope of any sign can appear at least somewhat relevant to any individual. The tightness or specificity with which smell-emotion correlations can be made remains to be determined. Much of the literature makes claims that are no more specific than a relationship to a simple dimension of arousal vs relaxation e.g., CNV research. Other authors add additional orthogonal dimensions, such as erotogenic vs sexually neutral (Carsch 1987). Of course, the more specific the prediction of a connection between a particular smell and a particular emotion, the easier will be the task of scientific test and the easier disproof becomes.

Providing clinical substantiation for a claim of effects of smells on mood, or of therapeutic benefits to essential oils is no simple matter. The classic experimental designs involving double-blind testing with placebos are difficult to achieve with fragrances. One problem concerns the nature of a placebo. Traditionally, it is a treatment which is sensorially indistinguishable from the treatment containing the pharmacologically active agent e.g., a pill with the same appearance, taste, etc., but is lacking in the active ingredients. A placebo with no fragrance makes no sense, as subjects will know when they are being treated and hence the test is no longer blind. Producing a placebo which is sensorially indistinguishable from the effective fragrance may be a formidable exercise in perfumery. Furthermore, how is activity to be defined as an independent variable? If the mode of action is through inhalation of a substance that enters the bloodstream, it makes sense to try to construct a sensorially equivalent placebo which merely lacks the "active" principles. If, however, the mode of action is through the psychological effects of associative memory, then two sensorially equivalent smells should evoke the same memory. In this case, activity is not defined by the physical constituents of the essential oil, but by some mediating perceptual consequences. If the hypothesis is that rose odors make someone depressed because they are reminded of funerals, activity cannot be defined on the basis of chemical constituents like phenethyl ethanol or beta damascenone, but on the subjective event of having evoked memories of funerals. This requires a new definition of what constitutes a placebo, or at least a theoretically satisfying idea of what constitutes a reasonable control condition against which to compare results.

Given the commercial interest in these phenomena, additional concerns surround the possibility of reduction to practice. Even if physiological and mood changes are demonstrated in the scientific laboratory, this may not mean the consumer will perceive benefits. Certainly there are many bodily changes and patterns of neural activity that may be unavailable to conscious inspection, and the causal influence of such changes on behavior is a matter for experimental demonstration. But what is the place of perception in substantiation? Suppose a 4 mm Hg change in blood pressure occurs in a stressful situation. Will a subject be aware of that change or aware of any change in mood or emotion that accompanies it? What if consumers show EEG changes to odors below their individual thresholds (Lorig et al. 1989)? It may be difficult to establish consumer perception of a psychic benefit for a product which they do not know they are using, and if statistically significant and reliable changes are observed, are they meaningful in any practical sense? How large are the effects compared to say, the arousing effect of a cup of coffee or the sedating effect of a few ounces of gin? Statistically reliable effects, which are seen through a narrow parametric window under controlled conditions, may be swamped by other overriding factors when examined outside the laboratory. Along these lines, one can question whether the variations in one time epoch of one particular EEG effect (the CNV) seen under the laboratory conditions of a vigilance/reaction time task with a warning signal, implies anything of practical importance regarding human arousal or sedation, moods, emotions or overt behavior in other situations.

The effects of ambient odors on behavior, however, were studied recently and involved cognitive function in four areas; a verbal memory task, a semantic reasoning task, an analogical reasoning task and an arithmetic reasoning task, as well as self-reports of mood and general affective response to the experiment (Ludvigson and Rottman 1989). Aromas of lavender or clove were dispersed throughout the room and the results compared to an odorless control. Statistical analysis revealed a detrimental effect of lavender aroma on arithmetic reasoning performance, and perhaps paradoxically, a more positive affective reaction to the experiment when lavender was the ambient odor. No other effects of odors on performance or mood were seen. Once again, there were limited albeit provocative findings in the face of a large number of statistical tests. Whatever effects are present during this sort of odor presentation and behavioral assessment, appear to be subtle.

One opportunity for the therapeutic use of aromas to affect moods might be in psychotherapies which use relaxation (King 1988). For example, in systematic desensitization, patients are trained to produce relaxation responses, which are incompatible with the anxiety evoked from some problem stimulus or situation (Wolpe 1982). Systematic desensitization can be used to treat phobic neuroses such as a fear of riding in elevators. Relaxation training precedes a series of encounters with anxiety producing stimuli which approximate, to a greater and greater extent, the actual or maximally fear-producing experience. The patient is trained to elicit a relaxed state of body (focussing on various muscle groups, breathing, etc.) and mind when encountering the anxiety-producing situation, which is incompatible with the fearful or panic-stricken moods previously felt. At least one practising psychiatrist has tried to use fragrances to support relaxation training in therapy (King 1988), and reported data from one patient which showed a reduction in EMG activity during odor presentation.

If odors can enhance relaxation, they could function through a variety of mechanisms to promote this process. In terms of classical conditioning, they could function either as a conditioned or unconditioned stimulus. In classical conditioning, a previously neutral stimulus is paired on repeated occasions with a stimulus which automatically evokes some behavior or physiological response e.g., a simple reflex. To the extent that some odors are physiologically relaxing, they could function as the unconditioned stimulus, and thus to directly enhance or support the relaxation that the patient is trained to produce during systematic desensitization. For example, with a patient who has a fear of flying in airplanes, a cologne could be constructed with relaxing constituents. This cologne would have been introduced during the relaxation training, and could also be smelled before or during a flight. As conditioned stimuli, other components of the fragrance with no intrinsic sedative properties would come to be associated with the relaxation experienced during training (therapy) sessions. According to King (1988), the odor works through this associational mechanism, since near-threshold levels of odors were used which are assumed to have no physiological/pharmacological action. Thus there are two potential mechanisms by which fragrances might support a therapeutic application in systematic desensitization.

There may be multiple effects and multiple modes of action involved in aromatherapy, as currently practiced. For example, Tisserand (1987) devotes an entire Chapter to baths, and also discusses the direct (oral) consumption of essential oils in preparations such as teas, infusions, etc., much like his predecessors in herbal medicine. At least two major routes of action were suggested, one through transdermal absorption that would occur during application of oils during massage, and a second mediated by the psychological effects of smelling the pleasant odor. Some proponents of aromatherapy decline to distinguish between the effects of the oil as achieved during massage or bathing, from the effect of the smell in isolation (Carsch 1987). For a recognizable benefit, it may be necessary to produce some synergistic effect of smelling combined with the relaxing or invigorating effects of tactile or thermal stimulation obtained from massage or bathing, that is responsible for the overall psychic result. If true, then laboratory tests of the effects of smells in isolation may miss the point. Of course, if it is claimed that the odor of the essential oil provides a relaxation benefit beyond that of a hot bath by itself, such additional benefit should be measurable. However, given the problems mentioned above regarding blind testing and placebos, establishing the reality of synergistic effects with scientifically acceptable documentation may prove very difficult. Such measurement problems may relegate aromatherapy to the permanent status of an art.

One potential mode of action that has largely been ignored in the literature on aromatherapy concerns trigeminal irritation. The awakening of sleeping persons by airborne chemical stimuli could be considered an extreme form of mood-alteration, or at least of CNS activity. In a study of chemicals for use as gas warning agents, Fieldner et al. (1931) found irritants like airborne capsaicin to be much more effective in awakening sleeping persons than were odorants. A related example of aggressive "aromatherapy" in the general sense is the administration of "smelling salts" to arouse the unconscious. These agents often consisted of ammonium carbonate (giving off ammonia) with an aromatic oil like lavender. The trigeminal reflex from the irritative ammonia provides the "kick" necessary to activate the brainstem arousal mechanisms of the unconscious individual. Perhaps the trigeminal effects of camphor and related terpenes in lavender oil also contribute to this rather extreme form of arousal. Given the knowledge that many common and even non-irritating odorants have some trigeminal impact (Silver 1987), there may be a trigeminal as well as an olfactory component in the mood altering effects of some essential oils.

As noted above, the degree to which reactions to smells are uniform among people may depend upon the mode of action. An effect which is based on associative memory will depend upon the evocation of common memories among individuals. Psychological differences may also be important. For example, individuals with "repressive coping styles", who are prone to hide their emotions, show higher bodily activity as measured in cardiovascular and respiratory changes under stress, but show poorer correlations with self-report data (Jamner and Schwartz 1986, Schwartz and Jamner 1986). The repressive-coping individuals put on a deceptively stoic face under stress and internalize their emotional reactions at the cost of heightened autonomic arousal. Investigations of smell/mood interactions should be sensitive to the potential influence of personality factors on the results that are observed (Ludvigson and Rottman 1989). Individual differences in odor preferences are another complicating factor. Tisserand (1977) noted that in order for the beneficial effects of a fragrance to be realized via the mode of action of smelling, the essential oil cannot have an odor that individuals find objectionable. Pleasantness of the stimulus was also important in the evaluation of respiratory parameters associated with lavender odor (Schwartz et al. 1986a). Subjects who disliked lavender inhibited their respiration and showed a pattern different from those who liked lavender and increased respiration, an action associated with alertness.

Finally, students of Proustian nostalgia need to transcend gratuitous statements about smell and the limbic system and to search for mechanisms, either on a psychological or physiological level, which will help predict the circumstances under which nostalgic recollections occur. Much has been said concerning the uniqueness of smell in these remembrances, but one can question whether or not emotion-laden recollections are strictly the province of odor stimuli. For example, perusing a family photo album or an old school yearbook can evoke nostalgic feelings in most people. Not all smells and all pictures evoke such memories. Perhaps the question should address why some stimuli are capable of stirring emotions associated with past times and places, while others do not. In this regard, the temporal parameters of learning and subsequent cuing could be studied further. A common sensation among people familiar with popular American music is to hear an old "top-40" radio hit from years ago, and have some nostalgic recollection of some other time and place associated with the frequent radio broadcast of that song e.g., of a teenage summer spent at the beach years ago. What do such recollections have in common with Laird's reports of nostalgic odor-cued memories? Radio stations have a practice of giving very frequent air play to new songs until they are displaced in popularity by yet newer songs. Then they are "retired" for a period of time until they reappear, sometimes years later, as "oldies." This temporal pattern, that of intensive exposure associated with a variety of people and places in a particular epoch of one's life, followed by a hiatus and then later re-stimulation with the cue after many years, is also characteristic of the sorts of odor experiences capable of later evoking emotional remembrances. The coffee aroma I smell every morning does not give me the Proustian jolt, but the aroma of fruitcake consumed only at Christmas is more likely to bring up scenes of dowager aunts at family gatherings and childhood holidays. Focussing on an experimental variable such as temporal patterning takes away some of the fascination with the uniqueness of odors, and places the question, perhaps more appropriately, into the realm of investigating the nature of nostalgia per se.

ACKNOWLEDGMENTS

The author thanks Tyler Lorig PhD for assistance with the literature review and for helpful discussions.

REFERENCES

Alsop R (1986) Firms push 'aroma therapy' to treat flat fragrance sales. Wall Street J, 20th March, p 33

Ashton H, Millman JE, Telford R, Thompson JW (1974) The effect of caffeine, nitrazapam and cigarette smoking on the contingent negative variation in man. Electroencephal Clin Neurophysiol 37: 59-71

Carsch G (1987) Aromatherapy, a status review and commentary. Soap Cosmet Chem Specialties August p 36-42, 88

Doty RL, Deems DA, Frye RE, Pelberg R, Shapiro A (1988) Olfactory sensitivity, nasal resistance and autonomic function in patients with multiple chemical sensitivities. Arch Otolaryngol Head Neck Surg 114: 1422-1427

Engen T (1982) The perception of odors. Academic Press, New York

Fieldner AC, Sayers RR, Yant WP, Katz SH, Shohan JB, Leitch RD (1931) Warning agents for fuel gases. (Monogr 4), US Dept of Commerce, Bureau of Mines

Filsinger EE, Braun JJ, Monte WC (1985) An examination of the effects of putative pheromones on human judgments. Ethol Sociobiol 6: 227-236

Filsinger EE, Braun JJ, Monte WC, Lindner DE (1984) Human (Homo sapiens) responses to the pig (Sus scrofa) Sex pheromone 5-Alpha-androst-16-en-3-one. J Comp Psychol 95: 219-222

Freedman AM (1988) Search is on for emotion-eliciting scents. Wall Street J, 13th October, B1

Grings WW, Dawson ME (1978) Emotions and bodily responses. Academic Press, New York

Jamner LD, Schwartz GE (1986) Integration of self report and physiological indices of affect: Interactions with repressive coping strategies. Psychophysiol 23: 444

Kanamura S, Kawasaki M, Indo M, Fukuda H, Torii S (1988) Effects of odors on the contingent negative variation and the skin potential level. Chem Sens 13: 327

Kanamura S, Kiyotas, S, Takashima Y, Kanisawa T, Indo M, Van Loveren G, Fukuda H, Torii S (1989) Effects of odours on CNV. Chem Sens 14: 303

King JR (1988) Anxiety reduction using fragrances. In: van Toller S, Dodd GH (eds) Perfumery: the psychology and biology of fragrance. Chapman and Hall, London, p 148

Laird DA (1935) What can you do with your nose? Sci Monthly 41: 126 -130

Lorig TS (1986) EEG and task performance: A comparison of three analytic techniques. Physiol Psychol 14: 651-662

Lorig TS, Huffman E, DeMartino AG, DeMarco J (1989) EEG and behavioral responses to low-level galaxolide adminstration. Assoc Chemoreception Sci Ann Meet, Sarasota, Florida. Abstr 98

Lorig TS, Schwartz GE (1987) Alteration of EEG by odor administration. Psychophysiol 24: 598

Lorig T, Schwartz GE (1988) Brain and odor: I. Alteration of human EEG by odor administration. Psychobiol 16: 281-284

Lorig T, Schwartz GE (1989-89) EEG activity during relaxation and food imagery. Imagin Cognit Personal 8: 201-208

Lorig T, Schwartz GE, Herman KB, Lane RD (1988) Brain and odor: II. EEG during nose and mouth breathing. Psychobiol 16: 285-287

Ludvigson HW, Rottman TR (1989) Effects of ambient odors of lavender and cloves on cognition, memory, affect and mood. Chem Sens 14: 525-536

McClintock MK (1971) Menstrual synchrony and suppression. Nature 229: 244-245

Muller J, Brauer H, Mensing J, Beck C (1984) The H&R book of perfume, vol I. Johnson, London, p 155

Price JL (1987) The central and accessory olfactory systems. In: Finger TE, Silver WL (eds) Neurobiology of taste and smell. John Wiley and Sons, New York, p 65

Rovesti P (1973) Aromatherapy and aerosols. Soap Perf Cosmet 46:475-477

Schacter S, Singer J (1962) Cognitive, emotional and physiological determinants of emotional states. Psychol Rev 69: 378-399

Schwartz GE, Jamner LD (1986) Subjective/respiratory dissociation and the repressive coping style. Psychophysiol 23: 459

Schwartz GE, Whitehorn D, Hernon JC, Jones M (1986a). Subjective and respiratory differentiation of fragrances: Interactions with hedonics. Psychophysiol 23: 460

Schwartz GE, Whitehorn D, Hernon JC, Jones M (1986b). The ARC method for averaging repetitive cycles: Application to respiration during stress and relaxation. Psychophysiol 23: 460

Silver WL (1987) The common chemical sense. In: Finger TE, Silver WL (eds) Neurobiology of taste and smell. John Wiley and Sons, New York, p 65

Sugano H (1989) Effects of odors on mental function. Chem Sens 14: 303

Tecce JJ, Savignano-Bowman J, Meinbresse D (1976) Contingent negative variation and the distraction-arousal hypothesis. Electroencephal Clin Neurophysiol 41: 277-286

Tisserand RB (1977) The art of aromatherapy. Destiny Books, Rochester, Vermont

Tisserand RB (1988) Essential oils as psychotherapeutic agents. In: van Toller S, Dodd GH (eds) Perfumery: the psychology and biology of fragrance. Chapman and Hall, London, p 167

Torii S, Fukuda H (1985) Effect of odors on contingent negative variation (CNV) no. 2. Takasago reprint. Takasago International Corp. Tokyo 144, Japan

Torii S, Fukuda H, Kanemoto H, Miyanchi R, Hamauzu Y, Kawasaki M (1988) Contingent negative variation (CNV) and the psychological effects of odour. In: van Toller S, Dodd GH (eds) Perfumery: the psychology and biology of fragrance. Chapman and Hall, London, p 107

van Toller S (1988) Emotion and the brain. In: van Toller S, Dodd GH (eds) Perfumery: the psychology and biology of fragrance. Chapman and Hall, London, p 121

Walter WG, Cooper R, Aldridge VJ, McCallum WC, Winter AL (1964) Contingent negative variation: An electrical sign of sensorimotor association and expectancy in the human brain. Nature 203: 380-384

Warren CB, Leight RS, Withycombe DA, Mookherjee BD, Trenkle RW, Munteanu MA, Benaim C, Walter HG, Schwartz GE (1986) Methods, compositions and uses thereof for reduction of stress. Europ. patent application 183,436 A2

Wolpe J (1982) The practice of behavior therapy. Pergamon Press, New York

INDEX

QMW LIBRARY
(MILE END)